ADVANCES IN
MEDICAL ONCOLOGY, RESEARCH
AND EDUCATION

Volume V

BASIS FOR CANCER THERAPY 1

ADVANCES IN MEDICAL ONCOLOGY, RESEARCH AND EDUCATION

Proceedings of the 12th International Cancer Congress,
Buenos Aires, 1978

General Editors: A. CANONICO, O. ESTEVEZ, R. CHACON and S. BARG, Buenos Aires

---

Volumes and Editors:

I - CARCINOGENESIS. *Editor:* G. P. Margison

II - CANCER CONTROL. *Editors:* A. Smith and C. Alvarez

III - EPIDEMIOLOGY. *Editor:* Jillian M. Birch

IV - BIOLOGICAL BASIS FOR CANCER DIAGNOSIS. *Editor:* Margaret Fox

V - BASIS FOR CANCER THERAPY 1. *Editor:* B. W. Fox

VI - BASIS FOR CANCER THERAPY 2. *Editor:* M. Moore

VII - LEUKEMIA AND NON-HODGKIN LYMPHOMA. *Editor:* D. G. Crowther

VIII - GYNECOLOGICAL CANCER. *Editor:* N. Thatcher

IX - DIGESTIVE CANCER. *Editor:* N. Thatcher

X - CLINICAL CANCER - PRINCIPAL SITES 1. *Editor:* S. Kumar

XI - CLINICAL CANCER - PRINCIPAL SITES 2. *Editor:* P. M. Wilkinson

XII - ABSTRACTS

(Each volume is available separately.)

Pergamon Journals of Related Interest

ADVANCES IN ENZYME REGULATION
COMPUTERIZED TOMOGRAPHY
EUROPEAN JOURNAL OF CANCER
INTERNATIONAL JOURNAL OF RADIATION ONCOLOGY, BIOLOGY, PHYSICS
LEUKEMIA RESEARCH

ADVANCES IN MEDICAL ONCOLOGY, RESEARCH AND EDUCATION

Proceedings of the 12th International Cancer Congress, Buenos Aires, 1978

Volume V

BASIS FOR CANCER THERAPY 1

Editor:

B. W. FOX

Department of Experimental Chemotherapy
Paterson Laboratories
Christie Hospital and Holt Radium Institute
Manchester

PERGAMON PRESS

OXFORD · NEW YORK · TORONTO · SYDNEY · PARIS · FRANKFURT

U.K.	Pergamon Press Ltd., Headington Hill Hall, Oxford OX3 0BW, England
U.S.A.	Pergamon Press Inc., Maxwell House, Fairview Park, Elmsford, New York 10523, U.S.A.
CANADA	Pergamon of Canada, Suite 104, 150 Consumers Road, Willowdale, Ontario M2J 1P9, Canada
AUSTRALIA	Pergamon Press (Aust.) Pty. Ltd., P.O. Box 544, Potts Point, N.S.W. 2011, Australia
FRANCE	Pergamon Press SARL, 24 rue des Ecoles, 75240 Paris, Cedex 05, France
FEDERAL REPUBLIC OF GERMANY	Pergamon Press GmbH, 6242 Kronberg-Taunus, Pferdstrasse 1, Federal Republic of Germany

First edition 1979

British Library Cataloguing in Publication Data

International Cancer Congress, *12th, Buenos Aires, 1978*
Advances in medical oncology, research and education.
Vol.5: Basis for cancer therapy, 1
1. Cancer - Congresses
I. Canonico, A II. Fox, Brian William
616.9'94 RC261.A1 79-40485

ISBN 0-08-024388-6
ISBN 0-08-023777-0 Set of 12 vols.

In order to make this volume available as economically and as rapidly as possible the authors' typescripts have been reproduced in their original forms. This method unfortunately has its typographical limitations but it is hoped that they in no way distract the reader.

Printed and bound at William Clowes & Sons Limited Beccles and London

Contents

Modern Trends and Prospectives in Cancer Surgery

Foreword

This book contains papers from the main meetings of the Scientific Programme presented during the 12th International Cancer Congress, which took place in Buenos Aires, Argentina, from 5 to 11 October 1978, and was sponsored by the International Union against Cancer (UICC).

This organisation, with headquarters in Geneva, gathers together from more than a hundred countries 250 medical associations which fight against Cancer and organizes every four years an International Congress which gives maximum coverage to oncological activity throughout the world.

The 11th Congress was held in Florence in 1974, where the General Assembly unanimously decided that Argentina would be the site of the 12th Congress. Argentina was chosen not only because of the beauty of its landscapes and the cordiality of its inhabitants, but also because of the high scientific level of its researchers and practitioners in the field of oncology.

From this Assembly a distinguished International Committee was appointed which undertook the preparation and execution of the Scientific Programme of the Congress.

The Programme was designed to be profitable for those professionals who wished to have a general view of the problem of Cancer, as well as those who were specifically orientated to an oncological subspeciality. It was also conceived as trying to cover the different subjects related to this discipline, emphasizing those with an actual and future gravitation on cancerology.

The scientific activity began every morning with a Special Lecture (5 in all), summarizing some of the subjects of prevailing interest in Oncology, such as Environmental Cancer, Immunology, Sub-clinical Cancer, Modern Cancer Therapy Concepts and Viral Oncogenesis. Within the 26 Symposia, new acquisitions in the technological area were incorporated; such acquisitions had not been exposed in previous Congresses.

15 Multidisciplinary Panels were held studying the more frequent sites in Cancer, with an approach to the problem that included biological and clinical aspects, and concentrating on the following areas: aetiology, epidemiology, pathology, prevention, early detection, education, treatment and results. Proferred Papers were presented as Workshops instead of the classical reading, as in this way they could be discussed fully by the participants. 66 Workshops were held, this being the first time that free communications were presented in this way in a UICC Congress.

The Programme also included 22 "Meet the Experts", 7 Informal Meetings and more than a hundred films.

METHODOLOGY

The methodology used for the development of the Meeting and to make the scientific works profitable, had some original features that we would like to mention.

The methodology used in Lectures, Panels and Symposia was the usual one utilized in previous Congresses and functions satisfactorily. Lectures lasted one hour each. Panels were seven hours long divided into two sessions, one in the morning and one in the afternoon. They had a Chairman and two Vice-chairmen (one for each session). Symposia were three hours long. They had a Chairman, a Vice-chairman and a Secretary

Of the 8164 registered members, many sent proferred papers of which over 2000 were presented. They were grouped in numbers of 20 or 25, according to the subject, and discussed in Workshops. The International Scientific Committee studied the abstracts of all the papers, and those which were finally approved were sent to the Chairman of the corresponding Workshop who, during the Workshop gave an introduction and commented on the more outstanding works. This was the first time such a method had been used in an UICC Cancer Congress.

"Meet the Experts" were two hours long, and facilitated the approach of young professionals to the most outstanding specialists. The congress was also the ideal place for an exchange of information between the specialists of different countries during the Informal Meetings. Also more than a hundred scientific films were shown.

The size of the task carried out in organising this Congress is reflected in some statistical data: More than 18,000 letters were sent to participants throughout the world; more than 2000 abstracts were published in the Proceedings of the Congress; more than 800 scientists were active participants of the various meetings.

There were 2246 papers presented at the Congress by 4620 authors from 80 countries.

The Programme lasted a total of 450 hours, and was divided into 170 scientific meetings where nearly all the subjects related to Oncology were discussed.

All the material gathered for the publication of these Proceedings has been taken from the original papers submitted by each author. The material has been arranged in 12 volumes, in various homogenous sections, which facilitates the reading of the most interesting individual chapters. Volume XII deals only with the abstracts of proffered papers submitted for Workshops and Special Meetings. The titles of each volume offer a clear view of the extended and multidisciplinary contents of this collection which we are sure will be frequently consulted in the scientific libraries

We are grateful to the individual authors for their valuable collaboration as they have enabled the publication of these Proceedings, and we are sure Pergamon Press was a perfect choice as the Publisher due to its responsibility and efficiency.

Argentina
March 1979

Dr Abel Canónico
Dr Roberto Estevez
Dr Reinaldo Chacon
Dr Solomon Barg

General Editors

Introduction

The use of cyto-active drugs continues to provide an alternative and effective therapy for a variety of different cancers. The basis of current chemotherapy has been derived from the empirical use of single agents, a combination of these agents with each other as well as with established methods of treatment such as surgery in humans. Any advance in this field, however, is dependent on a more detailed knowledge of their basic mode of action, the nature of a human tumour cell's intrinsic and acquired resistance to the drug, as well as to the relative sensitivity of the tumour to that of the important life-supporting systems of the bone marrow, gastro-intestinal tract, hepatic and renal function as well as to the central nervous system. The search for new synthetic drugs continues, but it is conspicuous from the papers presented at this conference that naturally occurring products maintain a dominant role. It is also clear that a much more concentrated effort is needed to unravel some of the fundamental mechanisms by which some cells survive and others succumb to the actions of these agents, before a really systematic approach to successful chemotherapy is possible.

B.W. FOX
March 1979

The Development of New Antitumour Agents

Modern Cancer Therapy Concepts

S. Eckhardt

National Institute of Oncology, Budapest, Hungary

ABSTRACT

Progress in clinical cancer therapy during the past four years can be analyzed by reviewing the following areas: specific cancer treatment methods such as surgery, radiotherapy, chemotherapy, immunotherapy and their combinations, monitoring of therapy by biological markers, supportive care of cancer patients, rehabilitation, clinical cancer data processing and cooperative trials. Development of cancer surgery is landmarked by the introduction of new techniques, progress in reconstruction surgery and revision of therapeutic principles. Advances in radiotherapy can be characterized by the following factors: availability of new radiation sources yielding heavy particles, improvement of techniques for tumour localization, progress in treatment planning, elaboration of methods for modification of radiosensitivity and application of radiotherapy within the frames of new multidisciplinary therapeutic concepts. Recent results achieved in cancer chemotherapy are: synthesis of new anticancer agents, revision of drugs not fully evaluated, procedures for facilitating drug uptake by target cell, methods for potentiation of antitumour drug effect, approaches for diminishing host toxicity. Monitoring of therapy by the use of biological markers became possible for several neoplastic diseases. Supportive care and rehabilitation of cancer patients is now in the focus of interest of many research groups. Data banks are established and cooperative trials are stimulated in order to increase the efficacy of the fight against cancer.

KEYWORDS

Surgery, radiotherapy, chemotherapy, immunotherapy, biological markers, supportive therapy, rehabilitation, data processing, cooperative trials.

Progress in clinical cancer therapy in the past twenty years is well reflected in the analysis of the topics of congress lectures of the international cancer congresses. In 1958 hormone therapy of breast cancer was in the focus of interest. In 1962 results obtained by the use of new high energy radiation sources landmarked the development. In 1966 cancer immunology offered a new therapeutic approach: immunotherapy. In 1970 data of experimental and clinical chemotherapy were reported at an explosion rate and the neccessity of a close link among basic and applied research was stressed.

In 1974, instead of congress lectures, ten conferences were organized out of which three were devoted to clinical therapy bearing the titles: /1/ Biological Concepts in Surgical Management of Cancer, /2/ Progress in Radiobiology and Radiotherapy, and /3/ Advances in Cancer Chemotherapy. In addition, the conference on tumour immunology offered one session to discuss the fundamentals of applied immunological control. Thus, at the last international cancer congress it became obvious, that clinical cancer therapy, as it is performed today, is based on a multidisciplinary approach consisting of several therapeutic modalities.

At the present congress, the critical review of the development in cancer therapy during the past four years has to consider the following areas for analysis: /1/ specific cancer treatment methods such as surgery, radiotherapy, chemotherapy, immunotherapy and their combinations, /2/ monitoring of therapy by biological markers, /3/ supportive care of cancer patients, /4/ rehabilitation, /5/ clinical cancer data processing and organizatory efforts for cooperative trials.

Among specific treatment modalities, surgery is the traditional approach, the basic goal of which is to remove as much tumour tissue as possible. Progress in cancer surgery in recent years is characterized by the introduction of new techniques, development of reconstruction surgery and revision of therapeutic principles. New techniques, such as criosurgery, became available not only for the excision of small and superficial neoplastic lesions, like premalignant and malignant tumours of certain regions of the skin or head and neck, but also for therapy of tumours more deeply located /e. g. prostate/. Hyperthermic local treatment of the neoplasm with simultaneous radiotherapy can make more tumours accessible for this type of therapy, especially in soft tissue sarcomas and melanomas of the extremities. Embolisation of highly vascularized tumours is another important development in cancer surgery. Its application proved to be successful in the removal of kidney tumours. Other fields of application are under evaluation in abdominal, thoracic and neuro-oncological areas. Laser beam surgery is a limited, but useful tool in the destruction of certain localized tumours, such as eye, skin, vulva, vagina etc.

Development in reconstruction surgery has set the goal to introduce organ transplantation into oncological surgery. Nevertheless, results so far achieved are purely investigational except for kidney. Some research groups became interested in the organ transplantation of liver, others in the replacement of pancreas or gut. Most of the findings are disappointing. On the other hand, techniques for substitution of gastrointestinal organs by neighbouring tissues /stomach, oesophagus, jejunum/ could be of substantial value to restore organ function. New plastic surgery approaches in the orofacial, breast and genital regions are greatly facilitated by the availability of new synthetic materials.

Far the most important development in cancer surgery seems to be the constant revision of traditional therapeutic principles. The concept of superradicality in removal of tumour mass is now reconsidered from the side of cost/benefit ratio. Many surgeons decline to perform aggressive surgical interventions which do not assure adequate quality of life for the patient. On the other hand, metastasis surgery, especially the removal of isolated tumour masses from brain, lung or liver seems to be justified to reduce tumour burden and increase the curative chances for radiation, chemo- and immunotherapy. Sometimes, surgical intervention after other treatment modalities might be curative as well, even in cases with disseminated cancer. A good example for such multidisciplinary approach is the successful "second look" operation of ovarian cancer after chemotherapy. In some instances, surgery might have a complex role within the multimodality therapeutic strategy. Staging laparotomy with or without splenectomy became integral part of many therapeutic programs proving its usefulness. In other instances, chemotherapy might have an important adjuvant role, especially in breast, testicular and osteogenic sarcoma patients. Positive findings, however, must be proved in the long-run and early overoptimistic views have to be strongly critized.

Progress in radiotherapy can be characterized by the following factors: /a/ availability of new radiation sources yielding heavy particles, /b/ improvement of techniques for tumour localization, /c/ progress in treatment planning, /d/ elaboration of methods for modification of radiosensitivity and /e/ application of radiotherapy within the frames of new multidisciplinary therapeutic concepts. Availability of new radiation sources yielding heavy particles have advantages which are physical and biological. The physical properties of high energy radiation are expressed in a better dose distribution by high linear energy transfer /LET/ while biologically the heavy particles possess lower oxygen enhancement ratio /OER/ than that of 250 kV photons traditionally used. The goal of any radiation therapy is to increase linear energy transfer and to have low oxygen enhancement ratio. Heavy particles under clinical or radiobiological investigation have the following characteristics:

TABLE 1 Progress in Radiotherapy of Cancer II.
Heavy Particles Under Investigation
/Data of Hall E., of Parker R., of Kaplan H./

Particles		LET /Ke V/μ/	OER
Photons /250 KV X-Ray/		2.0	$\approx$ 2.0 - 3.0
Neutrons		1. - 100.	$\approx$ 1.7
Helium ions		72.7	$\approx$ 1.7
Protons[x]		1. - 90.	$<$ 1.
Pi mesons[xx]		0.6 - 25.	1. - 5.
High energy nuclei	- boron ion	126.5	1.6 - 2.9
	- carbon ion	189.0	

[x]340 MeV: 0.3
2 MeV: 17.0
at Bragg peak: 90.0

[xx] $<$10 KeV/m = low
$>$10 KeV/m = high let

Neutron therapy possesses now a well established clinical past. Its dose distribution overpasses that of photon therapy. Proton and Pi meson therapy are in the initial phase of their clinical evaluation. High energy nuclei, such as boron or carbon ions seem to have extremely favourable radiobiological properties, however, for the time being, their clinical value is not assessed yet. One major task for the future development of radiotherapy is the critical review of the advantages of heavy particle radiation over those of telegamma or electron therapy. Another important field of application is the combination of radiations originated from various radiation sources. In this respect the simultaneous use of low and high energy radiation seems to be very promising.

Improvement of techniques for tumour localization was an important prerequisit for the development of radiotherapeutic programs. The elaboration of computer tomography opened a revolutionary era, especially for localizing brain, mediastinal, retroperitoneal, abdominal and pelvic tumour masses. Ultrasonography became another valuable aid yielding useful imaging of hepatic and gynecological tumours. The combination of traditional techniques, such as axial tomography, with the recently introduced methods substantially increased the possibility of exact localization of neoplastic processes.

Progress in treatment planning was highly facilitated by the availability of computer assisted dosimetry. This became especially essential when multiple radiation fields were planned and different radiation energy sources were used.

Methods for modification of radiosensitivity of tumour tissue are various. The potential use of protons and pi mesons with low OER is promising. The administration of radiosensitizer agents such as metronidazol, nitroimidazol, ICRF-159, nitrofurane, thiolated compounds or any antioxydants is an investigational approach of great interest. One major task for the future development of radiotherapy is the critical review of the advantages of heavy particle radiation over those of telegamma or electron therapy. The success of this multimodality treatment highly depends on the selectivity of the radiosensitizer agent. So far, none of the chemicals used is harmless to normal tissues and therefore, the research in this field continues. Nevertheless, compounds available at the present, can be clinically used with some success.

Increased oxygenation of tumour tissue by hyperbaric chambers is not a new radiosensitizing method, however, its combination with other techniques might increase therapeutic results. Local hyperthermia by microwave heating is another accessory method recommended by several authors for increasing radiosensitivity.

Although the application of compounds as radiosensitizers or synchronizing agents did not significantly improve the therapeutic results, however, it is still a promising field of investigation. The question whether effective radioprotection, by administration of chemicals, can be achieved, has still to be answered. Last but not least the rational use of radiotherapy in combination with other treatment methods brought outstanding clinical results. Therapeutic programs of stage III Hodgkin's disease or that of testicular tumours are favourable examples for such attempts.

Recent results achieved in cancer chemotherapy can be analyzed from the following aspects: /a/ synthesis of new anticancer agents, /b/ revision of drugs not fully evaluated, /c/ procedures for facilitating drug uptake by target cell, /d/ methods for potentiation of antitumour drug effect, /e/ approaches for diminishing host toxicity.

The synthesis of new anticancer agents was realized in both analog research and production of new molecules. Among analogs subjected to phase I-II trials they were nitrosoureas /chlorozotocin, GANU, RECNU, MENU/, hexitols /dibromodulcitol, galactitol/, and cyclophosphamid derivatives /ifosphamid, trophosphamid/. Most of them exerted antitumour activity upon solid tumours and some of them were also useful in case of resistance against other alkylating agents. In the group of antimetabolites Ftorafur /Furanydil-5-fluorouracil/ raised clinical interest due to its decreased toxicity. Moreover, a series of ARA-C analogs were subjected to clinical trials, among which cyclocytidine, anhydro-ara-5-fluorocytidine, N^4acyl-l-beta-d-arabinofuranosyl cytosine and deoxycoformicine showed antitumour activity. Plant alkaloid analogs were also synthetized and explored for antitumour effect in human malignant diseases. Epipodophyllotoxins /VM-26, VP-16213/ showed definite cytostatic activity and are now in clinical use. Among vinca derivatives vindesine and formyl-leurosine were intensively examined and found to be less toxic as the parent compounds. Trimethylcolchicine became also clinically studied in a limited extent. In order to decrease cardiotoxicity of Adriamycin a series of analogs was produced. AD-32 /N-Tri-fluoroacetyl-Adriamycin/ and Carminomycin were drugs of this category clinically used. Pilot studies are under way with some success. Bleomycin derivatives /Bleo-PEP, Bleo-5033/ also showed advantages in their toxicology data versus the parent compound.

New molecules are synthetized either on an empirical or a rational basis. In the former group many representatives of plant alkaloids can be found as a result of a special NCI program. Anguidine, Maytansine, Bruceantine and Homoharringtonine were drugs subjected to clinical trials showing antitumour activity in screening procedures. Among antibiotics a great amount of new products were developed. Most of them are in an investigational phase /Aclacinomycin, Neocarcinostatin/. Drugs having definite activity in a specific metabolic system were also synthetized. Examples are PALA /N-/phosphonacetyl/-L-Aspartate/ which is a transition state inhibitor of the reaction product in the de novo pyrimidine nucleotide synthetic pathway between carbamyl phosphate and L-aspartate. Similarly, pyrazofurin creates a profound block of de novo pyrimidine synthesis through inhibition of orotidylate decarboxylase. Thioguanosin is used as a purin analogue. Alanosin is an antagonist of l-aspartic acid and l-glutamic acid and its metabolite is inhibiting adenylsuccinate synthetase. Acetohydroxamic acid is a strong inhibitor of ribonucleotid reductase activity.

Other compounds with established clinical activity are platinum complexes and ICRF-159. A special class of substances is represented by the synthesis of estramustin phosphate and prednimustin. These drugs are hormones bearing alkylating radicals. Their selective antitumour activity and favourable toxicity data prove that clinically useful new molecules can be produced on a rational basis. A further interesting example of such type is the introduction of Tamoxyphen into clinical use. This drug, originally designed as an antioestrogen is mainly active if free estrogen receptors are present in the breast cancer cell. In general, it can be expected, that in the near future other types of cell receptors may specifically altered as well by newly synthetized molecules.

Revision of formerly rejected drugs is another important development in cancer chemotherapy. Such example is DDMP which due to its enormous toxicity was not accepted in the clinic, although it showed striking antitumour activity. Nevertheless this toxicity can completely be diminished by citrovorum factor rescue and therefore the drug is again subjected to clinical trial. Another example is cis--platinum-diethylene-diamin which had only limited use for a long time due to its serious nephrotoxicity. Since, however, this toxic manifestations might be overcome by forced diuresis, a renaissance of the application of this drug began, proving its high value in the treatment of solid tumours. A further approach is the inhibition of catabolization of the cytostatic agent. ARA-C for instance, is rapidly degradated by the enzyme pyrimidin nucleotid deaminase. Tetrahydrouridine, however, is capable to block the activity of this enzyme. Thus, potentiation of drug effect occurs due to prolongation of active ARA-C blood level. Another idea is the induction of enzyme activity responsible for producing active metabolite from the inactive parent compound. The effect of cyclophosphamide might be potentiated by liver microsomal enzyme inducers, since its cytostatic effect is dependent on the amount of active metabolite produced by these enzymes.

In many cases simply lack of interest is the reason for rejecting new drugs. This is usually the case when drugs show marginal activity in those tumours in which more effective compounds are still available. In this case, frequently, no further trials are initiated at a broad spectrum of tumours and important antitumour activity may escape for the observer. Therefore, some drugs are "revisited", especially those, the toxicity of which can be avoided by present new approaches. Sometimes, false positive and false negative interpretation of data hindered the objective evaluation of newly synthetized drugs. It is therefore essential to support recent efforts to complete the drug evaluation in all instances in which contradictory data are available.

Recently procedures for increasing drug uptake specifically by target cells were also designed. A series of liposome coated drugs were produced with the aim to "protect" the drug from metabolic degradation and to facilitate membrane penetration. Their clinical value has to be assessed. Hapten bound drugs were also synthetized and the conjugation of cytostatic agents to globulins reacting immunologically with target cell raised great hopes. This problem is, however, far from being solved.

Therapeutic synergism of cytostatics by appropriate timing is a further potentially useful method for increasing drug activity. It has been demonstrated by various authors, that 5-Fu and MTX may antagonize or potentiate their effect according to the order of their administration. Enzyme induction or inhibition have also peak values in timing which have to be elaborated for each cytostatic and potentiating drug. All these aspects have to be linked to cell kinetic considerations. Nevertheless, it should be mentioned, that despite of promising results in animal experiments, neither synchronization nor combination chemotherapy on a cell kinetic basis proved to be useful in clinical cancer therapy and most of the successful polychemotherapy programs were empirically designed.

Methods to overcome drug resistance offered very little to the clinician. Cross-over experiments with some drug combinations resulted in temporary improvement of the patient. It became also obvious, that pulse therapy rarely, maintenance regimens often cause drug resistance. True progress can only be seen in the treatment of acute leukemias, where the combination of antimetabolites on a biochemical basis brought promising results. Nucleotide pool determination in leukemic cells originated from untreated and treated leukemias may predict therapeutic response.

One of the most important efforts for potentiating drug effect is to design methods for diminishing drug toxicity. The realization of the "rescue" principle in clinical therapy is such an attempt of major importance. Although citrovorum factor "rescue" after high dose methotrexate is not a recent technique, nowadays its application has several new aspects. Optimum timing and dosage have been established, and 1.0 - 7.0 g of Methotrexate is now delivered to the patient without lethal toxicity in various chemotherapy resistant malignant tumours, such as head and neck cancer or osteogenic sarcoma. CF rescue of other compounds, like that of the above mentioned DDMP is now also under investigation. More recently thymidine rescue proved to be clinically successful. Since this compound is available in relatively limited amount, methods for potentiation of its effect were also initiated. Diazauracil, for instance, a compound capable to inhibit thymidine kinase activity, an enzyme responsible for thymidine degradation, potentiated thymidine effect. Attempts to reutilize thymidine from the urine brought also promising results.

Diminution of organ toxicity induced by cytostatic agents with the aid of various protective drugs was also studied carefully. The first step is to clarify the mechanism, by which toxicity develops. In case of Adriamycin induced cardiotoxicity this type of research resulted in the discovery that certain substances /coenzyme Q, carnitin/ might play a role in protecting heart tissue from anthracyclin toxicity. Similar studies are in progress to reveal the mechanism of intestinal toxicity caused by vinca alkaloids and some alkylating agents. If these investigations will be successful, a major breakthrough in cancer chemotherapy can be expected. It is also imperative to continue to study bone marrow protection by all possible means. In this respect, however, the results so far achieved did not meet the expectations. As a conclusion, in the future we have to concentrate our efforts in this field, perhaps by learning more on regulation of hemopoietic stem cells.

The role and usefulness of cancer immunotherapy became widely discussed in the past four years. Although most of the authors dealing with this therapeutic modality are in agreement that immune restoration of the immune deficiency - induced in cancer patient by either immunosuppressive therapy or the malignant disease itself - is an essential task, their efforts to do so offered controversial results.

Attempts for immune restoration might be classified as follows: /a/ non-specific /systemic or local/ methods, /b/ active, tumour-specific methods, /c/ adoptive immune therapy, /d/ combination of various approaches.

In recent years non-specific systemic immunotherapy of leukemias was extensively studied. Table 2 is summarizing the results obtained by various methods.

TABLE 2 Progress in Cancer Immunotherapy I. Leukemias

Disease	Therapy	No of studies with survival and/or remission prolongation	No effect
ALL	BCG Pasteur	-	1
	Tice	-	1
	Glaxo	-	1
	BCG Pasteur + Allog. cells	1	2
AML	BCG	3	1
	BCG + Allog. cells	4	1
	MER	2	-
	MER + Allog. cells	1	-
	Pseudom. vac.	1	-
	Viral oncolys.	-	1
	Poly IC	-	1
CML	BCG + Allog. cells	1	-

It can be seen that in AML both survival and remission prolongation could be achieved by various methods, while in ALL or CML results were not convincing yet.

Table 3 summarizes the findings obtained in lymphomas.

TABLE 3 Progress in Cancer Immunotherapy II. Lymphomas

Disease	Therapy	No of studies with survival and/or remission prolongation	No effect
Hodgkin D.			
I-III	BCG-Tice	1	-
IV		1	-
Non-H.D.	BCG-Tice	1	-
	Levamisol		1
Burkitt L.	BCG-Pasteur		1

Although there is certain evidence that in Hodgkin's disease immunotherapy might be useful, further trials are needed to establish its value. In other forms of lymphomas data are scarce and not promising yet.

For solid tumours results are controversial as it can be seen in Table 4.

TABLE 4 Progress in Cancer Immunotherapy III. Solid tumours

Disease	Therapy	Effect	No of trials
Melanoma	BCG	−	4
	BCG + Allog. cell	∓	2
	Local BCG	+	3
Breast cc.	BCG	−	2
	Corynebact. parvum	−	2
	Local BCG	∓	2
Lung cc.	BCG aerosol	?	2
	BCG intrapleur.	?	2
	BCG	±	5
Osteosarcomas	Viral oncolysates	−	1
	BCG	?	2

Local BCG treatment is the only method which has definite clinical value. Unfortunately, hematogenous dissemination cannot be controlled. Various methods in several other malignancies did not yield satisfactory results.

Passive, adoptive immunotherapy has new and traditional ways. Administration of transfer factor or immune RNA are spectaculous techniques, however, without convincing success. The immune restoration by syngeneic cells is a logical effort with limited results.

Some progress could be achieved in the field of synthesis of new immune modulatory substances. Molecules, from the class of antibiotics, like Bestatin or the synthetic glycolipids are new chemical structures. A series of analogs of Levamisol were also produced and several bacterial and viral products were tested. The main task in the research of microbiological products remained their chemical identification and standardization of their biological activity.

New concepts for correct dosage and timing of immunotherapy became available. It is now evident, that immunotherapy has its limitations. Among them far the most important is that its efficacy becomes manifest, only in the presence of residual tumour masses. Another significant statement is that immunotherapy should always be used as an adjuvant method. In these cases, however, the timing of its optimum application has to be clearly defined. It is also of paramount importance to know whether the dosage in a given case would not cause enhancement phenomena. Research in this field seems to be very promising.

Attempts for elaborating methods to monitor success of therapy proved to be only partly successful. Detection of various types of tumour markers, follow-up of metabolic changes by biochemical and biophysical methods during therapy as well as the application of immunological techniques for detecting therapeutic response were the major efforts to be enumerated here. Tumour markers presently proposed for clinical use can be classified as /a/ oncofetal antigens, /b/ hormones associated with endocrine tumours, /c/ hormones associated with non-endocrine tumours, and /d/ enzymes produced in excess by tumour cells.

Oncofetal antigens clinically measured are enumerated in Table 5. Among them, only AFP and CEA proved their value in monitoring therapeutic results.

TABLE 5 Progress in Monitoring Therapy Tumour Markers
I. Oncofetal Antigens

AFP	liver
CEA	colon, testis, breast
Sulfoglycoproteids	stomach
Placental alk. ph.	prostate, ovary
Unnamed	lung, urothel., melanoma, sarcomas

It is evident from table 5 that most of the oncofetal antigens are under investigation. Methods for their standardization and specificity are urgently needed.

Hormones associated with endocrine tumours are commonly used for estimating success of therapy. The main types of hormones suitable for this purpose are listed in Table 6.

TABLE 6 Progress in Monitoring Therapy Tumour Markers
II. Hormones Associated with Endocrine Tumours

HCG	Trophoblastic tumours
Katecholamine	Neuroblastoma
HIAA	Carcinoid
Insulin	Insuloma

More recently hormones associated with non-endocrine tumours became subjected to detailed studies. A list of major research trends in this field is produced in Table 7.

TABLE 7 Progress in Monitoring Therapy Tumour Markers
III. Hormones Associated with Non-Endocrine Tumours

ACTH	lung
PTH	lung
Gastrin	stomach
Calcitonin	lung, breast
Glucagon	pancreas
Erythropoietin	kidney
Prostaglandin	kidney

Obviously most of these hormones are non-specific to tumour types, but specific to individuals. It is necessary to underline that these products might be responsible for the great majority of paraneoplastic syndromes. As an example, prostaglandin E production can be mentioned as a possible inducer of hypercalcemia, in tumour patients. Lung tumour patients frequently show a wide variety of paraneoplastic symptoms due to ectopic hormone production. In all these cases alterations in hormone levels may be considered as consequences of therapy.

Great interest is attached to the possible application of enzyme determinations for monitoring therapy. Enzymes produced in excess by tumour cells is a possible approach, while qualitative changes in the isozyme pattern is another. Until now, none of the proposed methods is sufficiently enough to ensure diagnostic accuracy, but many procedures seem to be suitable for predict prognosis and follow up success of therapy. A rather arbitrary list of actually proposed enzyme determinations can be seen in Table 8.

TABLE 8 Progress in Monitoring Therapy Tumour Markers IV. Enzymes

Muramidase	leukemia /AMOL/
Acid phosphatase	prostate
LDH isozyme pattern	stomach
RNASE	colon, prostate, ovary
β-hexosaminidase	colon
Arylsulphatase B	colon
Sialyltransferase	breast
Fucosyl transferase	breast

Detection of metabolic changes in the tumorous organism for monitoring results of therapy was recently also proposed. Measurement of katabolic products such as polyamines /spermidin, spermin, putrescin, cadaverin/ was initiated by various groups. In some leukemia patients positive correlation was observed among the amount of polyamines excreted by the urine and the leukocyte count. Other data did not support these observations. Several purin and pyrimidin metabolites were also subjects of study. None of them proved to be of interest for monitoring therapy so far. Metabolic products derived from collagen tissue destroyed by tumour cell proliferation were studied with great expectation. In this group of molecules hydroxyprolin was found to be of interest for monitoring therapy. Non--specific enzyme alterations as indicators of cell destruction were also investigated. LDH isozymes, GGTP or LAP are such examples for detection changes in hepatic tumour involvement. It is now obvious that such enzyme alterations are not sufficiently specific for follow up of therapeutic results. On the other hand alterations in plasma and urinary protein content seem to be highly useful for determining drug activity. Coeruloplasmin, haptoglobin, transferrin are plasma reactants clearly indicating therapeutic response in stage III-IV Hodgkin's disease. Determination of immunoglobulins is not only important for monitoring therapeutic success in myeloma patients, but it is of great use in other malignant diseases as well.

The possibility to achieve progress in monitoring therapy by detection of changes at cellular level was widely recognized in the past four years. Great interest was focused to the research of hormone receptors. Especially breast cancer patients were subjected to estrogen receptor determinations and it became soon probable that patients with free estrogen receptors have better prognosis, than those having no free binding capacity. It is still not clear, however, whether changes in estrogen binding capacity might be useful for monitoring therapy. More recently, hormone receptors were detected in ovarian and kidney cancer and in some patients with testicular tumours. The concept that receptor activity might predict and monitor success of therapy has to be verified in these tumours as well.

Vast experience was collected in regard of immunological cell surface characteristics of various tumour cells, especially of those of leukemias and lymphomas. As a result of the development of sophisticated techniques it is now possible to make clear distinction among various types of lymphomas and leukemias on a basis of their cellular immunological properties. It is, however, far from being clear whether these methods are useful for predicting and monitoring therapeutic success. This applies for melanoma and ovarian cancer as well.

Immunological reactivity of various blood cells, such as leukocytes or macrophages originated from tumorous organism arose great enthusiasm among research workers engaged in tumour diagnostics. The leukocyte adherence test, however, although capable to yield quantitative evaluation of immunological reactivity, seems not to be suitable for monitoring therapy. Migration inhibitory tests are also without significance in this respect. As research trends, some biophysical methods have to be also mentioned here. Among them, proton magnetic resonance spectral characteristics of tumour cells are under study to reveal specific differences which can be used as diagnostic, prognostic and monitoring factors. For the time being, this type of research has purely theoretical but no practical value yet.

Progress in supportive care of cancer patients is explained by the great number of important discoveries in both technical and biological research. Cell separators, developed earlier but brought to perfection more recently, are available for subtituting cell loss due to the malignant disease or its therapy. Development in criopreservative techniques facilitated such efforts. Advances in anti-infectious therapy by the synthesis of antibiotics and antimycotics with broad spectrum of activity were most important in controlling otherwise lethal infections. It has to be stressed, that according to various authors, 30-40 % of cancer patients have fatal outcome due to infections. Sterile or semisterile environment also improved the chances of survival. Study of the nutritional support of cancer patient was not so long time ago a fairly neglected area for research. It is, however, of great importance to reveal factors responsible for the development of cachexia. More recently, some progress was made in this field. The assumption that anorexigenic peptides derive from the pituitary gland became supported by experimental evidence. Studies related to hyperalimentation of cancer patients were also indicating to the presence of hormonal factors responsible for the metabolic imbalance. Caloric substitution became now possible, however, restoring the metabolic balance in a disseminated cancer patient can not be accomplished yet. Paraneoplastic manifestations of various type might be normalized. Prostaglandin induced hypercalcemia can be inhibited by mithramycin or indomethacin. Prolactin production in excess might be controlled by bromoergocriptine. Last but not least analgesia and psychic care is now assured by the great variety of modern analgesics and psychostimulant drugs.

Progress in rehabilitation of cancer patient is due to the fact that physical rehabilitation became an integral part of therapy facilitated by availability of new synthetic materials, advances in reconstruction surgery and construction of up-to-date technical devices substituting organ functions. Psychological rehabilitation became also integrated with the long-term follow-up of patients. Attention was directed to the patient's individual, family and professional life. The urgent need for social rehabilitation was recognized.

After reviewing the enormous amount of valuable clinical research data emerging from the institutes, hospitals and centers from all over the world it becomes evident that improvement of clinical cancer data processing is a basic need for further progress. Among requirements to be fulfilled in this field, the first task was standardisation of the most important input data. WHO by establishing International Classification of Diseases in Oncology and standardizing histology of tumours, or UICC creating the TNM system greatly contributed to solving this problem. In order to assure up-to-date storage of data and their quality UICC designed and implemented a program to collect patient data on the basis of an internationally standardized data collection system. For this reason data collecting centers were established in Bruxelles and Houston which are closely linked to eachother. The completion of this program will assure a regular, open data service for users for evaluation of end results and therapeutic strategies and will assist in program planning. It will also strongly promote clinical cancer research in the whole international scientific community.

Despite of all advances cancer is still a dreadful disease destroying individual's life and representing a heavy burden for mankind. It is therefore an urgent need for unlimited international collaboration for exchanging observations and concepts and to learn from eachother. Therefore, at the end of this lecture I feel privileged to ask you to devote great appreciation to those thousands of colleagues who contributed to the success in clinical cancer research so far achieved, and to those millions of patients whose suffrance, cure or death taught us how to make progress in therapy.

Perspectives in the Research of New Anticancer Agents

A. Di Marco

Division of Experimental Oncology B, Istituto Nazionale Tumori, Milano, Italy

The empirical approach to the pharmacological therapy of cancer is certainly the most ancient; south american and african sciamani employed in fact plant extracts for antitumor therapy long before the western science acknowledge an imbalance in regulation of cellular proliferative activity as the starting point of cancer. Some of these plant extracts have been recently reconsidered as sources of powerful antitumor agents such as the Maytansine (Kupchan, 1976) to be added to the list of drugs, such as Colchicine and Vinca alkaloids with similar origin. Dependent from the observation, the leucopenic effect, was also the introduction in cancer chemotherapy of nitrogen mustard (Gilman, 1946; Goodman, 1946) after developed to the modern alkilating agents. The origin of powerful antitumor drugs, such as 5-Fluorouracyl, Metotrexate, may be traced to the theory of the competition of synthetic substances with natural substracts, or coenzymes, of essential biochemical functions (Wood, 1940; Fildes, 1940). The drug kills the bacterial cells, and saves the mammalian ones owing to the lack in these of the biosynthetic pathway which is sensitive to the inhibition; this is unfortunately not replicated with the antitumor agents due to the identical pathways of nucleic acid synthesis from small precursors in normal and cancer cells.

When Actinomycin was discovered by Waksman and Woodruff (1940) and its antitumor effect recognized (Waksman, 1940) many had the feeling that this empirical approach (so fruitful in antibacterial therapy) could lead to agents active in cancer therapy. It was soon realized that this substance interferes with cellular activity in a very peculiar manner binding to DNA to inhibit its ability to function as a template in the just discovered process of transcription of the genetic information.

Many others microbial products were subsequently isolated and a few such as Mitomycin, the Bleomycins and the Anthracyclines recognized to have clinical usefulness.

Their antiproliferative effect is generally understood on the basis of the damage to DNA, or, as concerns the anthracyclines, to the impairment in his template activity in the process of replication and transcription (Di Marco, 1967, 1975, 1978). A selective effect on the genetic material could arise from a preferential binding of the antibiotic to a specific base sequence or as a consequence of a peculiar

feature in the structure of the chromatin. An high frequency of these base pairs in repetitive sequences with regulatory functions could give some hints to explain a preferential inhibition of some polymerase system such as the MSV polymerase (Zunino, 1974).
In the in vivo situation however nuclear proteins should considerably influence the interaction of the intercalating agents with the native chromatin structure. In parallel with the observed reduction in number of binding sites it is possible that histones, and most likely non histone nuclear proteins (NHNP) control a specific binding to DNA sequences with a regulatory function.
A question that arised soon after the discovery of the antimicrobial antibiotics was: what is their function in the producing organism? It appears now that some of them such as gramicidin (and tyrocidins) may have a specific regulatory function as inhibitors of the transcription initiation during the transition from vegetative growth to sporulation (Paulus, 1976).
The mechanism of the action would consist in the inhibition of the formation of the complex between RNA polymerase and DNA, and in the increase in its dissociation rate. A similar mechanism appears to work in the inhibiting effect of Actinomycin D on the transcription in the producing organism (Yoshida, 1966).
May these antibiotics have some specific effects on the regulation of gene expression in eukaryotic cells?
It is amazing that Actinomycin D, when administered to cultured cells in very low doses, has a paradoxycal enhancing effect on RNA synthesis. As suggested by Tomkins (1972) this may be due to the inhibitory effect of the antibiotic on the transcription of regulatory DNA sequences.
In the Britten and Davidson model (1969) regulatory RNA molecules, or their translation products, codyfied by integrator genes, are supposed to interact with receptor DNA sequences contiguous to each producer gene of a battery of functionally related genes (Fig. 1). Since the inhibited transcription by Actinomycin D of the regulatory gene causes an enhancement in total RNA synthesis it should be deduced that the physiological function of integrator RNA or the corresponding protein is to repress the transcription of a series of integrator genes (Fig. 2).
Following the developmental model of Davidson and Britten (1971) this effect of Actinomycin D may be understood as the suppression of a regulatory feedback loop (Fig. 3) between different integrator gene sites.
Since the effect on regulatory genes activity is observed at lower concentration than on structural ones the drug appears to have an higher affinity for the structure of the chromatin of the regulatory gene.
There is some evidence that a similar negative control is working in the regulation of the cell cycle; it was observed in fact that the number of mitotic figures in epydermal cells after wounding is reduced by translation inhibitors but increased by Actinomycin D (Gelfant, 1972). The antibiotic effect would consist in the inhibition in the synthesis of a regulatory RNA codyfing a specific repressor (chalone) of the transition between G_2 and M.
These effects of Actinomycin and of the anthracycline antibiotics suggest an interference with regulatory mechanisms of cellular proliferation also in eukaryotic cells. However, at variance of the effect reported previously on the producing organism, this effect is not specifically directed to the regulation of a physiological function and brings about a derangement in the macromolecular synthesis. This is certainly related with the antitumor activity but may also be so deep to

lead to an unreversible change of the genotype. The knowledge of the precise interference of the drug with the genomic material may therefore contribute to discover the nature of the changes in the mechanisms of the regulation of gene expression which are at the basis of the malignant transformation and eventually lead to agents selectively acting on these processes.

REFERENCES

Britten, R. J., and E. H. Davidson (1969). Genetic regulation in higher cells: a theory. Science, 165, 349-357.

Davidson, E. H., and R. J. Britten (1971). Note on the control of gene expression during development. J. Theor. Biol., 32, 123-130.

Di Marco, A. (1967). Daunomycin and related antibiotics. In D. Gottlieb and P. D. Shaw (Eds.), Antibiotics, Vol. 1, Springer Verlag, Berlin. pp. 190-210.

Di Marco, A., F. Arcamone, and F. Zunino (1975). Daunomycin (daunorubicin) and adriamycin and structural analogues activity and mechanism of action. In D. Gottlieb and P.D. Shaw (Eds.), Antibiotics, Vol. 3, Springer Verlag, Berlin. pp. 101-128.

Di Marco, A. (1978). Mechanism of action and mechanism of resistance to antineoplastic agents that bind to DNA. In F. M. Shabel (Ed.), Antibiotics and Chemotherapy, Vol. 23, Karger, Basel. pp. 216-227.

Fildes, F. (1940). A rational approach to research in chemotherapy. Lancet, 1, 955-957.

Gelfant, S., and G. Candelas (1972). Regulation of epydermal mitosis. J. Invest. Dermatol., 59, 7-12.

Gilman, A., and F. S. Philips (1946). The biological actions and therapeutic applications of the -chlorethylamines and sulfides. Science, 103, 409-415.

Goodman, L. S., M. M. Wintrobe, W. Dameshek, M. J. Goodman, A. Gilman, and M. McLennan (1946). Nitrogen mustard therapy: use of methylbis (-chlorethyl) amino hydrochloride for Hodgkin's disease, lymphosarcoma, leukemia and certain allied and miscellaneous disorders. J. Am. Med. Ass, 132, 126-132.

Kupchan, S. M. (1976). Novel plant-derived tumor inhibitors and their mechanisms of action. Cancer Treat. Rep., 60, 1115-1126.

Paulus, H., and H. Sarkar (1976). Molecular mechanisms in the control of gene expression. Academic Press, pp. 177-194.

Tomkins, G.M., B. B. Levinson, J. D. Baxter, and L. Dethlefsen (1972). Further evidence for posttranscriptional control of inducible tyrosine aminotransferase synthesis in cultured hepatoma cells. Nature New Biol., 239, 9-14.

Waksman, S. A., and H. B. Woodruff (1940). Bacteriostatic and bactericidal substances produced by a soil actinomyces. Proc. Soc. Exp. Biol. Med., 45, 609

Wood, D. D. (1940). Relation of p-aminobenzoic acid to mechanism of action of sulphanilamide. Br. J. Exp. Pathol., 21, 74-90.

Yoshida, T., H. Weissenbach, and E. Kate (1966). Inhibitory effect of actinomycin upon producing organism. Arch. Biochem. Biophys., 114, 252

Zunino, F., A. Di Marco, A. Zaccara, and G. Luoni (1974). The inhibition of RNA polymerase by daunomycin. Chem.-Biol. Interactions, 9, 25-36.

Anthracyclines: New Developments

Federico Arcamone

Farmitalia, Ricerca Chimica, Via dei Gracchi No. 35, 20146 Milano, Italy

ABSTRACT

The identification of DNA as the biological receptor for the anthracycline glycosides has been confirmed at different levels of biological organization and an irreversible structural damage of the receptor macromolecule rather than a reduction of the amount of DNA synthesized in the cell appears to be the basis of cytotoxic activity. Selectivity of action is likely related to differences in the uptake and in intracellular distribution of the drug, but other phenomena such as the binding to other biological polymers or the interaction with other cellular processes can also be invòlved in the pecualiar toxic effects of these drugs. The molecular requirements for biological activity of the anthracycline glycosides have been established. Stereochemistry of ring A is a strict requirement for bioactivity. On the other hand variations in the 9 substituent do not generally result in a loss of biological activity. Structural and/or stereochemical modifications in the sugar moiety may induce substantial variation in the biological efficacy of the drugs. The equatorial orientation of the 3'-aminogroup and the absence of hydroxyl substituent at C-2' and C-6' appear important conditions for the exhibition of high levels of activity, while different variations at C-4' do not affect the antitumor efficacy of the compounds. Modification at position 1 to 4 are also of interest, because the 4-demethoxyanalogues of daunomycin and adriamycin show outstanding antitumor activity at lower doses when compared with the parent compounds, and also because new analogs with methyl substituent at the said positionsretain substantial activity.

INTRODUCTION

Adriamycin (synonim doxorubicin) has been clinically used since 1969 and has contributed significantly to the effects achieved with cancer chemotherapy and to the present wide interest in this field. Major objectives of the research at the preclinical level are the understanding of the mode of action of the antitumor anthracyclines and the development of improved analogs endowed with more favourable pharmacological properties or with enlarged spectrum of antitumor activity.

MODE OF ACTION

Biological activity of antitumor anthracyclines has been related with their properties of complexing double stranded deoxyribonucleic acid (DNA) and with consequent inhibition of nucleic acid synthesis and function (Di Marco, Arcamone, and Zunino, 1974). These properties, indicating nuclear DNA as the biological receptor for the anthracycline glycosides, have been further investigated and confirmed in different systems and at different levels of biological organization. Adriamycin induced chromosome aberrations in human fibroblasts have been observed and related to the effects of the antibiotic on nucleic acid metabolism (Newsome and Gayle Littlefield, 1975). Degradation of preexisting DNA of murine L 1210 leukemia cells was induced by adriamycin (Lee and Byfield, 1976). Both daunorubicin (synonims daunomycin, rubidomycin) and adriamycin have been demonstrated to be highly potent in producing mutations and malignant transformation in mammalian cell systems in vitro in agreement with previous findings of animal studies (Marquardt, Philips, and Stenberg, 1976). DNA damage has been demonstrated as DNA breaks in P-288 ascite tumor cells in mice treated with daunorubicin and adriamycin (Schwartz, 1975), and nucleolar alterations of tumor cells were found upon treatment of rats bearing Novikoff hepatoma ascited tumor with adriamycin as well as with actinomyçin O (Smetana and others, 1977). A mechanism for strand scission of DNA by bound adriamycin and daunorubicin has been proposed (Lown and others, 1977). According to Di Marco (1977) the lethal event which is the basis of cytotoxicity is related to an irreversible damage of the DNA structure rather than to a quantitative reduction of the amount of DNA synthesized in the cell. However, action on DNA is not sufficient to explain the relative selectivity of the drugs against tumor cells, as well as the different efficacy of the drugs towards specific tumors and the development of specific resistance. These aspects of the biological activity are currently related to differences in the rate of uptake of drug in the target cell and to differences in the intracellular distribution of the drugs (Di Marco, 1977, and ref. cited herein; Chewinsky and Wang, 1976) but different immunodepressive activities could also play a role (Casazza and others, 1975).

It seems worthwhile to mention, at this point, that other biological macromolecules have been found to complex adriamycin. These are the acidic phospholipids such as cardiolipin (Duarte-Karim and others, 1976), the mucopolysaccharides heparin and chondroitin sulphate (Menozzi and Arcamone, 1978), and proteins such as nuclear nonhistone protein from rat liver (Kikuchi and Sato, 1976), tubulin from calf brain (Na and Timasheff, 1977), spectrin from human erythrocytes (Mikkelsen and others, 1977). The relevance of these phenomena as determinants of cytotoxic effects and uptake mechanism remains to be established. On the other hand other properties, such as an action of adriamycin at the membrane level of Sarcoma 180 ascites cells affecting the movement of molecules in the membrane itself and a consequent interference with cell proliferation (Murphree and others, 1976), free radical formation in the presence of rat liver microsomes or heart sarcosomes (Bachur and others, 1977), the redox cycling of the antibiotic in the presence of rat liver microsomes and the concomitant generation of free radicals and lipid peroxidation (Goodman and Hochstein, 1977), and the inhibition of Ca^{++} transport in mitochondria from Erlich ascites carcinoma cells (Anghileri, 1977), have been recently proposed as being possibly involved in the antitumor and/or toxic side-effects of the antitumor anthracyclines. Depression of DNA synthesis in cardiac tissue of mice treated with adriamycin was also demonstrated (Rosenoff and others, 1975). It is however not established if one or more of these phenomena are operating in practice and whether or not a strict relationship be present between the mechanism of the antitumor and cardiotoxic effects.

STRUCTURE-ACTIVITY RELATIONSHIP

For the purpose of a discussion of structure-activity relationship, as deduced from the studies carried out in our laboratory with the cooperation of the 'Istituto Nazionale per la Cura e lo Studio dei Tumori' of Milan and the National Cancer Institute, Drug Research and Development Division of Cancer Treatment, Bethesda, Md., USA, we shall first consider the results of modification on ring A and its substituents, and than those concerning modifications on ring D.

RING A CONFIGURATION

In the biosynthetic antitumor anthracyclines and their derivatives ring A is substituted at position 7 and 9, both representing centres of chilarity which are indicated as 7 (S), 9 (S). Table 1 shows the results obtained with analogs differing in the stereochemistry of ring A (Penco and others, 1977; Arcamone and others, 1976; Arcamone, 1977). The 9 (R) stereoisomer Ib is more toxic and less active than the 9 (S) analogue Ia, but when the absolute configuration ad C-7 is also inverted, as in the 7 (R), 9 (R) analogues, biological activity disappears. The pivotal nature of the C-7 asymmetric centre is clearly evident in the three dimensional representation of the anthracycline structure as derived by X-ray diffraction (Neidle and Taylor, 1977), stressing the importance of the absolute configuration at C-7 as a determinant of the topological relationships among the different atomic groups in the anthracycline glycosides.

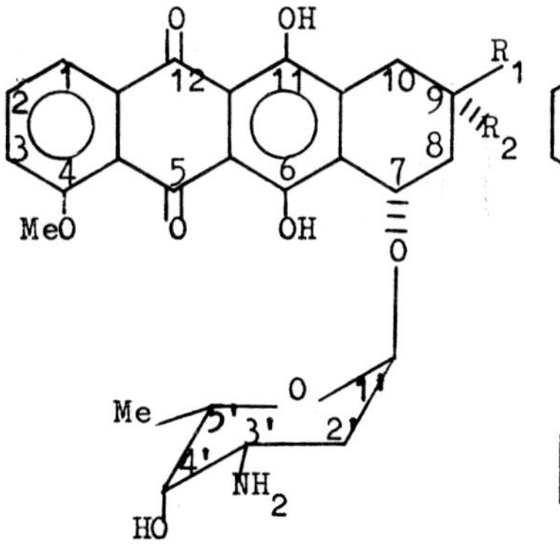

Ia,b

a : R^1 = H, R^2 = OH

b : R^1 = OH, R^2 = H

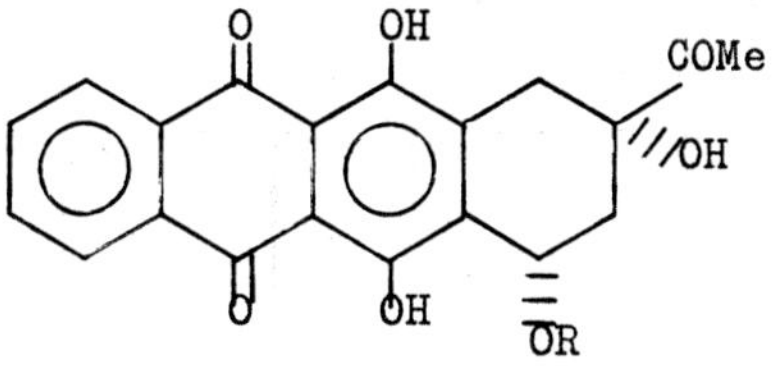

IIa,b

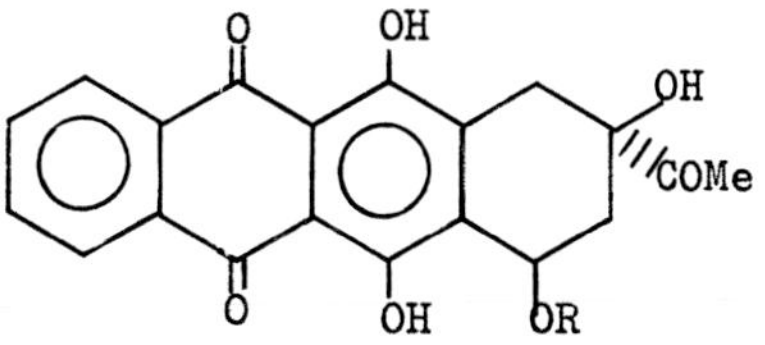

IIIa,b

a : R = α } – daunosaminyl

b : R = β }

TABLE 1. Affinity for DNA Antitumor activity of Analogues with Different Ring A Configurations

Compound	Absolute configuration	Kapp [a]	O.D. [b]	T/C % [c]
Ia	7(S), 9(S)	$2.2\ 10^5$	12.5	228 [d]
Ib	7(S), 9(R)	$0.8\ 10^5$	6.5	171 [d]
IIa	7(S), 9(S)		0.5	150 [e]
IIb	7(S), 9(S)		8.0	150 [e]
IIIa	7(R), 9(R)		8.0 [f]	100 [e]
IIIb	7(R), 9(R)		10.0 [f]	100 [e]

a) Stability constant of DNA complex.
b) Optimal dose mg/kg (ip, day 1).
c) Average survival time as percent of control mice.
d) P 388 experimental leukemia.
e) L 1210 experimental leukemia.
f) Max. dose tested.

THE 9β SUBSTITUENT

When the stereochemistry is not modified, variations in the 9β substituent do not generally result in a loss of biological activity. Variations shown in Table 2 correspond to a wide range of different structures for the 9β substituent and antitumor activity at reasonable dose levels is a characteristic of this group of analogues, although none appears clearly superior to adriamycin.

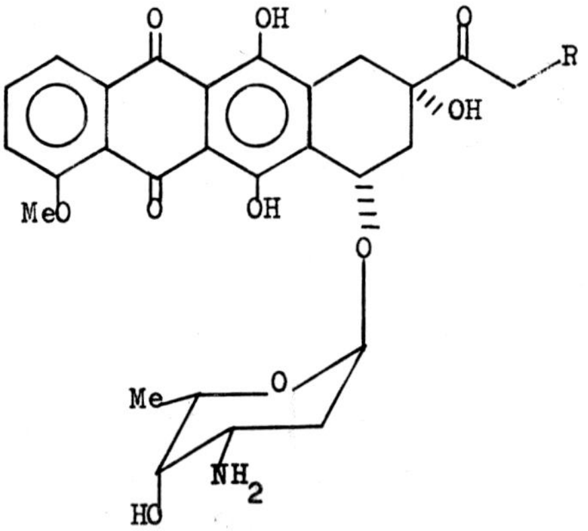

IV a-b

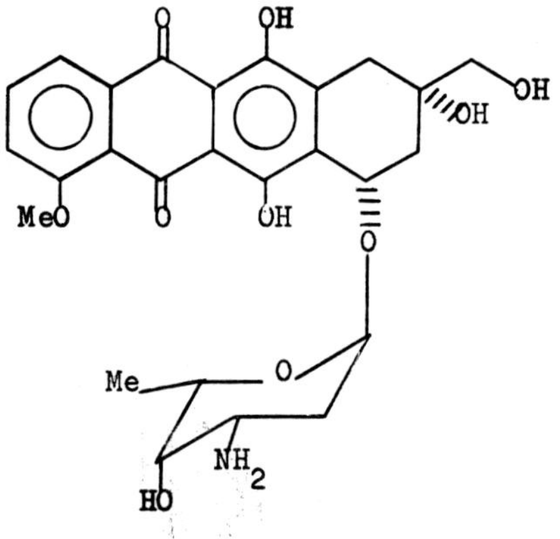

V

TABLE 2. Effect of the 9 Substituent on Antitumor Efficacy on P 388 Experimental Murine Leukemia [a,b].

Compound		QD 1-9		Q4 D 5,9,13	
		O.D.	T/C %	O.D.	T/C %
IVa	R = H (daunorubicin)	1.14	161	10.66	134
b	OH (adriamycin)	1.11	209	7.6	168
c	N(morpholino)O	10.4	170	-	-
d	$OCOCH_2OCO(CH_2)_{16}Me$	3.13	201	20.0	170
e	$OCOCH_2SCO(CH_2)_4Me$	3.13	201	37.5	160
f	SCOMe	9.4	145	-	-
g	SCOPh	3.13	141	-	-
h	OMe	-	-	31.25	125
i	OEt	3.13	160	-	-
j	OPh	6.25	180	-	-
V		1.56	175	10.5	136
Ia		12.5	231	-	-

a, b) NCI data. See footnotes to Table 1.

Substantial activity was already reported in the P 388 system for the simple 14--esters of adriamycin, among which the octanoate, the nicotinate, and the phenyl-acetate displayed outstanding antitumor activity (Goldin and Johnson, 1975). It should also be mentioned that the octanoate appeared to be able to form a complex with DNA (Arlandini and others, 1977) and also to be hydrolyzed in vivo by tissue esterases (Arcamone and others, 1974).

NEW SUGAR MOIETIES

Structural and/or stereochemical modifications in the sugar moiety have already been shown to involve substantial variation in the biological activity of the anti-tumor anthracyclines (Arcamone, 1977). The presence of an additional hydroxyl group at C-2' or at C-6' induces a moderate reduction of the affinity for DNA and of the inhibition of DNA dependent RNA and DNA polymerases in vitro, but the bio-logical activity appears to be strongly affected by the same modifications. A lower uptake of these derivatives in the cells, also accompanied by a lower activ-ity in cultured cell system, has been related with this effect (Di Marco and others, 1977). Table 3 shows the consequences of stereochemical modifications on affinity for DNA (Arlandini and others, 1977, and others results from Author's laboratory) and antitumor efficacy (Arcamone, 1977, Di Marco, personal communication). Within this series a strict relationship between the two properties can be observed. The axial orientation of the aminogroup is unfavourable for DNA complexation and for expression of activity, the L-xylo isomer appearing practically inactive. The lower affinity for DNA should however not be the only factor determining the loss of biological activity, because, as shown above for compounds Ia and Ib, the lower value of the association constant is not incompatible with the expression of anti-mitotic properties. It should be deduced, therefore, that the different absolute configuration of the sugar residue which destabilizes the DNA complex be also un-favourable for other interactions, as for instance with membrane components determining the rate of uptake into the target cells.

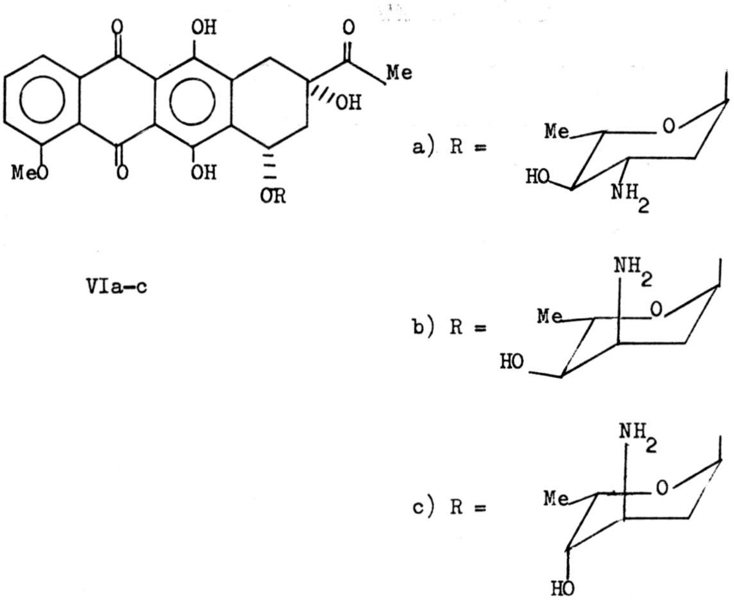

VIa-c

a) R =

b) R =

c) R =

TABLE 3. Configurational Analogues : Affinity for DNA and Antitumor activity on L 1210 leukemia in mice [a].

Compound	Stereochemistry	Kapp x 10^{-5}	O.D.	T/C %
IVa	L-lyxo	4.5	4	150
VIa	L-arabino	3.8	4	143
VIb	L-ribo	2.6	50	137
VIc	L-xylo	1.3	50	100

a) See footnotes to Table 1.

Full activity was retained in the L-arabino compound and for this reason other analogues showing a single modification at C-4', were synthetised and tested (Arcamone, 1977; Cassinelli, Ruggieri and Arcamone, 1978). Three new such derivatives in the adriamycin series are shown in Table 4 together with the stability constant of the DNA complex (Arlandini and others, 1977, and other results from Author's laboratory), and the activity on L 1210 leukemia. Compound VIIa, 4'-epiadriamycin, showed a behaviour very similar to that of adriamycin. This has been confirmed in different experimental tumors, and it has also been found that VIIa dispalys a lower toxicity when compared with I (Arcamone, Di Marco, and Casazza, 1977). Also remarkable are compounds VIIb and VIIc, the latter showing an outstanding efficacy on L 1210 leukemia, a tumor system which is somewhat resistant to adriamycin and its known derivatives. This finding opens the perspective that adriamycin analogs with enlarged spectrum of activity can be developed by performing further suitable chemical modifications in the said position of the aminosugar moiety.

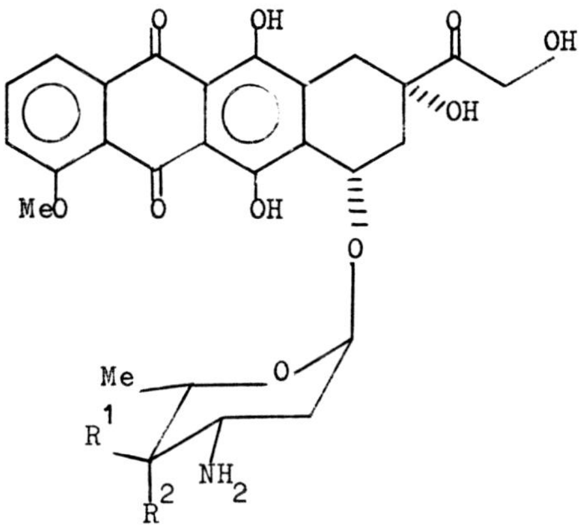

VIIa $R^1 = OH, R^2 = H$

VIIb $R^1 = R^2 = H$

VIIc $R^1 = H, R^2 = OMe$

TABLE 4. Effect of Modifications at C-4' on Affinity for DNA and on Antitumor Activity on L 1210 Leukemia in Mice[a].

Compound	Kapp x 10^{-5}	O.D.	T/C %
IVb	3.7	5	166
VIIa	3.6	5	150
VIIb	4.4	4	177
VIIc	3.0	4.4	287

a) See footnotes to Table 1.

RING D SUBSTITUTION

The 4-demethoxyanalogues of daunorubicin have already been mentioned. Further modifications of ring D involved substitution at C-1 and C-4, or at C-2 and C-3. As shown in Table 5, the methylated analogues VIIIa and VIIIc retain substantial *in vivo* activity and the same is found for the pentacyclic analog VIIIe (Arcamone and others, 1978). The 4-demethoxy series appears to be important also because of the interesting properties of 4-demethoxydaunomycin IIa and of the high antitumor efficacy of the corresponding adriamycin analogs. As shown in Table 5 the latter compounds retain the full activity of adriamycin at lower optimal doses (4-demethoxyadriamycin VIIIf and 4-demethoxy-4'-epiadriamycin VIIIh) or even a enhanced efficacy (4-demethoxy-2,3-dimethyladriamycin VIIIg) (Di Marco and others, 1978; Arcamone and others, 1978).

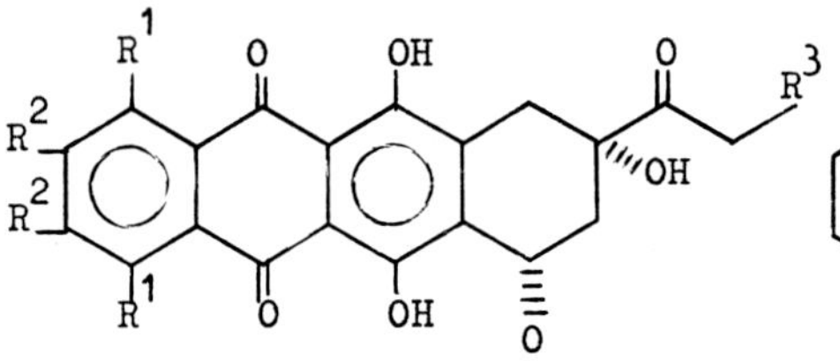

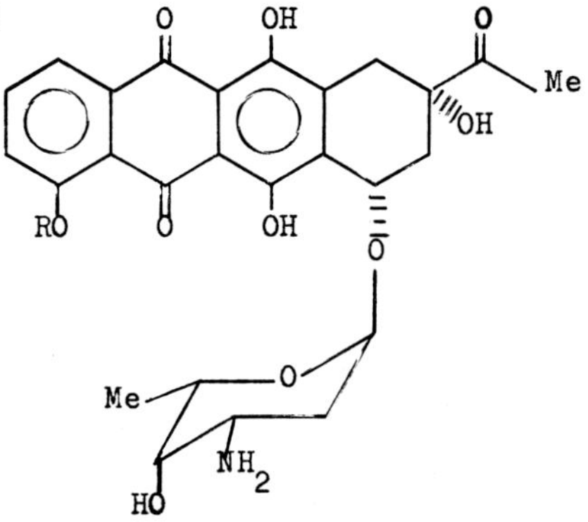

	R^1	R^2	R^3
VIIIa	H	Me	H
VIIIb	H	Cl	H
VIIIc	Me	H	H
VIIId	Cl	H	H
VIIIe	H	(*)	H
VIIIf	H	H	OH
VIIIg	H	Me	OH
VIIIh**	H	H	OH

(*) R_2,R_2

** 4'-epi

IXa : R = Et
IXb : R = iPr
IXc : R = $PhCH_2$
IXd : R = cyclohexyl

TABLE 5. Effect of Ring D Substitution on Antitumor Activity[a]

Compound	O.D.	T/C %
IIa	1.0	150[b]
VIIIa	1.25	131[b]
VIIIb	33.7	111[b]
VIIIc	6.6	147[b]
VIIId	20.0	116[b]
VIIIe	10.0	135[b]
VIIIf	0.5	166[b]
VIIIg	4.4	173[b]
VIIIh	0.5	166[b]
IXa	25.0	134[c]
IXb	18.75	161[c]
IXc	50.0	155[c]
IXd	50.0	122[c]

a) See footnotes to Table 1.
b) L 1210 murine experimental leukemia.
c) P 388 murine experimental leukemia.

The C-4 methoxyl group is a typical substituent of clinically useful anthracycline and their strictly related biosynthetic derivatives, but is absent in all other known anthracyclines of microbial origin which always display a C-4 hydroxyl (Thom son, 1971). The antibiotic carminomycin (4-O-demethyldaunomycin) has been found t exibit activity at lower doses than the related daunorubicin, from which it differs only for the presence of an hydroxyl group at C-4 in place of the methoxyl group in daunorubicin (Wani and others, 1975; Pettit and others, 1975). It should be noted that the presence at C-4 of an alkoxy group different than the methoxyl does not abolish the biological activity, and the antitumor efficacy of IXb and IXc in the P 388 system appears to be quite high (Bernardi and others, 1978).

CARDIOTOXICITY OF THE NEW ANALOGS

A particularly serious side-effect of daunomycin and doxorubicin therapy is the development of a cardiomyopathy leading to congestive heart failure in approximately 1 % of patients. This side-effect, being related with the total cumulative dose, limits the amount and duration of drug therapy (Carter, 1975). Although the cardiomyopathy experimentally produced in rabbits and monkeys be considered similar to that occurring in treated cancer patients and therefore use of the said animal species be recommended for a definitive assessment of cardiac toxicity at the preclinical level, a screening system which requires small quantities of drug and which is useful for the screening of new synthetic anthracycline derivatives has been developed in the rat by Zbinden and Brandle (1975). According to this model, ECG changes following repeated administration of the compounds are coincident with other known alterations of heart tissue consequent to anthracycline treatment. The Authors have decided to use the animal cumulative dose required to produce a prolongation of the QRS duration to 22 min s as screening parameter for the preliminary classification of the cardiotoxic properties of new analogs (Zbinden, Brochmann, and Holderegger, 1978). Table 6 lists the minimal cumulative doses together with the antitumor efficacy on P 388 leukemia for different anthracycline derivatives.

TABLE 6. Minimal Cumulative Cardiotoxic Doses (MCCD) in Rats and Activities on P 388 Murine Leukemia of Anthracycline Derivatives.

NSC No.	Compound	MCCD[a,b]	Antitumor activity[c,d]	
			O.D.[a]	T/C %
82151	Daunorubicin	12-16	11.66	140
268708	Daunorubicin,9-deacetyl	33	75.00	156
256439	Daunorubicin,4-demethoxy	1.75	2.40	163
123127	Adriamycin	10-12	9.33	170
149584	Adriamycin,14-octanoate	20	10.00	178
149583	Adriamycin,14-nicotinate	32	16.00	169
261057	Adriamycin,14-stearoyl-glycolate	40	20.00	170
256942	Adriamycin,4'-epi	16	7.50	160
267469	Adriamycin,4'-deoxy	15	3.13	166
256438	Adriamycin,4'-demethoxy	3	1.00	175

a) mg/kg.
b) Data from G. Zbinden, ETH, Zurich.
c) Data from NCI (Screener A.D. Little).
d) q4d 5,9,13 schedule, ip.

Optimal doses in the P 388 tests also vary considerably, and are found in the range 0.2 to 30 mg/kg for the qd 1-9 schedule and in the range 1.0 to 75 mg/kg for the q4d 5,9,13 schedule. Inspection of the data shows that the ratio of MCCD to optimal dose varies widely, indicating that different cardiotoxic properties such as, for instance, uptake in rat heart tissue, are associated to active doses of the different analogs. The compounds can therefore be ranked taking account for the above mentioned ratio and, accepting the comparison of data obtained in two different rodent species as inevitably dictated by the existing laboratory models, 4'-deoxyadriamycin, 4-demethoxyadriamycin, and 4'-epiadriamycin appear promising analogs with potentially reduced cardiotoxicity. The results also suggest that the clinically important side-effect represented by cardiotoxicity is likely to be overcome in the new chemically originated analogs.

Aknowledgements. Part of the investigations summarized in this report were supported by the National Cancer Institute, N.I.H. (Contracts NO1-CM57014 and NO1-CM 57036). Cardiotoxicity data were kindly provided by Dr. G. Zbinden, ETH, Zürich, and activity on L 1210 leukemia was tested by Prof. A. Di Marco, A.M. Casazza and their coworkers at Istituto Nazionale per la Cura e lo Studio dei Tumori, Milan.

REFERENCES

Anghilieri, L. J. (1977). Ca-2+ -Transport inhibition by antitumor agent adriamycin and daunomycin. Arzneim.-Forsch., 27, 1177-1180.

Arcamone, F. (1977). New antitumor anthracyclines. Lloydia, 40, 45-66.

Arcamone, F., L. Bernardi, B. Patelli, P. Giardino, A. Di Marco, A.M. Casazza, C. Soranzo, and G. Pratesi (1978). Synthesis and antitumour activity of new daunorubicin and adriamycin analogues. Experientia. Paper accepted for publication.

Arcamone, F., L. Bernardi, P. Giardino, B. Patelli, A. Di Marco, A. M. Casazza, G. Pratesi, and P. Reggiani (1976). Synthesis and antitumor activity of 4-demethoxydaunorubicin, 4-demethoxy-7,9-diepidaunorubicin, and their beta anomers. Cancer Treat. Rep., 60, 829-834.

Arcamone, F., A. Di Marco, and A. M. Casazza (1977). Chemistry and pharmacology of new antitumor anthracyclines. Paper presented at the 8th Symposium of the Princess Takamatsu, Cancer Research Foundation, Tokyo (Nov. 15-17, 1977).

Arcamone, F., G. Franceschi, A. Minghetti, S. Penco, S. Redaelli, A. Di Marco, A.M. Casazza, T. Dasdia, G. Di Fronzo, F. Giuliani, L. Lenaz, A. Necco, and C. Soranzo (1974). Synthesis and biological evaluation of some 14-O-acyl derivatives of adriamycin. J. Med. Chem., 17, 335-337.

Arlandini, E., A. Vigevani and F. Arcamone (1977). Interaction of new derivatives of daunorubicin and adriamycin with DNA. Il Farmaco, Ed. Sci., 32, 315-323.

Bachur, N. R., S. L. Gordon, and M. U. Gee (1977). Anthracycline antibiotic augmentation of microsomal electron transport and free radical formation. Mol. Pharmacol., 13, 901-910.

Bernardi, L., P. Masi, A. Suarato, and F. Arcamone (1978). Germ. Offen 2,750,812 (May 18, 1978).

Carter, S. K. (1975). Adriamycin. A review. J. Nat. Cancer Inst., 55, 1265-1274.

Casazza, A.M., A.M. Isetta; F. Giuliani, and A. Di Marco (1975). In M. Staquet and others (Eds.), Adriamycin Review; Second Internat. Sympos., Brussels (May 16-18, 1974).

Cassinelli, G., D. Ruggieri, and F. Arcamone (1978). Synthesis and antitumor activity of 4'-O-methyldaunomycin, 4'-O-methyladriamycin and their 4'-epi-analogues. J. Med. Chem. Paper accepted for publication.

Cherwinsky, D.S. and J. J. Wang (1976). Uptake of adriamycin and daunomycin in L 1210 and human leukemia cells: A comparative study. J. Medicine, 7, 63-79.

Di Marco, A. (1978). In E. Schabel (Ed.), Antibiotics and Chemotherapy, Vol. 23. S. Karger, Basel. pp. 216-227.

Di Marco, A., F. Arcamone, and F. Zunino (1974). In J.W. Corcoran and F.E. Hahn (Eds.), Antibiotics, Vol. 3. Springer Verlag, Berlin. pp. 101-128.

Di Marco, A., A.M. Casazza, T. Dasdia, A. Necco, G. Pratesi, P. Rivolta, A. Velcich, A. Zaccara, and F. Zunino (1977). Changes of activity of daunorubicin, adriamycin and stereoisomers following the introduction or removal of hydroxyl groups in the amino sugar moiety. Chem. Biol. Interactions,19, 291-302.

Di Marco, A., A.M. Casazza, F. Giuliani, G. Pratesi, F. Arcamone, L. Bernardi, G. Franchi, P. Giardino, B. Patelli and S. Penco (1978). Synthesis and antitumor activity of 4-demethoxyadriamycin and 4-demethoxy-4'-epiadriamycin. Cancer Treat. Rep., 62, 375-380.

Duarte-Karim, M., J.M. Ruysschaert, and J. Hildebrand (1976). Affinity of adriamycin to phospholipds: A possible explanation for cardiac mitochondrial lesions. Biochem. Biophys. Res. Comm., 71, 658-663.

Goldin, A. and R. K. Johnson (1975). In M. Staquet and others (Eds.), Adriamycin Review, European Press Medikon, Ghent, Part I, pp. 37-54.

Goodman, J. and P. Hochstein (1977). Generation of free radicals and lipid peroxidation by redox cycling of adriamycin and daunomycin. Biochem. Biophys. Res.Comm., 77, 797-803.

Kikuchi, H. and S. Sato (1976). Binding of daunomycin to nonhistone proteins from rat liver. Bioch. Biophys. Acta, 434, 509-512.

Lee, Y. C. and J. E. Byfiled (1976). Brief Communication : Induction of DNA degradation in vivo by adriamycin. J. Nat. Cancer Inst., 57, 221-224.

Lown, J.W., S. Sirn, K. C. Majumdar, and R. Chang (1977). Strand scission of DNA by bound adriamycin and daunorubicin in the presence of reducing agents. Biochem. Biophys. Res. Comm., 76, 705-710.

Marquardt, H., F. S. Philips, and S.S. Sternberg (1976). Tumorigenicity in vivo and induction of malignant transformation and mutagenesis in cell cultures by adriamycin and daunomycin. Cancer Res., 36, 2065-2069.

Menozzi, M. and F. Arcamone (1978). Binding of adriamycin to sulphated mucopolysaccharides. Biochem. Biophys. Res. Comm., 80, 313-318.

Mikkelsen, R. B., P. Lin, and D.F.H. Wallach (1977). Interaction of adriamycin with human red blood cells - Biochemical and morphological study. J. Mol. Med., 2, 33-40.

Murphree, S.A., L.S. Cunninghan, K.M. Hwang, and A.C. Sartorelli (1976). Effects of adriamycin on surface properties of sarcoma 180 ascites cells. Biochem. Pharmacol., 25, 1227-1231.

Na, C. and S.N. Timasheff (1977). Physical-chemical study of daunomycin-tubulin interactions. Arch. Biochem. Biophys., 182, 147-154.

Neidle, S. and G. Taylor (1977). Nucleic acid binding drugs. Part IV. The crystal structure of the anti-cancer agent daunomycin. Biochim. Biophys. Acta, 479, 450-459.

Newsome, Y.L. and L. Gaile Littlefield (1975). Adriamycin-induced chromosoma abberations in human fibroblasts. J. Nat. Cancer Inst., 55, 1061-1063.

Penco, S., F. Angelucci, A. Vigevani, E. Arlandini, and F. Arcamone (1977). Stereo-chemical requirements in the antitumor anthracyclines. J. Antibiotic, 30, 764--766.

Pettit, G. R., J.J. Einck, C.L. Herald, R. H. Ode, R. B. Von Dreele, P. Brown, M. C. Brazhnikova, and G. F. Gause (1975). The structure of carminomycin I. J. Am. Chem. Soc., 97, 7387-7388.

Rosenoff, S. H., E. Broocks, F. Bostick, and R. C. Young (1975). Alterations in DNA synthesis in cardiac tissue induced by adriamycin in vivo. Relationship to fatal toxicity. Biochem. Pharmacol., 24, 1898-1901.

Schwartz, H.S. (1975). DNA breaks in P-288 tumor cells in mice after treatment with daunorubicin and adriamycin. Res. Comm. Chem. Path. Pharmacol., 10, 51-64.

Smetana, K., J. Merski, Y. Daskai, R. K. Busch, and H. Busch (1977). Effects of adriamycin and actinomycin D on nucleolar morphology: A simple biologic assay. Cancer Treat. Rep., 61, 1253-1257.

Thompson, R.H. (1971). In R.H. Thompson (Ed.), Naturally Occurring Quinones, Academic Press Inc., London and New York, Chap. 6, pp. 536-575.

Wani, M. C., H. L. Taylor, M. E. Wall, A. T. McPhail, and K. D. Onan (1975). Antitumor agents. XIII.Isolation and absolute configuration of carminomycin I from Streptosporangium sp. J. Am. Chem. Soc., 97, 5955-5956.

Zbinden, G. and E. Brändle (1975). Toxicology screening of daunorubicin (NSC--82151), adriamycin (NSC-123127), and their derivatives in rats. Cancer Chemotherapy Rep., 59, Part I, 707-715.

Zbinden, G., E. Bochmann, and C. Holderegger (1978). Model systems for cardiotoxic effects of anthracyclines. Antibiot. Chemother., 23, 255-270.

Bleomycin: New Developments

Hamao Umezawa

Institute of Microbial Chemistry, Kamiosaki 3-Chome, Shinagawa-ku, Tokyo

ABSTRACT

The β-lactam part of bleomycin structure was recently revised to the open amide structure and the structure of bleomycin metal complex was proposed. On the basis of the structure of bleomycin ferrous complex the molecular mechanism of action of bleomycin to cause DNA strand was proposed. Behavior of bleomycin has been elucidated and therapeutic effect of bleomycin to squamous cell carcinoma is due to its low content of bleomycin hydrolase.

Derivatives and bleomycin analogs have been studied and on the basis of a low pulmonary toxicity pepleomycin was selected as the one worth clinical study. Bleomycin has many functional groups, and various types of active derivatives are being studied. Bleomycin is not immunosuppressive and the bleomycin treatment in combination with a small molecular immune-enhancing agent such as bestatin *etc.* is under study.

INTRODUCTION

Since 1969, bleomycin has been used in treatment of squamous cell carcinoma and Hodgkin's disease. Bleomycin treatment in combination with a vinca alkaloid or a platinum compound or in combination with both of them has shown therapeutic effect against testis tumors. Analogs of bleomycin which have a lower pulmonary toxicity than the bleomycin clinically used at present have been developed and suggested to be more useful in cancer treatment. Recently there was an important progress in chemistry and biochemical mechanism of action of bleomycin. In this paper, the author will report on recent studies in his institute and the Research Institute of Nihon Kayaku Co.

RECENT PROGRESS IN CHEMISTRY AND MECHANISM OF ACTION

As in the general case of the biosynthesis of peptide antibiotics, it is suggested that the peptide part of bleomycin is synthesized on a multienzyme system. Recently we succeeded in isolation of intermediate peptides produced in each step of biosynthesis of the peptide part of bleomycin. The first amino acid from which the biosynthesis is started can be called demethylpyrimidoblamic acid. This amino acid and pyrimidoblamic acid have the following structures:

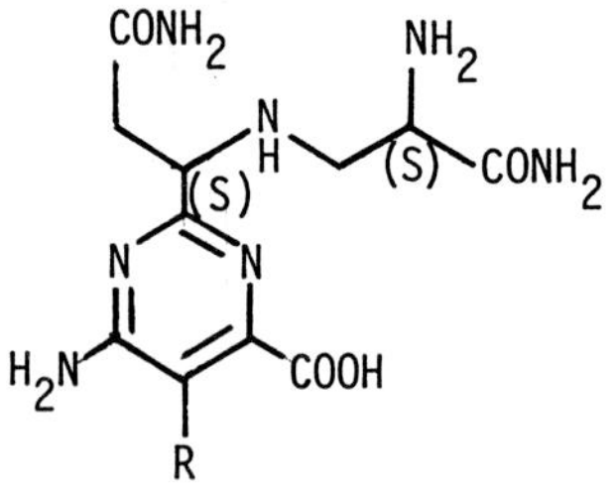

Demethylpyrimidoblamic acid: R=H

Pyrimidoblamic acid: $R=CH_3$

The following peptides were isolated from culture filtrates of a bleomycin-producing strain:

(1) Demethylpyrimidoblamyl-histidine

(2) Demethylpyrimidoblamyl-histidyl-alanine

(3) a) Demethylpyrimidoblamyl-histidyl-4-amino-3-hydroxy-2-methylpentanoic acid

b) Demethylpyrimidoblamyl-histidyl-2-amino-3-hydroxypentane

(4) Demethylpyrimidoblamyl-histidyl-(4-amino-3-hydroxy-2-methylpentanoyl)-threonine

(5) Pyrimidoblamyl-histidyl-(4-amino-3-hydroxy-2-methylpentanoyl)-threonine

(6) Pyrimidoblamyl-histidyl-(4-amino-3-hydroxy-2-methylpentanoyl)-2'-(2-aminoethyl)-2,4'-bithiazole-4-carboxylic acid

(7) Pyrimidoblamyl-β-hydroxyhistidyl-(4-amino-3-hydroxy-2-methylpentanoyl)-threonyl-2'-(2-aminoethyl)-2,4'-bithiazole-4-carboxylic acid

(8) Deglycobleomycin A2

Among these intermediate peptides, demethylpyrimidoblamylhistidylalanine was crystallized as its copper complex. The result of the X-ray crystal analysis of this copper complex by Iitaka and Nakamura, Pharmaceutical School of Tokyo University provided us very important information on the structure of bleomycin. We revised the structure of the β-lactam part of bleomycin molecule to the open amide structure as shown in Fig. 1 (Takita and others, 1978). On the basis of the crystal structure of a copper complex of a biosynthetic intermediate, our chemical study of bleomycin copper complex (Muraoka and others, 1976) and the ESR study of metal complexes of bleomycin by Sugiura _et al._, Science Faculty of Kyoto University, we also proposed the structure of bleomycin ferrous complex as shown in Fig. 2. Ferrous ion and oxygen are necessary for the reaction of bleomycin with DNA. Bleomycin binds to double strands of DNA (Asakura, 1978). In this binding, the bithiazole moiety of bleomycin molecule binds to guanine base in DNA. Oxygen molecule can bind to the ferrous ion in bleomycin complex. It is suggested that this activated oxygen reacts with the deoxyribose moiety of DNA and causes strand scission.

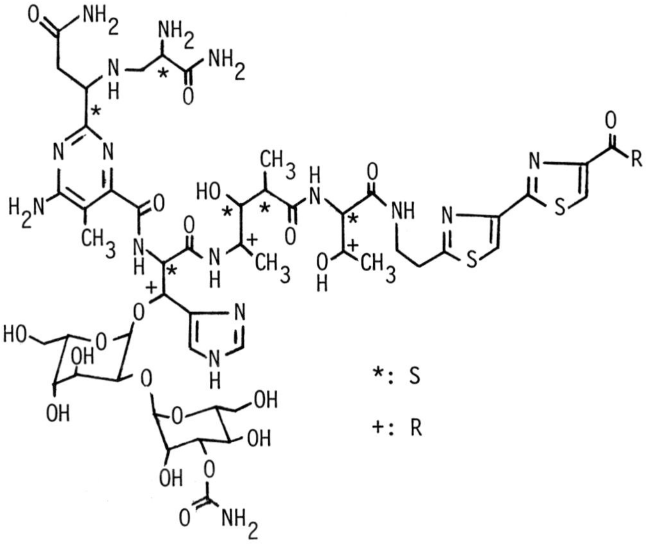

O NH2 NH2
N * NH2
H O
*
N N O
CH3 H O N R
HO N
H2N O * * NH N S
O + O * S
CH3 N N +
H * N CH3 O CH3
+ H H
O N
HO O
OH N
O H
OH OH
O OH
OH O
O NH2

*: S

+: R

Bleomycins:

A1: R = $NH-(CH_2)_3-SO-CH_3$; Demethyl-A2: R = $NH-(CH_2)_3-S-CH_3$;

A2: R = $NH-(CH_2)_3-S^{+}(CH_3)_2$: A2'-a: R = $NH-(CH_2)_4-NH_2$;

A2'-b: R = $NH-(CH_2)_3-NH_2$; A2'-c: R = $NH-(CH_2)_2$-(imidazolyl, N H)

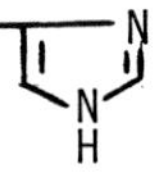

A5: R = $NH-(CH_2)_3-NH-(CH_2)_4-NH_2$;

A6: R = $NH-(CH_2)_3-NH-(CH_2)_4-NH-(CH_2)_3-NH_2$;

B1': R = NH_2; B2: R = $NH-(CH_2)_4-NH-C(=NH)-NH_2$;

B4: R = $NH-(CH_2)_4-NH-C(=NH)-NH-(CH_2)_4-NH-C(=NH)-NH_2$; Bleomycinic acid: R = OH

Fig. 1. Structures of bleomycins.

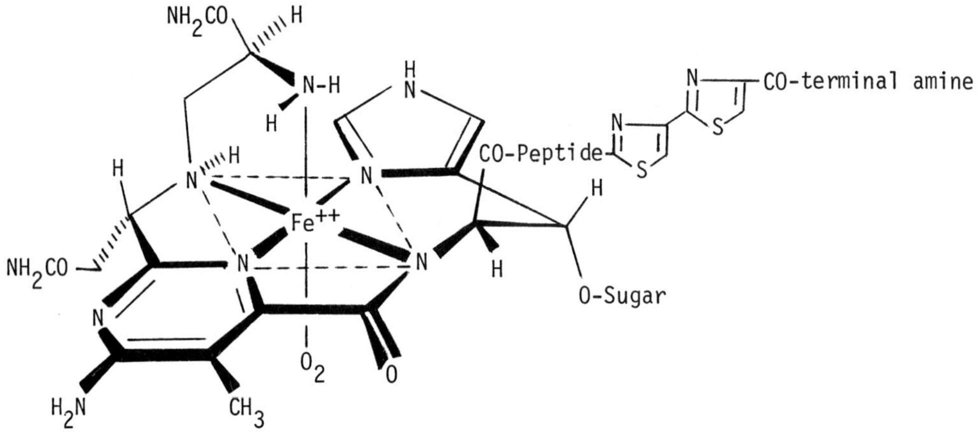

Fig. 2. Bleomycin ferrous complex in binding to O_2.

The behavior of bleomycin in vivo can be shown as follows (Umezawa, 1976): injected bleomycin binds with cupric ion in blood and forms bleomycin copper complex; after penetration into cells the cupric ion in bleomycin copper complex is reduced by intracellular small molecular reducing compounds and transferred to a cellular protein which can bind to cuprous ion (Takahashi and others, 1977); copper-free bleomycin undergoes the action of bleomycin hydrolase which cleaves the carboxamide bond of the α-aminocarboxamide moiety and is inactivated (Umezawa, 1976); copper free bleomycin which was not inactivated reaches nuclei, binds to DNA and ferrous ion is taken into bleomycin molecule binding to DNA; the reaction with DNA occurs and the growth of cells is inhibited; it is also possible that bleomycin copper complex from which the cupric ion was not reduced and removed reaches nuclei, binds to DNA, and the cupric ion in bleomycin copper complex binding to DNA is reduced, and replaced by ferrous ion. The mechanism of therapeutic action of bleomycin against squamous cell carcinoma is due to a significantly low content of bleomycin hydrolase and a high concentration of bleomycin in this tumor.

DEVELOPMENT OF NEW BLEOMYCINS

As already described, various bleomycins are different each other in their terminal amine moiety. If an amine is added to a fermentation medium and a bleomycin-producing strain is cultured, a bleomycin containing the amine added is produced and the production of other bleomycins are suppressed. By this fermentation method about 250 bleomycins have been prepared. The hydrolysis of bleomycin B2 by a Fusarium enzyme (Umezawa and others, 1973) which we named acylagmatine amidohydrolase or the application of the cyanogen bromide method to bleomycin demethyl-A2 (Takita and others, 1973) gives bleomycinic acid. Bleomycinic acid is the main molecular part common to all bleomycins. Various bleomycins can be synthesized from bleomycinic acid. Aminoalkyl esters of bleomycinic acid also causes a single strand scission of DNA. We have confirmed that various bleomycins are different in the degree of the renal and pulmonary toxicity (Umezawa, 1976).

Bleomycin does not show pulmonary toxicity in young mice, because this antibiotic is rapidly inactivated in lungs of young animals. This inactivation is much slower in old mice. A method to test the pulmonary toxicity using mice older than 15 weeks age has been established by Matsuda et al., Nihon Kayaku Co. Recently, a much more simple method where bleomycin is administered intratracheally was presented by Prof. S. Cohen. Among bleomycins which have lower pulmonary toxicity than present bleomycin, bleomycin PEP which is called pepleomycin and contains N-(3-aminopropyl)-α-phenethylamine as the terminal amine has been studied clinically in most detail. It inhibits Ehrlich carcinoma and squamous cell carcinoma induced by methylcholanthrene in mouse skin in the same or stronger degree than the present bleomycin. Its high dose inhibits also an adenocarcinoma in rat stomach induced by N-methyl-N'-nitro-N-nitrosoguanidine. The present bleomycin does not inhibit this adenocarcinoma in rat stomach. In pulmonary toxicity test using aged mice and dogs, pepleomycin has been shown to have a significantly lower pulmonary toxicity than the present bleomycin. The clinical study has suggested that this bleomycin has a significantly lower pulmonary toxicity than the present bleomycin and is more useful in treatment of bleomycin-sensitive tumors.

As reported by Oka et al. and Svanberg et al. (Carter and others, 1976), bleomycin is useful in treatment of squamous cell carcinoma in lung. But pulmonary toxicity occurs more frequently in lung cancer patients than in other cancer patients. Therefore, it is thought that a bleomycin which has a lower pulmonary toxicity than the present bleomycin is more useful in treatment of lung cancer.

Miyamoto and others (1977) reported that daily infusion of 5 mg of bleomycin for 7 days and 10 mg of mitomycin after the last injection of bleomycin and two to five times repetition of this treatment with 7 days interval exhibited a strong therapeutic effect on lung metastasis of cervix cancer. Daily 5 mg of bleomycin PEP for seven days and the repetition of the treatment with 7 days intervals seem to be an interesting administration schedule in treatment of tumor sensitive to bleomycin treatment. The clinical study of pepleomycin in the last 2 years has suggested its usefulness in cancer treatment. Bleomycin BAPP which contains N-(3-aminopropyl)-N'-butyldiaminopropane in the terminal amine was shown by the test in NCI, USA to exhibit an inhibition against B16 melanoma. It has about 5 times lower pulmonary toxicity than the present bleomycin. It is also one of new bleomycins worth clinical study. Tallisomycins A and B (Kawaguchi and others, 1977) are bleomycin group antibiotics. If they have a low pulmonary and renal toxicity, they may be interesting analogs of bleomycin. We are continuing the study of various types of derivatives of bleomycin and its analogs. This study will give more effective compounds useful in cancer treatment.

COMBINATION TREATMENT OF BLEOMYCIN WITH AN IMMUNE RESPONSE-ENHANCING AGENT

As described above, bleomycin has been used in combination of radiation or other cytotoxic chemotherapeutic agents. It is one of the characteristics of bleomycin that it does not suppress the bone marrow cells and immune responses. The bone stem cells are rich in bleomycin hydrolase and resistant to the action of bleomycin. On the basis of this characteristic, it can be imagined that therapeutic effect of bleomycin may be enhanced by an immune-enhancing agent.

Recently we found that microorganisms produce small molecular compounds enhancing immune response. Aminopeptidases are located on the surface of all kinds of animal cells. We found bestatin, [(2S,3R)-3-amino-2-hydroxy-4-phenylbutanoyl]-L-leucine, which inhibited aminopeptidase B and bound to macrophages, lymphocytes and other animal cells. Very low doses of bestatin such as 0.1-100 μg/mouse enhanced delayed-type hypersensitivity (Umezawa and others, 1976) and the daily administration of 10 or 100 μg started from 8 days after the inoculation of tumor cells inhibited slowly growing solid tumors such as Gardner lymphosarcoma and IMC carcinoma (Umezawa, 1977).

The cancer reduces the cellular immune activity of the host. Therefore, it is possible that a treatment with an immune-enhancing agent added to bleomycin treatment may enhance the curative effect. The study on the effect of bleomycin in combination with bestatin is going at present.

REFERENCES

Asakura, H., H. Umezawa and M. Hori (1978). DNA structures required for bleomycin binding. J. Antibiotics, 31, 156-158.

Carter, S. K. and others (Ed.) (1976). GANN Monograph on Cancer Research, No. 19, University of Tokyo Press.

Kawaguchi, H. and others (1977). Tallysomycin, a new antitumor antibiotic complex related to bleomycin. I. Production, isolation and properties. J. Antibiotics, 30, 779-788.

Miyamoto, T. and others (1977). Therapeutic effect of the continuous treatment of bleomycin in combination of mitomycin on advanced uterus cervix cancer. Gann (Cancer) and Kagakuryoho (Chemotherapy), 4, 273-276 (in Japanese).

Muraoka, Y. and others (1976). Chemistry of bleomycin. XVI. Epi-bleomycin. J. Antibiotics, 29, 853-856.

Takahashi, K. and others (1977). Intracellular reduction of the cupric ion of bleomycin copper complex and transfer of the cuprous ion to a cellular protein. J. Antibiotics, 30, 861-869.

Takita, T. and others (1973). Chemical cleavage of bleomycin to bleomycinic and synthesis of new bleomycins. J. Antibiotics, 26, 252-254.

Takita, T. and others (1978). Chemistry of bleomycin. XIX. Revised structures of bleomycin and phleomycin. J. Antibiotics, 31, 801-804.

Umezawa, H. (1976). In S. K. Carter and others (Ed.), GANN Monograph on Cancer Research, No. 19, University of Tokyo Press. pp.3-36.

Umezawa, H. (1977). Recent advances in bioactive microbial secondary metabolites. Japanese J. Antibiotics, 30 Suppl., S138-S163.

Umezawa, H. and others (1973). Preparation of bleomycinic acid: Hydrolysis of bleomycin B2 by a fusarium acylagmatine amidohydrolase. J. Antibiotics, 26, 117-119.

Umezawa, H. and others (1976). Enhancement of delayed-type hypersensitivity by bestatin, an inhibitor of aminopeptidase B and leucine aminopeptidase. J. Antibiotics, 29, 857-859.

The Special Position of Ifosfamide in the Series of Cytostatically Active Oxazaphosphorines

N. Brock

Pharmacological Department of Asta-Werke AG, D-4800 Bielefeld 14, Federal Republic of Germany

ABSTRACT

Ifosfamide is inactive in vitro and becomes activated in the liver of warm-blooded animals to cytotoxic metabolites. Its cancerotoxic selectivity depends on the formation of the primary metabolites 4-hydroxy/aldo-ifosfamide. In vivo ifosfamide shows a high cancerotoxic activity against various experimental tumours. Its margin of safety is better than that of cyclophosphamide, due to its cumulative behaviour. The leukotoxicity of ifosfamide in man is less pronounced than that of other alkylating compounds. In clinical use ifosfamide has markedly widened the spectrum of tumours amenable to chemotherapy. Even in cyclophosphamide-resistant patients and in cases which no longer respond to any other therapy it is possible to obtain remissions. The problem of urotoxic side-effects could be overcome by developing a specific antidote.

KEYWORDS

Chemotherapeutic properties, metabolism, clinical indications, urotoxic detoxification.

INTRODUCTION

Ifosfamide (Holoxan®) like cyclophosphamide (Endoxan®, Cytoxan®) belongs to the group of the N-2-chloroethyl-amino-oxazaphosphorines (Fig. 1). It was developed in 1967 in cooperation with Arnold, Bourseaux and Bekel, and after prolonged large-scale pharmacological and clinical trials it was introduced on the occasion of an International Symposium in 1977.

CHEMOTHERAPEUTIC PROPERTIES

From among the great number of results available, I have chosen some few characteristic examples of the drug's pharmacotherapeutic behaviour on which the specific advantages for its clinical use are based, especially in comparison with cyclophosphamide.

Just like cyclophosphamide, ifosfamide is inactive in vitro. In the liver of warm-blooded animals, it is activated to form cytotoxic metabolites. Ifosfamide is better tolerated than cyclophosphamide, particularly when

Cyclophosphamide

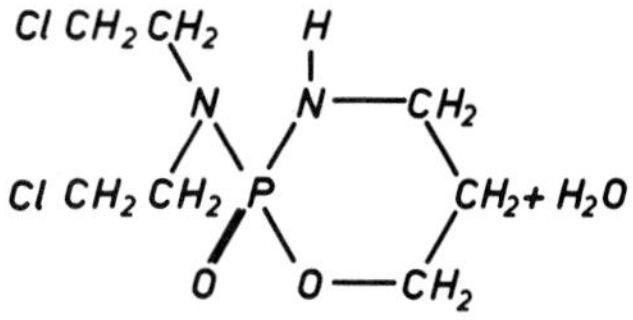

Trofosfamide

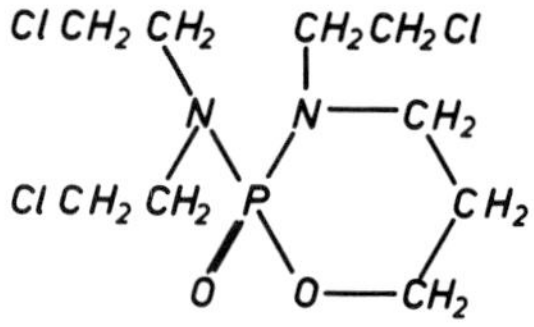

Ifosfamide

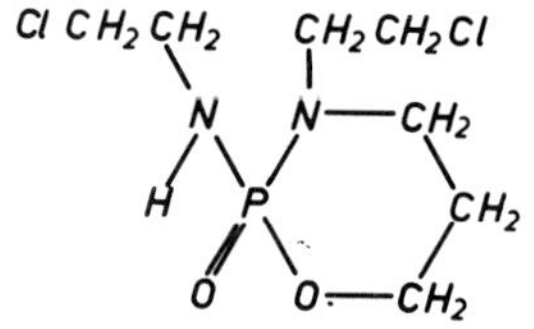

Fig. 1 Structural formula of the N-2-chloroethyl-amino-oxazaphosphorines cyclophosphamide, trofosfamide and ifosfamide.

administered in fractionated doses. Table 1 shows the LD 50 of cyclophosphamide in Sprague-Dawley rats to be practically the same after single-dose administration and after 4 divided doses, whereas ifosfamide is markedly better tolerated, the LD 50 rising from 430 mg/kg on single-dose administration to 519 mg/kg after the total dose is spread over 4 consecutive days.

TABLE 1 Lethal Doses of Cyclophosphamide and Ifosfamide in Sprague-Dawley Rats (i.v.). D = Total Dose, Administered on 1 Day or 4 Consecutive Days.

Compound	LD 50(1) [mg/kg]	LD 50(4) [mg/kg]
Ifosfamide	430 (393 – 476)	519 (469 – 636)
Cyclophosphamide	255 (234 – 278)	244 (215 – 266)

Like cyclophosphamide, ifosfamide also proved to be chemotherapeutically superior to direct alkylating agents in a great number of tumours of increasing chemoresistance. In comparison to cyclophosphamide fractionation of the total dose again proved to be of special benefit. For instance, in Yoshida's ascitic sarcoma AH 13 (Table 2) the therapeutic indices of cyclophosphamide and ifosfamide are largely the same on single-dose administration, whereas ifosfamide shows a considerably wider margin of safety, whenever the total dose is distributed over 4 consecutive days, its D 50 index being 82, and that of cyclophosphamide only 27.

TABLE 2 Curative and Lethal Doses as well as D 50 Indices of Ifosfamide and Cyclophosphamide with one Single Dose and 4 Separate Doses in the Yoshida Ascites Sarcoma AH 13 of the Rat

Compound	Number of adminis-trations	LD 50 [mg/kg]	CD 50 [mg/kg]	LD 50 / CD 50	Therapeutic units % LD 50
Ifosfamide	1	430	7,7	56	1,8
	4	519	6,3	82	1,2
Cyclophosphamide	1	255	3,9	65	1,5
	4	244	9,0	27	3,7

Due to its wider margin of safety, ifosfamide has also yielded better therapeutic results in chemoresistant tumours, for instance in the DS-carcinosarcoma and in the TA-nephroblastoma of the rat, which can be seen from the dose/action regression lines of ifosfamide and cyclophosphamide (Fig. 2). These results are of outstanding importance for the chemotherapy of human tumours, since they show that even chemoresistant tumours can be cured by adequate doses of a highly potent and well tolerated compound.

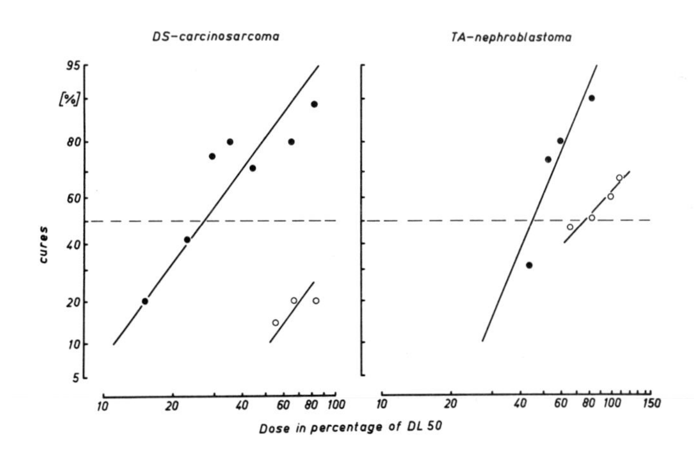

Fig. 2 Definite cures in DS-carcinosarcoma and TA-nephroblastoma in rats BD II after single i.v. administration of ifosfamide ● in comparison to cyclophosphamide ○. Administration 12 days after transplantation of the tumour (weight: 2 - 5 g)

(Druckrey, H. 1970, J. Kanser [Ankara], 1, 131 - 149).

The above described results chosen from among a great number of analogous results obtained in various tumour types of the rat and the mouse have repeatedly been confirmed. Special mention should be made of the publications of Goldin, Bethesda, and Larionov, Moscow, who also proved the therapeutic superiority of ifosfamide in their model tests.

CUMULATION AND TIME/ACTION RELATIONSHIP

Almost any antitumour agent produces cumulative effects. Therefore the pharmacologist is challenged to study the cumulation of the various actions so as to provide the doctor with a new drug which is clinically safe and of great therapeutic potency. This important aspect is often neglected. Although cumulation of toxic actions is dangerous and therefore undesirable, cumulation of desirable therapeutic effects is welcome. We assessed the individual actions of ifosfamide and cyclophosphamide according to Druckrey's method by quantitatively determining the cumulative action remaining after 24 hours, the cumulative rest C being the counterpart of the value R of the reversible part. The results are shown in Fig. 3. The cumulating toxicity of ifosfamide remaining after 24 hours was C = 83 %, with cyclophosphamide it was C = > 95 %. For the curative action the situation was vice versa, i.e. the cumulative rest being 45 % with cyclophosphamide, and approximately 100 % with ifosfamide. Thus the curative action of ifosfamide is definately more cumulated than that of cyclophosphamide, whereas the toxicity is less cumulated, which explains why in the case of ifosfamide, fractionated dosage exerts a considerably more pronounced therapeutic effect.

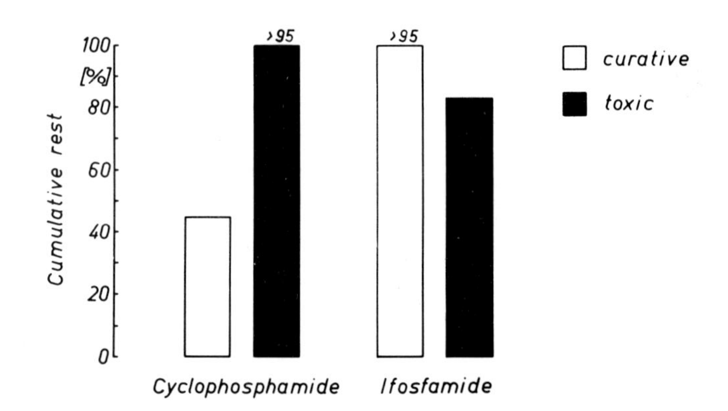

Fig. 3 Cumulative rest C of the curative and toxic effects of cyclophosphamide and ifosfamide persisting after 24 hours if the total dose is subdivided into 4 daily doses.

METABOLISM OF IFOSFAMIDE

As with cyclophosphamide the metabolism of ifosfamide (Fig. 4) starts in the liver. This initial metabolization stage, which we call "activation", is a mixed-function hydroxylation of ifosfamide by means of which the inert transport form is activated to form the primary metabolites 4-hydroxy-ifosfamide/aldo-ifosfamide. The higher reactivity caused by enzymatic

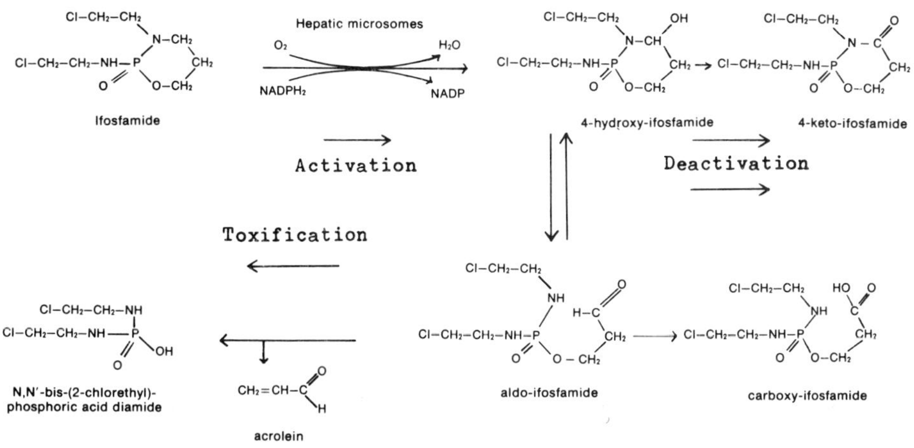

Ifosfamide

Hepatic microsomes

O_2 → H_2O

$NADPH_2$ → NADP

Activation

4-hydroxy-ifosfamide

4-keto-ifosfamide

Deactivation

Toxification

N,N'-bis-(2-chlorethyl)-phosphoric acid diamide

acrolein

aldo-ifosfamide

carboxy-ifosfamide

Fig. 4 Metabolism of ifosfamide.

hydroxylation leads to spontaneous liberation of a 3-carbon compound in the form of acroleine and formation of the direct alkylator N,N-bis-2-chloroethyl-phosphoric acid diamide, which reaction we have called "toxification". Spontaneous toxification subsequent to enzymatic activation allows also in vivo intermediary reactions in the oxazaphosphorine ring which partly or completely deactivate the activated metabolites, yielding the biologically inert excretion products 4-keto-ifosfamide and carboxy-ifosfamide. As partial deactivation, the non-enzymatic reaction of 4-hydroxy-ifosfamide with sulfhydryl compounds leading to 4-mercapto-ifosfamide plays an important biological role for understanding the drug's selectivity.

The proved enzymatic hydroxylation of the chloroethyl-side-chain is not an activation process in our sense and may be neglected with regard to the problem of cancerotoxic selectivity of ifosfamide.

SUMMARY OF CLINICAL RESULTS OBTAINED WITH IFOSFAMIDE

Ifosfamide is a new cytostatic agent whose pharmacological properties will ensure its firm place in the antitumour armamentarium. The leukotoxic effect in man, so often a limiting factor in the chemotherapeutic use of alkylators, proved to be definately less pronounced than that of the alkylators so far available, which is in conformity with the laboratory findings in animals. Thus the doses found to be effective in animals could be directly transferred to man. Decisively better results were seen on fractionated dosage, which did not only enhance the therapeutic potency and ensure higher remission rates, but also improve the drug's tolerance.

Ifosfamide has a wide spectrum of activity and is particularly indicated in the following conditions:

Oat cell bronchogenic carcinoma, ovarian carcinoma, breast carcinoma, testicular tumours of all histological types, soft-tissue sarcomas (chondro-, leiomyo- and rhabdomyosarcomas).

It also exerts a good effect in:

Endometrial carcinoma (corpus carcinoma), hypernephroma, pancreatic carcinoma, bronchial carcinoma of other than oat cell types, malignant lymphomas (lymphogranulomatosis, lymphosarcoma, reticulosarcoma).

In some isolated cases, ifosfamide is also effective in:

Gastric carcinoma (except the parvicellular and scirrhous forms), carcinoma of the cervix, osteosarcoma, malignant melanoma, chronic and acute leukaemia, multiple myeloma.

Two outstanding therapeutic features should be especially stressed:

a) Even in cyclophosphamide-resistant patients and in cases which no longer respond to any other treatment, it is possible to obtain remissions with ifosfamide therapy.

b) The introduction of ifosfamide has markedly widened the spectrum of tumours amenable to chemotherapy (cf. its great effectiveness even in testicular tumours not arising from the spermatogonial cells).

With regard to its clinical dosage two schemes of treatment are mainly recommended:

In tumours of relatively rapid growth, daily i.v. injections of 50 to 60 mg/kg of ifosfamide should be given on 5 consecutive days.

In tumours of a lower proliferation rate, e.g. in soft tissue sarcoma and endometrial carcinoma and in debilitated patients, 20 to 30 mg/kg of ifosfamide should be given by daily i.v. injection for 10 consecutive days.

Ifosfamide treatment has so far been limited by its urotoxicity, usually causing haemorrhagic cystitis which, due to the considerably higher dosage and metabolite excretion,is definitely more pronounced than with cyclophosphamide. It should be borne in mind that lesions may not only occur in the vesical mucosa, but also in the upper region of the efferent urinary tract. Such lesions can usually be prevented by careful prophylactic measures, such as liberal supply of fluid (about 4 litres daily), specific use of diuretics, alkalization of the urine and instillation of sulfhydryl-group-containing agents into the bladder.

DETOXIFICATION

The problem of haemorrhagic cystitis following the administration of oxazaphosphorines has long been known and in the past it has seriously worried many a doctor using cyclophosphamide. After many years of intensive experimental and clinical work we have succeeded in solving this problem just recently.

Basic Concept

Regional detoxification of the kidneys and of the efferent urinary tract would be possible by means of an antidote which should be readily soluble in water and physiologically inert. To maintain its full curative effect such a compound should not penetrate into the tissues, but it should rather be excreted rapidly and completely via the kidneys. In addition it should have a high affinity to the cytotoxic metabolites and detoxify them biologically in the kidneys and the efferent urinary passages.

Design of Biological Assays

Rats are treated with 68,1 mg/kg of ifosfamide intravenously. After 24 hours they are given trypan blue by i.v. injection, 10 minutes later they are sacrificed.

The top row of Fig. 5 shows normal bladders, the third row those with pathological lesions, recognizable by the blue coloration of the oedematously enlarged bladders.

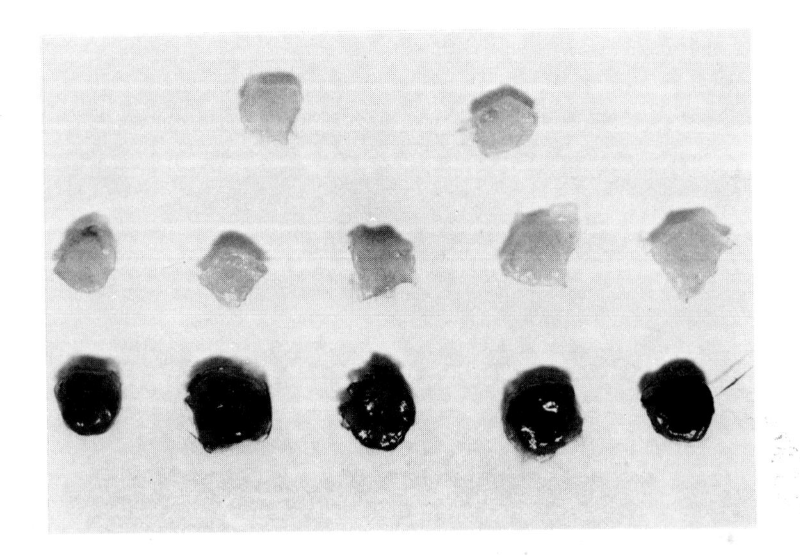

Fig. 5 Bladders of rats following i.v. administration of trypan blue.

First row: Controls without administration of ifosfamide, normal bladders

Second row: Treatment with 68,1 mg/kg of ifosfamide i.v., contemporary administration of the antidote ASTA D 7093

Third row: Administration of 68,1 mg/kg of ifosfamide without antidote.

Concurrent administration of ifosfamide and of sodium 2-mercapto-ethane sulphonate (Fig. 6) - 21,5 mg/kg i.v. - completely prevents such severe vesical lesions (cf. middle row).

A great number of model tests in various types of tumour have proved that the chemotherapeutic effect of ifosfamide is not impaired even by high doses of the antidote.

$$\begin{array}{l} \quad SH \\ CH_2 \\ | \\ CH_2 - SO_3^{\ominus}\ Na^{\oplus} \end{array}$$

Sodium-2-mercaptoethane sulphonate

Fig. 6 Structural formula of ASTA D 7093.

The macroscopic findings are strongly confirmed by the histological results. Fig. 7 shows the results of ifosfamide treatment in the rat: purulent necrotic cystitis with oedema of the submucosa, ulcerative decay of the mucosa overlayered by tissue exudate.

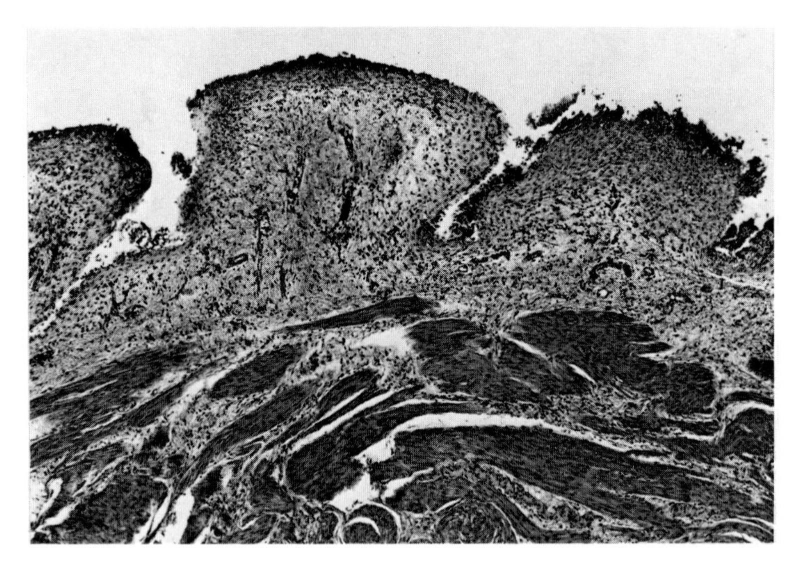

Fig. 7 Bladder of a rat, treated with 68,1 mg/kg of ifosfamide.

(The relative severity of the appearance depends on experimental modalities: no drinking water for 24 hours before treatment)

Fig. 8 (ifosfamide + antidote) shows a completely normal bladder.

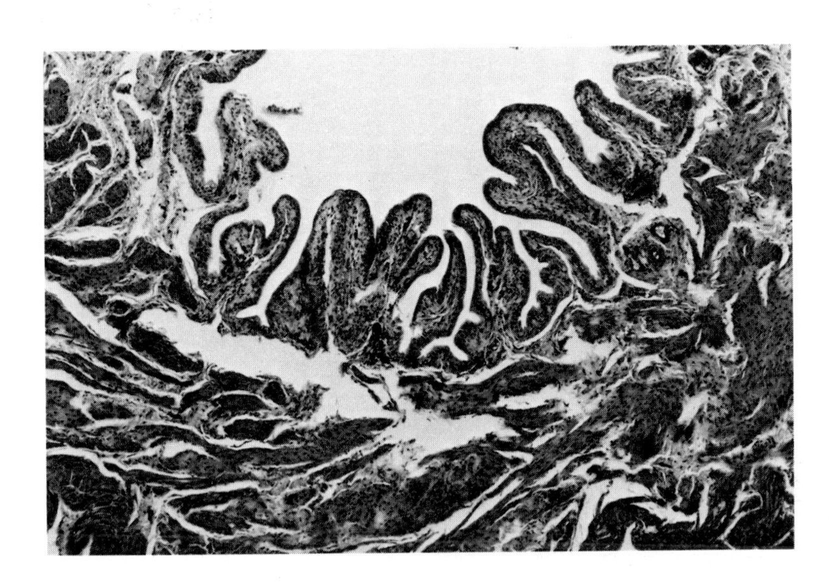

Fig. 8 Bladder of a rat following treatment of ifosfamide and ASTA D 7093.

Clinical Trials

In randomised studies made in the Radiological Clinic Janker, Bonn, and in the Medical Unit of Cologne University, the above detoxification method was compared with the standard prophylactic procedures so far taken in high-dose treatment. After standard prophylaxis nearly all patients showed at least microhaematuria, whereas the antidote-treated cases did not reveal any signs of urotoxic side-effects.

Even large-scale use of this new method in several hospitals has invariably confirmed these positive results and thus we hope soon to be able to provide all clinical oncologists with this new compound to the benefit of their patients.

Triazenylimidazoles and Related Compounds

Y. Fulmer Shealy

Kettering-Meyer Laboratory, Southern Research Institute, Birmingham, Alabama 35243, U.S.A.

ABSTRACT

After we had synthesized the initial group of triazenyl derivatives of imidazole-4-carboxamide, including DTIC, and had observed the antineoplastic activity of DTIC, we prepared a large number of heteroaryl and aryl triazenes for anticancer evaluation. The compounds prepared and tested against leukemia L1210 and, sometimes, against other transplantable neoplasma included triazenyl derivatives of the following aromatic groups: imidazole-, v-triazole-, and pyrazole-carboxamides; phenylalkanoic, benzoic, imidazolecarboxylic, and pyrazolecarboxylic acid esters and hydrazides; benzamides, benzamidine, and benzenesulfonamides; imidazolecarbonitrile; and miscellaneous heterocycles and benzene derivatives. Many of these triazenes that contained at least one methyl group in the triazenyl moiety were active against leukemia L1210, whereas few of those lacking such a methyl group showed significant activity. A rather consistent pattern of activity was shown by imidazolecarboxamides, pyrazolecarboxamides, and o-benzamides containing at least one methyl group in the triazenyl moiety; by 3-(2-hydroxyethyl)-3-methyl)-1-triazenyl derivatives; and by bis(2-fluoroethyl)triazenes. None of the triazenes were significantly more active than DTIC except for certain chloroethyltriazenes. Like DTIC, the highly active bis(2-chloroethyl)triazenyl analog of DTIC is enzymatically dealkylated, apparently, to the monochloroethyltriazene, which also effects some cures. Other chloroethyltriazenes were also studied and prepared, but most compounds of this type are too unstable to show high activity. However, the 3-benzyl-3-(2-chloroethyl)-1-triazenyl analog of DTIC is highly active.

KEYWORDS

Anticancer agents. Cancer chemotherapy. Leukemia L1210. DTIC. Triazenes. DTIC analogs. Chloroethyltriazenes. Triazenylbenzenes. Triazenylheterocycles.

INTRODUCTION

A rather large number of triazenes has been synthesized in our Institute for evaluation against experimental neoplasma. This group of compounds includes triazenyl derivatives of imidazole-4-carboxamide, of related heterocycles, and of benzene derivatives. One of the first of the triazenes that we prepared was DTIC (4) (Shealy, Krauth, and Montgomery, 1962), which is designated in the literature by chemical names or as DTIC, DIC, dacarbazine, imidazolecarboxamide, or NSC-45388. The story of this compound and of our work on related structures begins with 5-diazoimidazole-4-carboxamide (2) (Shealy and co-workers, 1961), which we were able to isolate after diazotizing 5-aminoimidazole-4-carboxamide (1, AIC). Traditionally, diazotization of ortho-amino carboxamides, such as the ortho-aminobenzamides, produced v-triazinones. Indeed, Woolley and Shaw (1951) had reported earlier that they had obtained 2-azahypoxanthine (3), a v-triazinone, by diazotizing AIC. Studies of our diazotization product left no doubt that it was, in fact, the diazoimidazole (quite similar in character to aromatic diazonium salts) and that it does cyclize readily to 2-azahypoxanthine in acidic or basic solutions. These studies also showed that the diazoimidazole is cytotoxic to cells in culture and has modest antitumor activity (Shealy and co-workers, 1961). These results were the principal motivation, at that time, for the preparation of triazene derivatives of this interesting diazoheterocycle.

SYNTHESIS

Soon after DTIC was synthesized, its antineoplastic activity was observed (Shealy, Montgomery, and Laster, 1962; Tables 1 and 2). Consequently, additional heteroaryl and aryl triazenes were prepared. During the first phase of our synthesis work we prepared 5-(3,3-dialkyl-1-triazenyl)imidazole-4-carboxamides (5) (Shealy, Krauth, and Montgomery, 1962), 3-monoalkyl or 3-monoaryl triazenylimidazoles (6) (Krauth, Shealy, and Montgomery, 1962; Shealy and Krauth, 1966a), and analogous dialkyltriazenyl triazolecarboxamides (8) (Shealy and Krauth, 1966b).

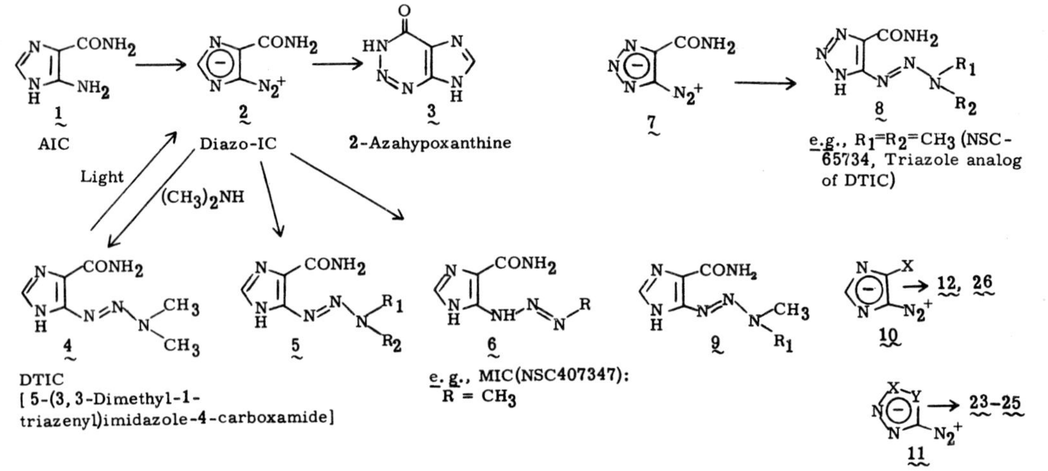

1 AIC → 2 Diazo-IC → 3 2-Azahypoxanthine

Light; $(CH_3)_2NH$

7 → 8

e.g., $R_1=R_2=CH_3$ (NSC-65734, Triazole analog of DTIC)

4 DTIC [5-(3,3-Dimethyl-1-triazenyl)imidazole-4-carboxamide]

5

6 e.g., MIC(NSC407347): R = CH_3

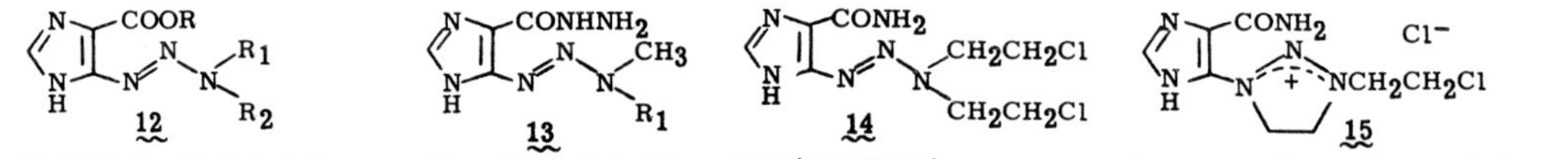

9

10 → 12, 26

11 → 23–25

12 R=Methyl, ethyl, butyl, octyl, diethylaminoethyl e.g., $R_1=R_2=R_3$(NSC87982). Antileukemia, antibacterial, antifungal activity

13 R_1 = Methyl, butyl cyclohexyl

14 BIC (NSC82196)

15 Ionic transformation product (a 1,2,3-triazolinium chloride)

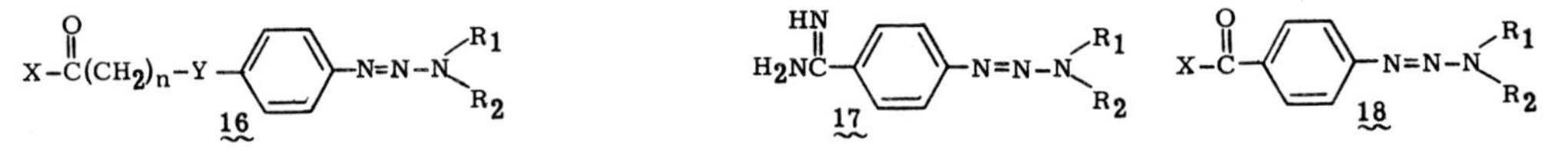

16 X=RO– or H_2NNH–
n=2, Y=CH_2: Phenylbutyric acid esters and hydrazides (chlorambucil analogs)
n=1, Y=CH_2: Hydrocinnamic acid esters and hydrazides
n=1, Y=O: Phenoxyacetic acid esters and hydrazides

17 p-Benzamidines

18 X=RO–: p-Benzoic acid esters
X=H_2N–: p-Benzamides
X=H_2NNH–: p-Benzoic acid hydrazides

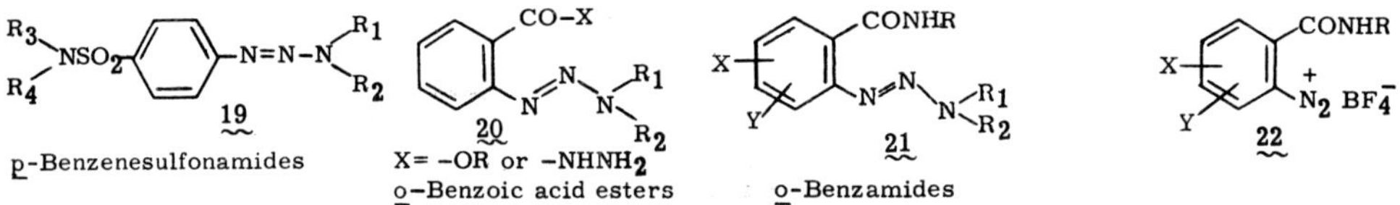

19 p-Benzenesulfonamides

20 X= –OR or –$NHNH_2$ o-Benzoic acid esters and hydrazides

21 o-Benzamides

22

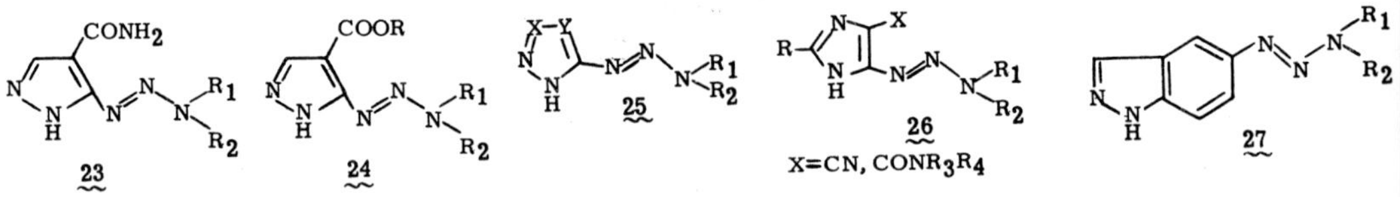

23 24 25

26 X=CN, $CONR_3R_4$

27

The triazenyl-triazoles (8) were prepared from 5-diazo-1,2,3-triazole-4-carboxamide (7), which we had obtained from the triazole analog of AIC soon after we had isolated diazo-IC (Shealy and co-workers, 1961).

During the second phase of our synthesis work, we effected further modifications of the triazenyl group and replaced the imidazolecarboxamide moiety with other structural units. Among the imidazoles, these structural variants consisted of a series of methyl alkyl triazenylimidazolecarboxamides (5, $R_1=CH_3$) (Shealy and co-workers, 1968a), in addition to DTIC and the butyl methyl triazene prepared during the first phase; imidazolecarboxylic acid esters (12) (Shealy and co-workers, 1967, 1971b; Krauth, Shealy, and O'Dell, 1972a, 1972b) and hydrazides (13); and the bis(2-chloroethyl)triazene (14) (Shealy and Krauth, 1966c; Shealy and co-workers, 1968b), which is very active against leukemia L1210 in mice, but cyclizes to a triazolinium salt (15). During the same period, we were also replacing the imidazole group by substituted benzene-ring groups (Shealy and co-workers, 1971c, 1971d; Tables 5, 6). These triazenes are summarized by the general structures (16–20) and may be broadly classified as triazenyl phenylalkanoic-acid (16) and triazenylbenzoic-acid (17–20) derivatives and analogs. As the NSC numbers in the tables show, many of the benzenoid triazenes preceded, or were contemporary with, BIC (Shealy and Krauth, 1966c) and the imidazolecarboxylic acid esters (Shealy and co-workers, 1967). Still another type that we prepared during the second phase are triazenylpyrazoles (23, 24) (Shealy and O'Dell, 1970, 1971a), analogous to the earlier imidazole and triazole triazenes.

Since about 1970, our synthesis work has consisted of further structural variations of both the triazenyl moiety and the heterocyclic or aryl group (e. g., 25–27) with considerable emphasis on chloroethyltriazenes and substituted o-benzamides (21) (Shealy, O'Dell, and Krauth, 1975; Shealy and O'Dell, 1975; Shealy and co-workers, unpublished). In general, the triazenylheterocycles were prepared from isolated diazoheterocycles (e. g., 2, 7, 10, 11), whereas the triazenylbenzene derivatives were usually prepared from aqueous diazonium salt solutions. The o-benzamides (21), however, were prepared from isolated diazonium fluoborates (22) (Shealy and co-workers, 1971d) in organic solvents because of the facility of the cyclization of o-carbamoyldiazonium salts to benzo-v-triazinones. Because of limitations of time and space, this report deals almost entirely with syntheses and biological evaluations carried out in our Institute and does not attempt to review subsequent work on triazenes by others (e. g., Lin and Loo, 1972; Audette and co-workers, 1973; Connors and co-workers, 1976).

BIOLOGICAL EVALUATION

Some of the results of numerous tests of DTIC against leukemia L1210 are summarized in Table 1. The doses listed are not necessarily optimal; but, in general, the highest increases in lifespan observed for a given dose, route, and schedule are shown. These data show that DTIC is active against leukemia L1210 when administered by various routes and schedules and that the highest ILS values were about 95–100%. The results of some of our tests of DTIC against other leukemias and solid tumors are summarized in Table 2. This sampling of results shows that DTIC exhibited activity against a spectrum of mouse neoplasma; note, however, that it was not active against intracerebral leukemia L1210. The broad spectrum of activity of DTIC coupled with the novelty of its structure in comparison with known anticancer agents were the reasons that it was chosen for clinical trial by the National Cancer Institute (U. S. A.).

TABLE 1 DTIC versus L1210 Leukemia

Dose mg/kg	Route*	Schedule	Average Lifespan, Days Treated/Control	%Increase in Lifespan (ILS)
100	IP	qd 1-death	17.7/9.0	+96
100			15.0/8.9	+68
100			17.2/10.0	+72
50			14.7/9.0	+63
50			12.8/8.2	+56
100		qd 1-9	13.2/7.9	+67
200		qd 1-5	13.3/7.9	+68
100			12.0/7.9	+51
600		Day 1 only	12.0/7.9	+51
317			14.5/8.0	+81**
200	SQ	qd 1-9	15.7/7.9	+98
100			12.8/7.9	+62
600	Oral	Days 1, 5, 9	15.3/7.9	+93
300			13.9/7.9	+75
200		qd 1-9	15.5/7.9	+96
100			13.9/7.9	+75
50			13.2/7.9	+67
100	IM	qd 1-death	14.5/9.0	+61

*IP, intraperitoneal; SQ, subcutaneous; IM, intramuscular.
** Median increase in lifespan.

TABLE 2 DTIC versus Other Murine Leukemias and Solid Tumors

Neoplasm	Dose mg/kg	Schedule IP	Lifespan Treated/Control	% ILS or MLS* (30-Day Survivors)
P815	100	qd 1-15	>20.8/14.3	>+45 (2/10)
P815/FU	100	qd 1-15	>28.6/14.7	>+95 (5/10)
L5178y	75	qd 1-15	12.2/9.2	+32
L4946	75	qd 1-15	>21.8/11.4	>+91 (2/10)
L1210/MTX, 10^6 cells	87	qd 1-15	14.0/7.0	+100*
	50	qd 1-death	12.5/7.0	+78*
L1210/MP, 10^6 cells	125	qd 1-15	11.0/7.0	+57*
L1210/TG, 10^6 cells	75	qd 1-15	18.5/10.0	+85*
IC L1210, 10^4 cells**	317	Day 1	9.0/9.0	0*
				% of Control
Sarcoma 180	100	qd 1-7	7 tests	23–50
	50			43
Adenocarcinoma 755	125	qd 1-11	9 tests (2 toxic)	3–44
	100			22
P1798 Lymphosarcoma	100	qd 1-11		0 (10/10)***
	75	qd 1-11		0 (7/10)***

* Median Increase in lifespan. ** Intracerebral L1210. *** Without tumors.

Almost all of our triazenes were tested against the standard mouse leukemia L1210, in which 10^5 cells are inoculated intraperitoneally on day zero; some were also tested against solid tumors. The following tables summarize a rather large number of compounds and results of tests against leukemia L1210, but not all of the triazenes that we prepared and not all of the tests can be summarized here. The purpose of these tables is two-fold: first, to show many of the structural modifications that we have effected and, secondly, to convey some information on the activity of structural types and of individual compounds against leukemia L1210 in mice. In these tables, 1-d means daily treatment on days 1 to the day of death of the animal; 1-9, treatment on days 1 through 9; D1 or D2, single-dose treatment on day 1 or day 2, respectively; 1, 5, 9 means treatment on days 1, 5, and 9. The data in Table 3 show that several imidazolecarboxamide and triazolecarboxamide analogs of DTIC displayed little, if any, activity against L1210 leukemia. All of these compounds are disubstituted triazenyl derivatives without a methyl group; and, as the NSC numbers indicate, all were among the first analogs of DTIC that we prepared. The absence of activity among these early DTIC analogs lacking a methyl group strongly influenced the choice of other triazenes for synthesis.

All of the imidazolecarboxamide analogs of DTIC shown in Table 4 contain a methyl group in the triazenyl moiety; and, is contrast to those in Table 3,

TABLE 3 Early Imidazole and v-Triazole Analogs of DTIC Lacking a Methyl Group

(Structure: X-containing ring with N, NH; substituents $CONH_2$ and $N=N-N(R_1)-R_2$)

Triazene		NSC*	% ILS, L1210**	Rating	IP Doses, mg/kg	Schedule
			X = CH (Imidazolecarboxamides)			
R1	R2					
n-Butyl	n-Butyl	45396	+1	–	250	1-d
Ethyl	Ethyl	52371	+13, +1, +4	– ?	400, 200, 100	D1
			+19, +8, +2		400, 200, 100	1, 5, 9
			t, t, +14, +27, +11, +17		400, 200, 100, 100, 50, 25	1-9
n-Octyl	n-Octyl	404182+	–2	–	400	1-d
Benzyl	Benzyl	52372	–7	–	250	1-d
R_1R_2N =	Piperidinyl	406800+	t, –11	–	500, 250	1-d
R_1R_2N =	Pyrrolidinyl	406801+	–14, +5, +26, +3	– ?	500, 375, 250, 125	1-d
			X = N (v-Triazolecarboxamides)			
R1	R2					
n-Butyl	n-Butyl	65735	t, –3	–	500, 250	1-d
R_1R_2N =	Pyrrolidinyl	406450+	–10, –7	–	300, 175	1-d

*Numbers of the 400,000 series were contemporary with those of the 40,000 – 65,000 series.
** t = Toxic. The listing of % ILS values corresponds to the listing of IP doses; e.g., both of the ILS values of +14 and +27 for NSC 52371 resulted from a dose of 100 mg/kg/day in separate tests, qd 1-9.

TABLE 4 5-(3-Methyl-3-alkyl-1-triazenyl)imidazole-4-carboxamides (Structure 9)

R	NSC	%ILS, L1210	Rating	IP Dose, mg/kg	Schedule
H (MIC)	407347+	+83, +77	+	30	1-d
		+54		20	1-d
		+54		15	1-d
Propyl	76418	+58	+	50	1-d
		+51		38	1-d
n-Butyl	70874	+86, +58	+	80	1-d
		+64		40	1-d
		+68		36	1-9
Isobutyl	83113	+67, +48	+	75	1-d
		+70		37	1-d
sec.-Butyl	144216	+86, +50	+	88	1-9
		+72		264	1, 5, 9
n-Pentyl	87981	+59	+	100	1-d
		+70, +61		400	D2
Cyclohexyl	83111	+66	+	75	1-d
		+64		200	1, 5, 9
		+42, +21, +29		400	D2
Benzyl	83695	+76	+	200	1-9
		+55		400	1, 5, 9
Phenethyl	146371	+55	+	150	1-9
		+79		600	1, 5, 9
2-Hydroxyethyl	83112	+69	+	400	1-9
Phenyl	405287+	– 7	–	75	1-d
$-CH_2CN$	84961	+16	–	100	1-d
$-C(CH_3)-CH(C_6H_5)OH$	152644	+ 6	–	400	1-9
$-CH_2CH_2N(CH_3)_2$	152645	– 5	–	400	1-9

most of them are active against L1210 leukemia. Activity was retained when one of the methyl groups of DTIC was replaced by a propyl, n-butyl, sec.-butyl, isobutyl, pentyl, cyclohexyl, benzyl, phenethyl, or hydroxyethyl group.

Benzenoid triazenes were being prepared simultaneously with imidazole and other heterocyclic triazenes shown in Tables 4 and 7. All of the benzenoid triazenes shown in Tables 5 and 6 contain a triazenyl grouping that produced active compounds when it was attached to the imidazolecarboxamide moiety (Tables 1 and 4). All contain at least one methyl group in the triazene portion. However, many of these compounds were inactive or showed only modest activity in our tests. One of the best members of this group was the methyl butyl triazenyl derivative of hydrocinnamic acid ethyl ester (NSC 77587), shown in the lower part of Table 6. Several other esters and hydrazides (e.g. NSC 80637, 84957, 83693 of Table 5 and NSC 93192 and 80344 of Table 7) displayed modest, reproducible activity, as did the p-dimethyltriazenyl derivatives of benzamide (NSC 86441) and benzamidine (NSC 112486). The p-(dimethyltriazenyl)benzamide (Lin and Loo, 1972; Shealy and co-workers, 1971d) has subsequently received further study (Abel, Connors, and Giraldi, 1977; Connors and co-workers, 1976; Farquhar, 1976). o-Benzamides were prepared later (Shealy and co-workers, 1971d, unpublished) and showed a better pattern of activity against leukemia L1210 than did the p-benzamides.

Most of the methyl alkyl triazenyl heterocycles shown in Table 7 are also analogs of active imidazolecarboxamide triazenes (Table 4). Overall, the pattern of activity appears to be somewhat better than that of the benzenoid triazenes shown in Tables 5 and 6; note, especially, the consistent pattern of activity shown by the pyrazole triazenes.

Most of the ortho-triazenylbenzamides summarized in Table 8 are active against leukemia L1210. Thus, they are similar to the imidazolecarboxamide and pyrazolecarboxamide triazenes containing a methyl substituent; and, as a group, they were superior, in our tests, to other benzenoid and heterocyclic triazenes shown in Tables 5–7. Both the o- and p-triazenylbenzamide analogs of DTIC are more stable to light than is DTIC (Shealy and co-workers, 1971d), but they are less stable in acidic media.

TABLE 5 3,3-Dimethyl-1-triazenylbenzene Derivatives

A: X–C_6H_4–N=N–N$(CH_3)_2$ (para)   B: X–C_6H_4–N=N–N$(CH_3)_2$ (ortho)

Structure A or B	X	NSC	% ILS, L1210	Rating	IP Dose (mg/kg)	Schedule
A	EtO_2CCH_2O-	75945	+15	–	187	1-d
A	$EtO_2C(CH_2)_2-$	76422	+25	–	83	1-9
			+5		250	1-d
A	$EtO_2C(CH_2)_3-$	80637	+34	±	75	1-9
			+33		50	1-9
A	$EtO_2C(CH_2)_2CH(CO_2Et)NH(CO-)$	74252	–4	–	500	1-d
A	$H_2NNHC(O)CH_2O-$	92162	+11	–	50	1-d
A	$H_2NNHC(O)(CH_2)_2-$	84957	+32, +41	+	45	1-d
A	$H_2NNHC(O)(CH_2)_3-$	83693	+42	+	45	1-d
			+46		63	1-9
			+59		400	D2
A	EtO_2C-	86445	+10, +24	–	90	1-d
A	$H_2NNHC(O)-$	88728	+29	±	100	1-d
			+34		75	1-d
A	$H_2NC(O)-$	86441	+46	+	100	1-d
			+44		75	1-d
A	$H_2NC(=NH)-$	112486	+33	±	35	1-9
			+28		53	1-9
A	$(ClCH_2CH_2)_2NSO_2-$	95962	–8	–	25	1-9
			toxic		50	1-9
B	$-COOCH_3$	102245	–11	–	70	1-15
			toxic		140	1-15
B	$-CONHNH_2$	102247	+20	–	100	1-d

The compounds shown in Table 9 represent a return to the imidazolecarboxamide structure of DTIC. The structural modifications consisted of the introduction of a substituent at position 2 of the imidazole ring and at the nitrogen atom of the carboxamide group. Activity was generally retained when a substituent was introduced at position 2; one of the carboxamide derivatives showed good activity. Table 10 includes results of tests of several 3-[2-(hydroxyethyl)-3-methyltriazenes. Most of the triazenes bearing this polar triazene moiety were active at high doses.

The first group of our imidazole and triazole carboxamides included several with disubstituted triazene groups lacking a methyl substituent, and these compounds showed little, if any, activity against L1210 leukemia (Table 3). We continued to prepare a few of this type of triazenyl derivatives of various heterocyclic and benzene nuclei to test this non-activity correlation (Table 11). All of these later triazenes were devoid of significant activity except for the diethyltriazenyl pyrazolecarboxamide (NSC 145927), which showed modest, reproducible activity.

2-HALOETHYLTRIAZENES

The preparation of BIC, the bis(2-chloroethyl)-triazenyl derivative of imidazole-4-carboxamide, and the fact that it cyclizes readily to a triazolinium chloride were mentioned above. Despite its instability, this derivative showed excellent activity against L1210 leukemia, effecting cures of the majority of treated animals at optimal doses (Hoffman and co-workers, 1968; Shealy and Krauth, 1966c). It was, by far, the best of the triazenes that we prepared through

TABLE 6 3-Methyl-3-alkyl-1-triazenyl Benzene Derivatives vs. Leukemia L1210*

ArN=NN(R)–CH_3

R	Ar = EtO_2C–C_6H_4– NSC	Rating	$H_2NC(O)$–C_6H_4– NSC	Rating	$H_2NNHC(O)$–C_6H_4– NSC	Rating	$H_2NC(=NH)$–C_6H_4– NSC	Rating	o-$COOCH_3$–C_6H_4– NSC	Rating	o-$CONHNH_2$–C_6H_4– NSC	Rating
n-Butyl	87980	+	87429	–	92161	–	112485	–	102246	–	102248	–
Isobutyl	93191	±	93189	–			112472	–	103540	–	103539	–
sec.-Butyl	93192	+	93194	–					103536	–	103535	±
Cyclohexyl	93190	–	93188	–			119653	–	103538	–	103537	–

R	Ar = CH_2O(CO_2Et)–C_6H_4– NSC	Rating	$OC(OEt)(CH_2)_2$–C_6H_4– NSC	Rating	$OC(OEt)(CH_2)_3$–C_6H_4– NSC	Rating	$OC(NHNH_2)(CH_2)_2$–C_6H_4– NSC	Rating	$(R)_2NSO_2$–C_6H_4–, R= H NSC	Rating	R= $ClCH_2CH_2-$ NSC	Rating
n-Butyl	77586	–	77587	+	80344	±	86446	–	84959	–	92163	–
Isobutyl					93496	–						
sec.-Butyl			93505	±	93502	±						
Cyclohexyl	93187	–			93504	–						

CH_3–N(n-C_4H_9)–N=N–C_6H_4–$(CH_2)_2COOEt$

NSC 77587

% ILS, Leukemia L1210	Dose	Schedule	% ILS, Leukemia P388	Dose	Schedule
+79	600	1-9	+84	600	1, 5, 9
+61, +30, +62	400	1-9	+46	400	1, 5, 9
+42	250	1-d	+34	266	1, 5, 9

* +, ILS>35%; ±, ILS 24–35%; –, ILS<24%.

TABLE 7 3-Methyl-3-alkyl-1-triazenyl Heterocycles

Heterocyclyl–N=N–N(R)CH_3

Heterocyclyl	R	NSC	%ILS, L1210	Rating	IP Dose mg/kg	Schedule
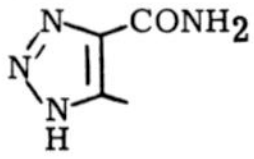 4-carbamoyl-1,2,3-triazol-5-yl ($CONH_2$)	Methyl	65734	+62	+	500	1-d
			+54		400	1-9
			+47		200	1-9
	n-Butyl	70878	+18	–	37	1-d
	Isopropyl	91014	+36, +15	±	100	1-d
	Benzyl	70877	+16	–	175	1-d
	Phenyl	65736	–6	–	375	1-d
imidazol-5-yl, X = CO_2CH_3	Methyl	87982	+71, +65, +61	+	31	1-d
	n-Butyl	90338	+36, +32	+	48	1-d
			+25		75	1-d
	Cyclohexyl	95438	+9	–	150	1-d
			+17		54	1-d
	H	105530	+23	–	50	1-9
CO_2Et	Methyl	98662	+47, +46	+	32	1-d
	n-Butyl	98663	+25, +21	±	38	1-d
	Cyclohexyl	98664	–2	–	400	1-15
	n-Octyl	105767	+23	–	400	1-9
$CO_2C_4H_9$	Methyl	112477	+32	+	42	1-15
			+46		400	D2
	n-Butyl	112481	+12	–	80	1-9
$CO_2C_8H_{17}$	Methyl	100863	+51	+	300	1-9
			+28, +27		200	1-15
	n-Butyl	100864	+17	–	200	1-15
	Cyclohexyl	100865	+11	–	295	1-15
$CO_2CH_2CH_2NEt_2$	Methyl	106628	+17	–	200	1-15
	Cyclohexyl	107380	–2	–	140	1-15
$CONHNH_2$	Methyl	101346	+17	–	400	1-15
	n-Butyl	101344	+4	–	400	1-15
	Cyclohexyl	101345	+27, +12	±	170	1-15
CN	Methyl	138430	+15	–	200	1-9
	n-Butyl	136898	+9	–	300	1-9
pyrazol-4-yl, X = $CONH_2$	Methyl	114924	+44	+	300	D1
			+76*		109	1-9
	n-Butyl	121239	+84	+	66	1-9
			+65, +34		44	1-9
	n-Propyl	145929	+44, +88	+	100	1-9
			+78		66	1-9
	Allyl	153186	+54	+	100	1-9
			+57, +38		50	1-9
	Phenethyl	145930	+34, +35	+	400	1-9
			+48		266	1-9
CO_2CH_3	Methyl	117122	+75, +63	+	88	1-9
	n-Butyl	118320	+60, +36	+	400	1-9
			+50		268	1-7
CO_2Et	Methyl	115757	+34	+	120	1-9
			+41, +28		400	D2
	n-Butyl	118319	+70, +53	+	400	1-9
			+68		266	1-9
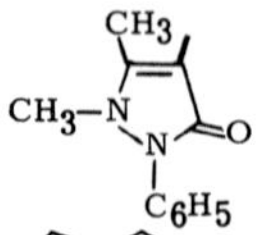 1-phenyl-2,3-dimethyl-5-oxopyrazolin-4-yl (CH_3, CH_3–N, C_6H_5)	Methyl	152646	+13	–	50	1-9
			+19		75	D1
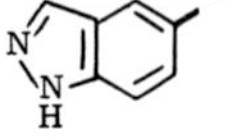 indazol-5-yl	Methyl	168824	+52, +49	+	33	1-9
			+28		22	1-9

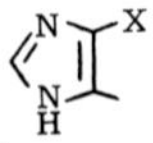

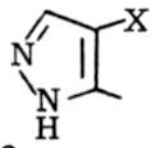

most of our synthesis program; and, obviously, it is an exception to the generalization that activity is associat[ed] with the presence of at least one methyl group in the triazenyl moiety.

Although bis(2-fluoroethyl)triazenes also undergo intramolecular cyclization, these triazenes are more readi[ly] isolated than the 2-chloroethyl triaze[nes] (Shealy and O'Dell, 1970). Except fo[r] the imidazole nitrile (NSC138432) and the pyrazole ester (NSC119651), this group of compounds was active agains[t] L1210 leukemia (Table 10). These bi[s] (2-fluoroethyl)triazenes were more stable, more toxic, and much less active than BIC.

The literature contained some examp[les] of triazolinium salts obtained from several phenyldiazonium salts and bis (2-chloroethyl)amine. We also attemp[ted] to prepare a few simple bis(2-chloroethyl)triazenyl-benzene derivatives without much success. When there is an adjacent carboxamide group, both reactants (30, 31), as well as the product (32), are subject to competing intramolecular cyclizations (30→28, 31→29, 32→33). However, when we tried to prepare heterocyclic derivati[ves] closely related to BIC, we were able to isolate specimens of the bis(2-chloroethyl)triazenyl derivatives of the triazolecarboxamide (34) and pyrazol[e]carboxamide (35), but we were able t[o] isolate only the triazolinium chloride derivatives of the imidazole (36) and py[ra]zole (37) esters. Thus, we obtained bi[...]

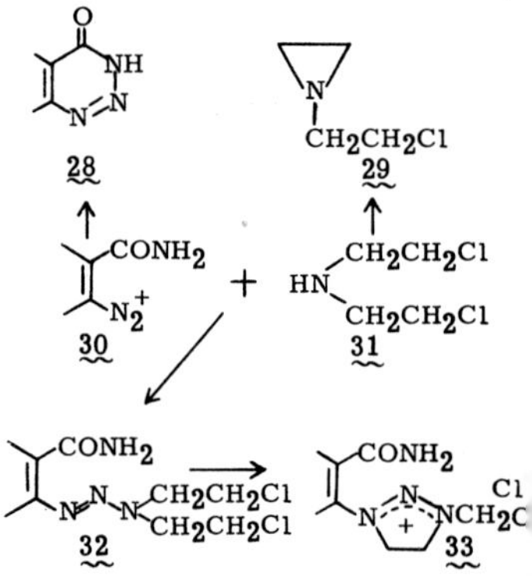

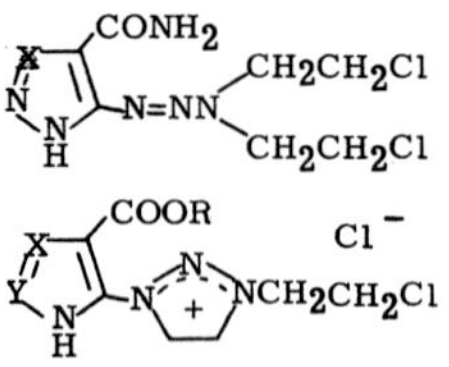

28 29

30 + HN(CH_2CH_2Cl)$_2$ 31

32 → 33 (Cl^-)

34: X=N
35: X=CH

36: X=N, Y=CH, R=CH_3, Et
37: X=CH, Y=N, R=CH_3

TABLE 7 (continued)

		R	NSC	%ILS, L1210	Rating	IP Dose mg/kg	Schedule
	X=CH	Methyl	170321	+103	+	25	1-9
				+51, +47		12	1-9
	X=N	n-Butyl	189821	+13	-	400	D2, 6**

* Noell and Cheng (1969). ** Only schedule

TABLE 8 o-(3,3-Dialkyl-1-triazenyl)benzamides and Related Aryl Triazenes

Ar	R	X	Y	R_1	R_2	NSC	% ILS, L1210	Rating	IP Dose mg/kg	Schedule
A	H	H	H	Methyl	Methyl	136896	+62, +50, +27	+	150	1-9
A	H	H	H	Methyl	n-Butyl	136892	+50, +44, +37	+	200	1-9
A	H	H	H	Methyl	Allyl	145123	+41	+	133	1, 5, 9
							+48, +58		88	1-9
A	H	H	H	Methyl	n-Propyl	145137	+53	+	150	1-9
							+34, +31		100	1-9
A	H	H	H	Methyl	n-Pentyl	145134	+15, +12, +6	-	400, 200, 100	D1
							+12, +1, +9		400, 200, 100	1, 5, 9
							t, +15, +14		400, 200, 100	1-9
A	H	H	H	Methyl	2-Hydroxyethyl	142025	+76	+	600	1-9
							+77, +62		400	1-9
A	H	H	H	n-Propyl	n-Propyl	145136	+3, +2, +7	-	400, 200, 100	D1
							−2, −6, +4		400, 200, 100	1, 5, 9
							t, +2, −10		400, 200, 100	1-9
A	H	Cl	H	Methyl	Methyl	143907	+46	+	400	1-9
A	H	Cl	H	Methyl	n-Butyl	147766	+5, +5, +10	-	400, 200, 100	1-9
A	H	Cl	Cl	Methyl	Methyl	146372	+65, +41	+	200	1-9
							+38, +40		400	1, 5, 9
A	H	Cl	Cl	Methyl	n-Butyl	145138	+43	+	100	1-9
							+81, +34		400	1, 5, 9
A	H	NO_2	H	Methyl	Methyl	143908	+70	+	180	1-9
							+67		268	1-9
							+54		400	1-7
A	H	NO_2	H	Methyl	n-Butyl	143149	+100	+	600	1-9
							+70, +65		400	1-9
							+60		266	1-9
B				Methyl	Methyl	153183	+70, +18	+	400	1-9
							+33		266	1-9
C				Methyl	Methyl	155952	+38, +15	±	200	1-9
A	C_4H_9	H	H	Methyl	n-Butyl	239735	+36*	+	300	1-9

Only test

TABLE 9 Derivatives of DTIC

	R_1	R_2	NSC	% ILS, L1210	Rating	IP Dose mg/kg	Schedule
	H	Propyl	127837	+80, +62	+	150	1-9
	H	Isobutyl	127836	+75	+	150	1-9
				+44		75	1-9
	H	Benzyl	127838	+36, +38	+	400	1-9
				+95*		600	1-9
				+50,* +45*		400	1-9
	H	Methyl	140406	+47	+	400	1-9
	H	Heptyl	145130	+30	±	100	1-9
	H	3-Phenylpropyl	145129	+10		400	1, 5, 9
				Not tested			1-9
enethyl	H	H	87983	+22	-	140	1-9
clohexyl	H	H	157948	+ 1	-	300	1-9
tyl	Butyl	H	145931	+17	-	50	1-9
R_1N=Pyrrolidinyl		H	145924	+50, >+94	+	200	1-9

Leukemia P388

(2-chloroethyl)-triazenyl derivatives of imidazole, triazole, and pyrazole carboxamides, but we did not succeed in isolating the analogous esters. Also, the bis(2-fluoroethyl)triazenyl derivatives of the imidazole and pyrazole carboxamides cyclize more slowly than the corresponding esters. We postulated that the adjacent carboxamide group may exert a stabilizing effect on the triazenyl moiety through hydrogen bonding. For this reason, and despite the three competing intramolecular cyclizations, we investigated the preparation of ortho-bis(2-chloroethyl)triazenyl benzamides. Furthermore, we believed that the presence of electron-attracting groups on the benzene ring should deter cyclization of the triazene to the triazolinium chloride by decreasing electron density at the triazenyl nitrogens. Some of our attempts to prepare additional bis(2-chloroethyl)triazenes are summarized in Table 12. The results shown, in terms of isolated products, indicate that both the presence of an adjacent carboxamide group and strongly electron-attracting aryl moieties may have a stabilizing effect on the triazene, although this picture is certainly not complete. Of those products tested, all of which were predominantly the triazene, the 5-nitrobenzamide and the pyrazolecarboxamide, gave the best results against L1210 leukemia, but they did not approach BIC in effectiveness.

Early in our work, we showed that the monoalkyltriazenyl imidazolecarboxamides dissociate in solution, even in the dark, to AIC;

TABLE 10 3-(2-Hydroxyethyl)-3-methyltriazenes and 3,3-Bis(2-fluoroethyl)triazenes

	X	$ArN{=}NN(CH_3)CH_2CH_2OH$ NSC	% ILS L1210	Rating	IP Dose, mg/kg	Schedule	$ArN{=}NN(CH_2CH_2F)_2$ NSC	% ILS, L1210	Rating	IP Dose, mg/kg	Sched
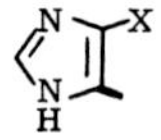	$CONH_2$	83112	+69	+	400	1-9	118317	+58	+	30	D1
			+48		200	1-9		+90, +65, +48		7.5	1-9
	CO_2CH_3	98661	+57	+	600	1-15	115758	+69	+	15	1-9
			+51		400	1-15		+118*	+	66	1,5,9
	CO_2Et	105766	+52, +37	+	400	1-9	---				
			+36		266	1-9					
	CN	142026	+13	–	400	1-9	138432	+21	–	18	1-9
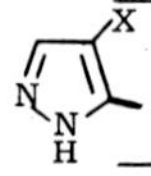	$CONH_2$	133729	+122, +89	+	600	1-9	128730	+24	+	19	D1
			+106, +84		400	1-9		+36		15	1-9
	CO_2R	153179	+36	+	400	1-9	119651	+11	–	19	D1
		($R = CH_3$)	+63		200	1-9	(R = Et)	+11		9.5	1-9
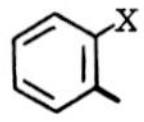	$CONH_2$	142025	+76	+	600	1-9	145121	+13	+	25	D1
			+77, +62		400	1-9		+56, +53		12	1,5,9
								+20		6	1-9
$EtO_2C(CH_2)_2CH(CONH\text{–})–CO_2Et$	CONH	95442	–2	–	400	1-d	---				

* Leukemia P388

TABLE 11 Later Triazenes Lacking a Methyl Group

Ar ($ArN{=}N{-}NR_1R_2$)	NSC	% ILS L1210	Rating	IP Dose mg/kg	Schedule
$EtO_2C(CH_2)_3$–C6H4– ; $R_1 = R_2 = n$-Butyl 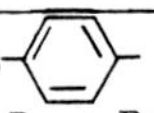	81036	–3	–	500	1-d
		+1		250	1-d
$EtO_2C(CH_2)_2$–C6H4– ; R_1R_2N=Pyrrolidinyl 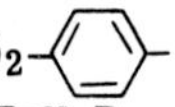	76423	t	–	500	1-d
		–3		250	1-d
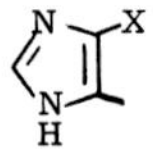 $R_1 = R_2 = n$-Butyl, $X = CO_2CH_3$	108860	t, –3, –3	–	400, 200, 100	D2
		–3		200	1-15
$R_1 = R_2$=Allyl, $X = CO_2CH_3$	150034	–7 to +9	–	400, 200, 100*	
$R_1 = R_2$=Cyclohexyl, $X = CO_2Et$	105770	–8 to +2	–	400, 200, 100	D2
		–18 to –11		400, 200, 100	1-9
$R_1 = R_2$=Propyl, $X = CONH_2$	146370	–5 to +13	–	400, 200, 100**	
		t, +6, –3, –3		400, 200, 100, 50	1-9
R_1=Ethyl, $R_2 = n$-Butyl, $X = CONH_2$	145928	–6, to +16	– ?	400, 200, 100**	
		–12, +24, +17		400, 200, 100	1-9
$R_1 = R_2 = -CH_2CH_2CN$, $X = CONH_2$	92157	–7 to +5	–	400, 200, 100*	
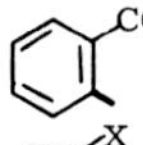 $CONH_2$; $R_1 = R_2$=Propyl	145136	–6 to +7	–	400, 200, 100**	
		t, +2, –10		400, 200, 100	1-9
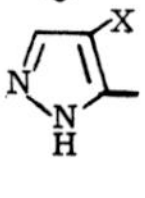 $R_1 = R_2$=Allyl, $X = CO_2Et$	152643	–8 to +11	–	400, 200, 100***	
$R_1 = R_2$=Ethyl, $X = CONH_2$	145927	+26, +44	+	200	1-9
		+32		133	1-9
		+26		100	1-9
$R_1 = R_2$=Isopropyl, $X = CONH_2$	147765	–10 to +14		400, 200, 100*	

* All three doses on schedules D1; 1,5,9; 1-9. ** All three doses on schedules D1 and 1,5,9.
*** All three doses on schedules D1 and 1-9.

and we postulated the f mation of alkylating ag such as methylating ag (42–44, X=H) from MI (Shealy and Krauth, 19 We found that the half-life of MIC in phosphat buffer is about 8 minut but, if administered quickly, it is active against leukemia L121 These properties of M are relevant to later studies of the metaboli of DTIC by Skibba and co-workers (e.g., 1970a, 1970b) and Hous and Loo (e.g., 1971), who showed that it is d methylated by liver mi crosomal oxidases to AIC and that a methylating agent is generate by this process. Presumably, MIC is form by oxidative demethyla via 38 and undergoes t kind of decomposition sented by 39–44 (X=H). Although the markedly greater activity of BI and the presence of th nitrogen mustard-like group suggested a different mechanism of action, we postulated a similar mechanism of activation of BIC (3 44, $X = CH_2Cl$). We pr pared the monochloro ethyltriazene (41), fou it to be highly active, and showed that it dissociates in aqueous media, two of the products being AIC and 2-chloroethanol (Shealy, O'Dell, and Krauth, 1975). Hill (1975) confirmed that AIC is, indeed, formed by liver microsomal oxidases.

Tests of a few monochloroethyltriazenes are summarized in Table 13. The first compound is the postulated intermedi in the metabolic activation of BIC. There were high proportions of 30-day survivors after treatment with this compound at doses much lower than doses of BIC that caused long-term survival. The next two compounds also showed good act

TABLE 12 Bis(2-chloroethyl)triazenes

$ArN_2^+ \rightarrow Ar-N=N-N(CH_2CH_2Cl)_2 \rightarrow Ar-N$ (triazolinium, Cl^-) $N-CH_2CH_2Cl$

A → B

Ar	Product Isolated*	NSC	% ILS** L1210	IP Dose mg/kg	Schedule
5-X-imidazol-4-yl					
X=$CONH_2$	A (BIC)	82196			
X=$COOCH_3$	B				
X=$COOC_2H_5$	B				
X=CN	A (5–10% B)	140405	+7	200	1-9
$CONH_2$-triazolyl	A(20% B)	132927	+50	100	1-9
X-pyrazolyl					
X=$CONH_2$	A (0–5% B)	128729	+96, +78	300	1-9
X=CO_2CH_3	B				
tetrazolyl	A (2 Days, −10°) → A (25% B)	178030	+4	200	D1***
			-7	400	D1
2-$CONH_2$-phenyl	B				
$H_2NC(=O)$-phenyl	B				
Cl, $CONH_2$-phenyl	A (5–10% B)	157952	+5	300	1, 5, 9***
Cl, Cl, $CONH_2$-phenyl	A (12–15% B)	153184	+6	400	1, 5, 9***
Cl, Cl, Cl, $CONH_2$-phenyl	A (5–10% B) → (4 Days, −10°) → A (30–35% B)				
O_2N, $CONH_2$-phenyl	A	150026	+65	400	1, 5, 9
			+40	100	1-9
O_2N, Cl-phenyl	B (25% A)				

* % of B in isolated A determined by NMRanalysis. **Highest observed values. *** Only schedule.

TABLE 13 Monochloroethyltriazenes

5-[$N=N-N(R)CH_2CH_2Cl$]imidazole-4-$CONH_2$

R	NSC	30-Day Survivors	% ILS of Non-Survivors	Dose mg/kg	Schedule
H	157949	4/6	+46	37	D1
		6/6	–	25	D1
		5/6	+51	25	D1
		3/6	+44	17	D1
		3/6	+80	12	D1
		2/6	+66	9	1-9
		3/6	+103	6	1-9
		1/6	+126	6	1-9
		0/6	+56	4	1-9
Methyl	196556	1/10	+42	150	D1
		0/3	+88	100	D2, 6
		1/10	+96	150	1, 5, 9
		1/10	+117	100	1, 5, 9
		0/10	+68	66	1, 5, 9
Benzyl	245436	0/6, 0/6	+95, +74	400	D1
		2/6	+174	400	1, 5, 9
		0/6	+121	200	1, 5, 9
		2/6	+127	100	1-9
X-C6H3(CONH2)-$N=N-N(R)CH_2CH_2Cl$					
R=CH_3 X=H	238112	0/6	+23	200	1, 5, 9*
R=H X=NO_2	284220	0/6	+83	200	D1*

* Only schedule

ity. The ILS values for the benzyl derivative (NSC 245436) were the highest that we have observed except for BIC and the monochloroethyltriazene (NSC 157949); it is also modestly active against Lewis Lung carcinoma.

It is difficult to draw unequivocal conclusions about the antineoplastic activity of triazenes. Although leukemia L1210 was the neoplasm used to compare the activity of our triazenes, some of them were also tested against various solid tumors. These tests cannot be summarized here, but certain of the triazenes tested were active against Walker carcinosarcoma 256, Adenocarcinoma 755, B16 Melanoma, or Lewis Lung carcinoma. Our results with leukemia L1210 indicate that activity is almost always associated with the presence of a methyl group in the triazenyl moiety. Other alkyl groups, including the 2-hydroxyethyl group, may be present in combination with the methyl group. Bis(2-chloroethyl)triazenes and mono(2-chloroethyl)triazenes, if they are sufficiently stable, are highly active, and bis(2-fluoroethyl)triazenes are also active. Chemical, biochemical, and biological studies of DTIC and BIC are consistent with metabolic activation by oxidative dealkylation. An advantage of DTIC and similar imidazole-4-carboxamide triazenes over benzenoid and other heterocyclic triazenes is that the arylamine formed by metabolic activation is a natural metabolite, AIC.

DTIC (4): X=H; BIC (14): X=CH_2Cl ($N=N-N(CH_2-X)_2$) → 38 ($N=N-N(CH(OH)-X)CH_2-X$) → 39 ($N=N-NH-CH_2X$) + X-CHO (40)

MIC: X=H; 41: X=CH_2Cl ($NH-N=N-CH_2X$) ⇌ 39; → 1 AIC (NH_2) + $X-CH_2N=NOH$ (42) → $X-CH_2N_2^+$ (43) → $N_2 + X-CH_2^+$ (44)

ACKNOWLEDGEMENT

The chemical and biological studies described in this report were supported by contracts with the National Cancer Institute, U.S.A. The biological tests were carried out under the supervision of Dr. F. M. Schabel, Jr., and Dr. W. R. Laster, Jr., of this Institute.

REFERENCES

Abel, G., T. A. Connors, and T. Giraldi (1977). In vitro metabolic activation of 1-p-carboxamidophenyl-3,3-dimethyl triazene to cytotoxic products. Cancer Lett., 3, 259-264.

Audette, R. C. S., T. A. Connors, H. G. Mandel, K. Merai, and W. C. J. Ross (1973). Studies on the mechanism of action of the tumour inhibitory triazenes. Biochem. Pharmacol., 22, 1855-1864.

Connors, T. A., P. M. Goddard, K. Merai, W. C. J. Ross, and D. E. V. Wilman (1976). Tumour inhibitory triazenes: structural requirements for an active metabolite. Biochem. Pharmacol., 25, 241-246.

Farquhar, D. (1976). p-(3,3-Dimethyl-1-triazeno)benzamide (DTB): a potential central nervous system (CNS) acti analogue of dacarbazine (DTIC). Proc. Am. Assoc. Cancer Res., 17, 176.

Hill, D. L. (1975). Microsomal metabolism of triazenylimidazoles. Cancer Res., 35, 3106-3110.

Hoffman, G., I. Kline, M. Gang, D. D. Tyrer, J. M. Venditti, and A. Goldin (1968). Influence of treatment sched and route of aministration on the chemotherapy of murine leukemia L1210 with 5(or 4)-[3,3-bis(2-chloroethyl)-1-triazeno]imidazole-4(or 5)-carboxamide (NSC 82196). Cancer Chemother. Rep. (Part 1), 52, 715-724.

Housholder, G. E., and T. L. Loo (1971). Disposition of 5-(3,3-Dimethyl-1-triazeno)imidazole-4-carboxamide, a new antitumor agent. J. Pharmacol. Exp. Ther., 179, 386-395.

Krauth, C. A., Y. F. Shealy, and J. A. Montgomery (1962). Coupling reactions of 5-diazoimidazole-4-carboxamic J. Alabama Academy Sci., 33, 37-38.

Krauth, C. A., Y. F. Shealy, and C. A. O'Dell (1972a). Triazeno compounds. U. S. Patent 3,649,613, March 14.

Krauth, C. A., Y. F. Shealy, and C. A. O'Dell (1972b). Diazo derivative of imidazole carboxylic acid esters. U. Patent 3,654,257, April 4.

Lin, Y.-T., and T. L. Loo (1972). Preparation and antitumor activity of derivatives of 1-phenyl-3,3-dimethyl-triazene. J. Med. Chem., 15, 201-203.

Noell, C. W. and C. C. Cheng (1969). Pyrazoles. III. Antileukemic activity of 3-(3,3-dimethyl-1-triazeno)pyrazol 4-carboxamide. J. Med. Chem., 12, 545-546.

Shealy, Y. F., R. F. Struck, L. B. Holum, and J. A. Montgomery (1961). Synthesis of potential anticancer agents. XXIX. 5-Diazoimidazole-4-carboxamide and 5-diazo-v-triazole-4-carboxamide. J. Org. Chem., 26, 2396-2401.

Shealy, Y. F., C. A. Krauth, and J. A. Montgomery (1962). Imidazoles. I. Coupling reactions of 5-diazoimidazole-4-carboxamide. J. Org. Chem., 27, 2150-2154.

Shealy, Y. F., J. A. Montgomery, and W. R. Laster, Jr. (1962). Antitumor activity of triazenoimidazoles. Bioc Pharmacol., 11, 674-676.

Shealy, Y. F., and C. A. Krauth (1966a). Imidazoles. II. 5(or 4)-(Monosubstituted triazeno)imidazole-4(or 5)-carboxamides. J. Med. Chem., 9, 34-38.

Shealy, Y. F., and C. A. Krauth (1966b). Triazeno-v-triazole-4-carboxamides. Synthesis and antitumor evaluatic J. Med. Chem., 9, 733-737.

Shealy, Y. F., and C. A. Krauth (1966c). Complete inhibition of mouse leukemia L1210 by 5(or 4)-[3,3-Bis(2-chlo ethyl)-1-triazeno]imidazole-4(or 5)-carboxamide (NSC 82196). Nature, 210, 208-209.

Shealy, Y. F., C. A. Krauth, R. F. Pittillo, and D. E. Hunt (1967). A new antifungal and antibacterial agent, met 5(or 4)-(3,3-dimethyl-1-triazeno)imidazole-4(or 5)-carboxylate. J. Pharm. Sci., 56, 147-148.

Shealy, Y. F., C. A. Krauth, S. J. Clayton, A. T. Shortnacy, and W. R. Laster, Jr., (1968a). 5(or 4)-(3-Alkyl-methyl-1-triazeno)imidazole-4(or 5)-carboxamides. J. Pharm. Sci., 57, 1562-1568.

Shealy, Y. F., C. A. Krauth, L. B. Holum, and W. E. Fitzgibbon (1968b). Synthesis and properties of the antileukemic agent 5(or 4)-[3,3-bis(2-chloroethyl)-1-triazeno]imidazole-4(or 5)-carboxamide. J. Pharm. Sci., 57, 83-86.

Shealy, Y. F., and C. A. O'Dell (1970). Imidazole and pyrazole bis(2-fluoroethyl)triazenes. J. Pharm. Sci., 59, 1358-1360.

Shealy, Y. F., and C. A. O'Dell (1971a). Synthesis, antileukemic activity, and stability of 3-(substituted-triazeno) pyrazole-4-carboxylic acid esters and 3-(substituted-triazeno)pyrazole-4-carboxamides. J. Pharm. Sci., 60, 554-

Shealy, Y. F., C. A. O'Dell, and C. A. Krauth (1971b). Imidazole sanitizing compounds and process of making same. Canadian Patent 871,699, May 25.

Shealy, Y. F., C. A. Krauth, C. E. Opliger, H. W. Guin, and W. R. Laster, Jr. (1971c). Triazenes of phenylbutyric, hydrocinnamic, phenoxyacetic, and benzoylglutamic acid derivatives. J. Pharm. Sci., 60, 1192-1198.

Shealy, Y. F., C. A. O'Dell, J. D. Clayton, and C. A. Krauth (1971d). Benzene analogs of triazenoimidazoles. J. Pharm. Sci., 60, 1426-1428.

Shealy, Y. F., C. A. O'Dell, and C. A. Krauth (1975). 5-[3-(2-Chloroethyl)-1-triazenyl]imidazole-4-carboxamide and a possible mechanism of action of 5-[3,3-bis(2-chloroethyl)-1-triazenyl]imidazole-4-carboxamide. J. Phar Sci., 64, 177-180.

Shealy, Y. F., and C. A. O'Dell (1975). Synthesis of 5-(3,3-disubstituted-1-triazenyl)imidazole-4-carbonitriles. J. Pharm. Sci., 64, 954-956.

Skibba, J. L., D. D. Beal, G. Ramirez, and G. T. Bryan (1970a). Demethylation of the antineoplastic agent 4(5)-(3, dimethyl-1-triazeno)imidazole-5(4)-carboxamide by rats and man. Cancer Res., 30, 147-150.

Skibba, J. L., G. Ramirez, D. D. Beal, and G. T. Bryan (1970b). Metabolism of 4(5)-(3,3-dimethyl-1-triazeno)-imidazole-5(4)-carboxamide to 4(5)-aminoimidazole-5(4)-carboxamide in man. Biochem. Pharmacol., 19, 2043-2

Woolley, D. W., and E. Shaw (1951). Some imidazo-1,2,3-triazines and their biological relationship to purines. J. Biol. Chem., 189, 401-410.

Maytansine

J. Douros*, M. Suffness*, D. Chiuten and R. Adamson*****

**Natural Products Branch, National Cancer Institute, Silver Spring, Maryland, U.S.A.*
***Investigational Drug Branch, National Cancer Institute, Bethesda, Maryland, U.S.A.*
****Laboratory of Chemical Pharmacology, National Cancer Institute, Bethesda, Maryland, U.S.A.*

ABSTRACT

Maytansine, an antitumor agent, was the first ansa macrolide isolated from higher plants. This compound was discovered by S. M. Kupchan and co-workers (1972) in the plant *Maytenus ovatus* which has been collected in Kenya. Since that time, a complex of similar compounds called ansamitocin was isolated from *Nocardia* fermentations (Higashide and co-workers, 1977). Several groups in the United States have been attempting to synthesize maytansine and E. J. Corey and co-workers (1978) have recently reported the total synthesis of N-methyl maysenine which contains the maytansinoid ring system and all except two of the functional groups present in maytansine. Maytansine showed sufficient antineoplastic activity against murine tumors in the tumor panel of the NCI to be evaluated in clinical trials conducted under the auspices of the National Cancer Institute. This paper summarizes the preclinical and clinical results so far obtained with maytansine.

ISOLATION

S. M. Kupchan and his co-workers (1972) reported that extracts from *Maytenus ovatus* (family Celastraceae) possessed significant antitumor activity and identified maytansine as one of the active compounds from the plant. The isolation of maytansine was a difficult task since the concentration in the dried plant was only 0.0002% (0.2 mg per kg of dried plant). This group subsequently isolated maytanbutine, maytanprine, maytanvaline, maytanbutacine, maytanacine, maytansinol, maysine, normaysine and maysenine during their work with the *Maytenus* plants (Kupchan and co-workers, 1972a, 1975, 1977). Maytansine (Fig. 1) is an ansa macrolide (an aromatic nucleus to which a macrocyclic aliphatic bridge is attached at two non-adjacent positions). This was the first ansa macrolide isolated from higher plants and this was the first compound of this class to show well defined antitumor activity.

Many ansa macrolides have been isolated from microbial fermentations (geldanomycin, rifamycins, streptovaricins, tolypomycin). On this assumption, in 1974, the NCI had one of its contractors visit Kenya to obtain soil samples from the root of *Maytenus* in hopes of isolating an organism which produced maytansine. To carry out this study, a highly sensitive quantitative assay was developed (Hanka and Barnett, 1974). The microbial assay used was *Penicillium avellaneum* UC-4376. Maytansine was screened against hundreds of organisms and only showed activity against this fungal strain. After the assay was developed, eighteen hundred organisms were

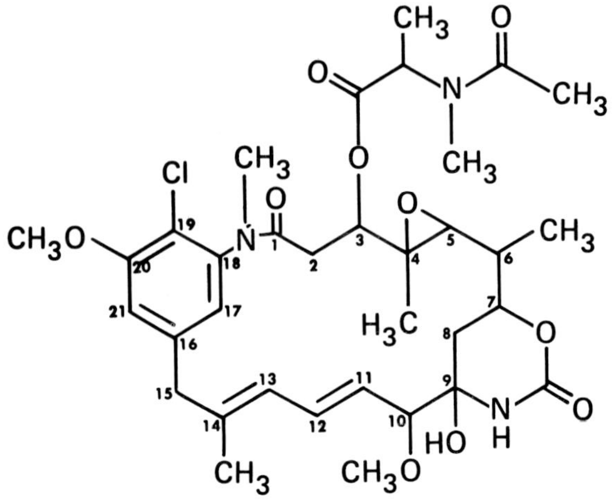

Fig. 1. Maytansine

isolated from the Kenyan soil samples and were fermented but none produced maytansine. However, in December 1977, novel maytansinoid antibiotics were isolated from a Nocardia fermentation (Higashide and co-workers, 1977). This antitumor antibioti complex was named ansamitocin and two of the components, ansamitocin P-3 (TAM-330) and ansamitocin P-4 (TAM-340) were evaluated at NCI and have activity similar to maytansine. TAM-330 is undergoing a direct comparison in the NCI tumor panel with maytansine. The possibility still exists that maytansine might be a transformation product in the plant of a substance of microbial origin (ansamitocins or related compounds)(Fig. 2)

Since maytansine costs approximately $75,000/g to produce by isolation from the plant, the discovery of ansamitocin was a significant breakthrough as analogs of maytansine can now be produced in higher yields at much lower costs. The capabilit to convert ansamitocin P-3 to maytansine has been demonstrated by Takeda Co. which supplied NCI with this compound. The yields of maytansine from various plants and plant parts are given on Table 1.

BIOLOGICAL ACTIVITY

Maytansine demonstrated potent cytotoxicity against cells derived from human carcinoma of the nasopharynx (KB). Maytansine, in low concentrations, inhibits mitosis in sea urchin eggs and inhibits the polymerization of brain tubulin. This drug shows no significant inhibition of cellular DNA and RNA polymerases and viral reverse transcriptase.[1]

Maytansine demonstrated activity in a variety of murine tumor systems. This compound was effective against B16 melanoma in mice giving an average increase in life span of 57%. The drug was effective against L1210 leukemia and showed excellent activity against P388 leukemia in mice (Table 2).

---

[1]Personal Communication - Dr. Sethi, Bowman Gray Medical School, Winston Salem, North Carolina.

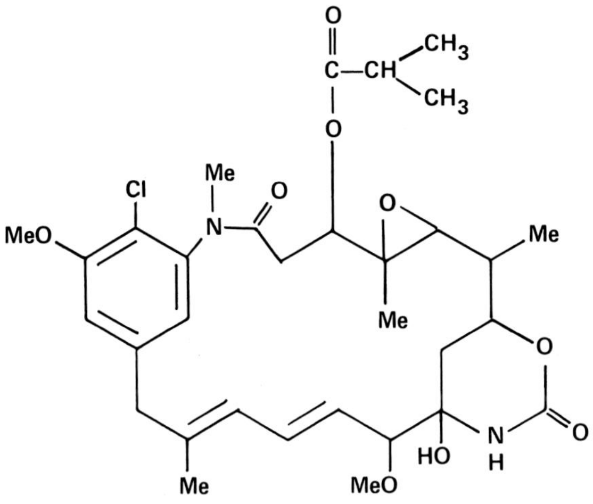

TAM-330 (Ansamitocin P-3)

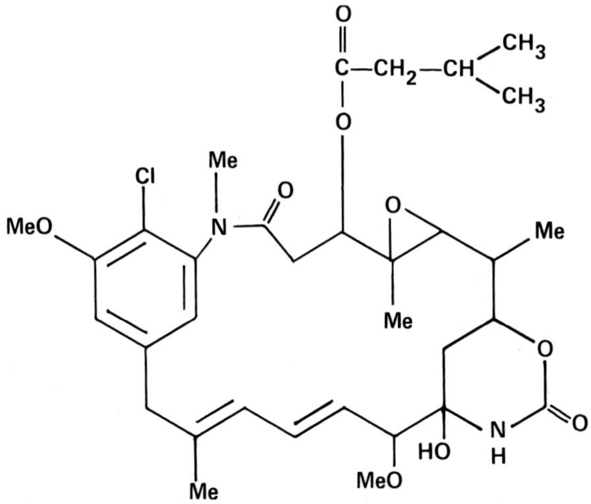

TAM-340 (Ansamitocin P-4)

Fig. 2. Ansamitocins of microbial origin

TABLE 1 Yield of Maytansine from Maytenus and Putterlickia ssp.

Plant Name	Plane Part*	Content** (mg/kg)	Isolated Yield (mg/kg)
Maytenus acuminata	ws	0.13	
Maytenus boaria	ws-sb	<0.1	
Maytenus buchananii	ws-sb-tw		2.5
	ws-sb		0.9
	ws-sb	0.5	
Maytenus emarginata	rt-ws-sb-lf		0.5
Maytenus heterophylla	ws-sb	0.3	
Maytenus mossambicensis	ws-sb		0.05
Maytenus nemorosa	tw-lf	0.01	
Maytenus polycantha	ws-sb-tw	0.08	
Maytenus senegalensis	ws-sb	0.05	
	tw-lf		0
Maytenus texana	tw-lf		0
Putterlickia pyracantha	tw-lf	1.8	
	ws-sb-tw	4.5	
	tw-lf	1.1	
	fr-sd	5.5	
	st	1.3	
Putterlickia verrucosa	ws-sb	12.0	8.5
	tw		6.6
	ws-sb		7.5
Maytenus rothiana	rt-st-lf	5.6	
	fr	6.0	

* as determined by chromatographic data (HPLC)

** ws = woody stems, sb = stembark, lf = leaf, fr = fruit, sd = seed, rt = root, tw = twigs

TABLE 2 Activity of Maytansine in Mouse Tumor Systems

Tumor	Dose *	Increased Life Span
B16 melanoma	16 μg/kg	57%
Colon 26	64 μg/kg	31%
L1210 Leukemia	180 μg/kg	49%
Lewis Lung Carcinoma	32 μg/kg	32%
P388 Leukemia	256 μg/kg	142%

* daily injection 1-9 days IP

Maytansine was inactive against the P388 leukemia subline resistant to vincristine and had little activity against mice inoculated intracerebrally with the P388 tumor. Maytansine was effective in increasing the life span of mice bearing the mast cell tumor P815 and in the plasma cell tumor YPC-1. Maytansine was also active against the Walker 256 carcinosarcoma in rats.

ANALOGS

Kupchan and co-workers (1978) prepared 21 analogs of maytansine to determine the structural requirements for significant antileukemic, cytotoxic, antitubulin and antimitotic activity. They found that the presence of the C-3 ester and the free hydroxyl group at C-9 was necessary for significant antileukemic activity. This was paralleled by the results in the 9KB cell cytotoxicity system, the brain tubulin polymerization assay and the sea urchin egg division assay except that the sea urchin assay showed no decrease in activity by modification at C-9.

However, thus far none of the analogs have shown any marked superiority over maytansine in the testing so far conducted. The activity of some of the compounds screened in the P388 leukemia is shown in Table 3, and the cytotoxicity data in the 9KB system is shown in Table 4. The structures of those compounds are found in Fig. 3. Since many derivatives are highly active in the P388 leukemia, other tumor systems will have to be used to differentiate among the analogs. Now that a better source of maytansinoids is available by fermentation, it is the hope of NCI to evaluate many more analogs of maytansine in order to better understand what are the critical positions in this molecule for activity and toxicity.

TABLE 3 Antitumor Activity of Maytansinoids in the P-388 Leukemia in Mice

Compound	Dose Range (μg/kg/inj.)	Optimal Dose (μg/kg/inj.)	ILS (%)[1]	Comments
Maytansine	100 - 0.4	25	120	highly active
Maytanprine	200 - 3.1	3.1	60	active
Maytanbutine	50 - 0.1	0.8	90	highly active
Maytanvaline	50 - 0.1	6.2	101	highly active
Maytansine ethyl ether	400 - 1.6	25	16	inactive,non-toxic
Maysine	100 - 0.2	100	10	inactive,non-toxic
Normaysine	100 - 0.2	3.1	15	inactive,non-toxic
Maytansinol	400 - 3.1	50	12	inactive,non-toxic
Maytanacine	200 - 0.8	100	130	highly active
Maytansinol, propionate ester	200 - 0.8	25	158	highly active
Maytansinol, bromoacetate ester	200 - 0.9	100	132	highly active
Maytansinol, crotonate ester	200 - 0.8	50	110	highly active
TAM-330	50 - 3.1	50[2]	72	moderately active
TAM-340	50 - 3.1	50[2]	78	moderately active

[1] ILS% = percentage increase in life span of test over control tumored animals.

[2] best activity at highest dose tested, further testing at higher doses indicated

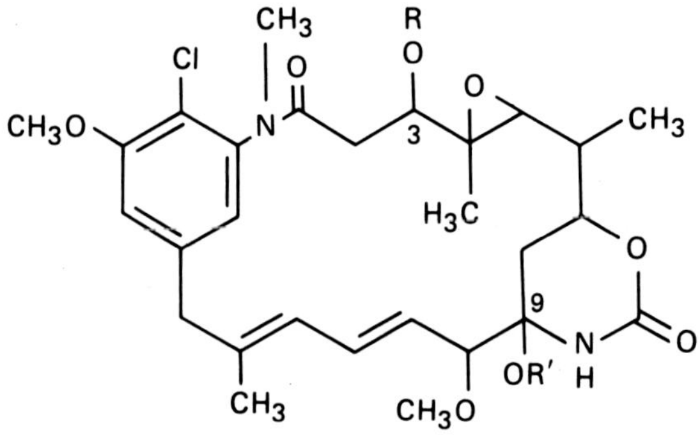

Maytansine	$R=CO \cdot CHCH_3 \cdot NCH_3 \cdot COCH_3, R'=H$
Maytanprine	$R=CO \cdot CHCH_3 \cdot NCH_3 \cdot CHCH_2CH_3, R'=H$
Maytanbutine	$R=CO \cdot CHCH_3 \cdot NCH_3 \cdot COCH(CH_3)_2, R'=H$
Maytanvaline	$R=CO \cdot CHCH_3 \cdot NCH_3 \cdot COCH_2CH(CH_3)_2, R'=H$
Maytansinol	$R=R'=H$
Maytanacine	$R=CO \cdot CH_3, R'=H$
Maytansinol, propionate ester	$R=CO \cdot CH_2CH_3, R'=H$
Maytansinol, bromoacetate ester	$R=CO \cdot CH_2Br, R'=H$
Maytansinol, crotonate ester	$R=CO \cdot CH=CHCH_3, R'=H$
Maytansine ethyl ether	$R=CO \cdot CHCH_3 \cdot NCH_3 \cdot COCH_3, R'=C_2H_5$

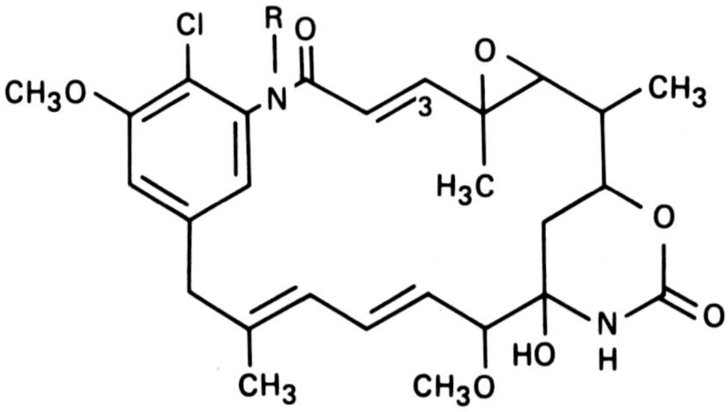

Maysine $R=CH_3$
Normaysine $R=H$

Fig. 3. Structures of maytansinoids

TABLE 4 Cytotoxicity of Maytansinoids in the KB Cell Line

Compound	Number of Experiments	Average ED_{50} (μg/ml)
Maytansine	2	2.5×10^{-5}
Maytanbutine	1	3.6×10^{-6}
Maytanvaline	1	2.1×10^{-6}
Maytanacine	4	1.7×10^{-4}
Maytansinol, propionate ester	2	2.2×10^{-4}
Maytansinol, bromoacetate ester	2	4.7×10^{-4}
Maytansinol, crotonate ester	2	3.5×10^{-4}
Maytansinol	4	1.8×10^{0}
Maytansine ethyl ether	1	1.1×10^{-1}
Maysine	3	2.4×10^{-1}
Normaysine	7	1.9×10^{-1}

PHARMACOLOGY

Maytansine has been found to produce a mitotic arrest similar to that seen with agents such as vincristine and colchicine (Wolpert and co-workers, 1975) (Bhattacharyya and Wolff, 1977). Maytansine, at doses as low as 10^{-8} m, increases the mitotic index of L1210 cells. Maytansine arrests mitosis in metaphase, suggesting that maytansine impairs the function of mitotic spindles (Wolpert and co-workers, 1975) (Johns and co-workers, 1978). Maytansine binds to tubulin at the vincristine binding site and is about 100 times more potent than vincristine (Mandelbaum-Shavit and colleagues, 1976). The C-3 position of the maytansine molecule is important for the binding of tubulin (Kupchan and co-workers, 1978). Maytansine, unlike vincristine, does not protect the independent colchicine binding site (Bhattacharyya and Wolff, 1977). Results indicate an additional binding site for maytansine not shared by vincristine.

Recent work (Luduena, 1978) indicates that maytansine not only operates through the capping mechanism of microtubule inhibition, but also actively stimulates rapid microtubule disassembly in a process actually antagonistic to vinblastine. Recent studies (Sieber and co-workers, 1978) compared the embryotoxic and teratogenic effects of maytansine and vincristine in pregnant Swiss albino mice that received a single i.p. injection of drug on day 6, 7, or 8. Vincristine had greater embryotoxic and teratogenic activity than maytansine at equimolar doses with the peak effects appearing after injection on day 7 of gestation. The teratogenic dose of vincristine was comparable to its effective antitumor dose in transplantable rodent tumor systems; in contrast, the teratogenic dose of maytansine was approximately 10-fold higher than its antitumor dose.

Studies on the pharmacological disposition of maytansine have been initiated. Rats given an iv injection of ^{3}H-maytansine excreted 61% and 6% of the administered radioactivity in 24-hour bile and urine, respectively. Less than 1% of the ^{3}H-activity was recovered in 24-hour thoracic duct lymph. Tritium activity disappeared from plasma with half-times of 54 minutes and more than 24 hours for the alpha and beta phases, respectively. Preliminary results, using thin-layer chromatography, suggest that approximately 18% of the radio-activity in bile was attributable to unchanged maytansine. ^{3}H-maytansine was given as a 20 minute infusion to a patient with pancreatic carcinoma in whom a biliary fistula had been established.

Radioactivity in samples of blood, bile and urine was monitored at intervals for 20 hours after the end of the infusion. Bile contained 28% of the administered tritium activity and 45% of the radioactivity was recovered in urine. Tritium activity disappeared from plasma with half-times of 200 minutes and 16 hours for the alpha and beta phases, respectively.[2]

CLINICAL STUDIES

Phase I clinical studies of maytansine were first begun in the United States in late 1975, with the following objectives: 1) to establish the qualitative and quantitative toxicity of maytansine, 2) to establish the maximum tolerated dose (MTD) with predictable and reversible toxicity, 3) to evaluate any potential therapeutic effects. These studies were carried out in patients with far advanced disease who had become refractory to conventional chemotherapy or had histologic proof of malignancy for which there was no available form of therapy of proven clinical benefits. Table 5 illustrates the different dose schedules, starting and highest escalation dosage, and MTD for each phase I study. The maximum tolerated dose on a daily times three schedule has been .5 mg/m^2/day every 2 to 3 weeks, depending on recovery from toxicity. Patients with liver disease were more susceptible to myelosuppression. In this group of patients, the recommended dose is .25 mg/m^2/day for 3 days. The maximum tolerated dose has been .4 to .5 mg/m^2/day for the daily times five schedule, .75 mg/m^2 for the every other day times three schedule, and 2 mg/m^2 for the single dose schedule. There seemed to be no significant differences in the maximum tolerated dose per course among the 4 different dose schedules explored.

TOXICITY

The major toxicities of maytansine in the phase I-II clinical trials are summarized in Table 6. Gastrointestinal toxicity (i.e. nausea, vomiting, and diarrhea) was dose-related and dose-limiting. It usually occurs on day 3 and lasts for 3-7 days after drug administration (Blum and Kahlert, 1978; Chabner and co-workers, 1978; Eagan and co-workers, 1978). The overall incidence of GI toxicity is 22% (Eagan and co-workers, 1978).

Hepatic toxicity which was reported in all series of patients occurred in 11% of the courses of therapy (Eagan and co-workers, 1978). The incidence increased with increasing doses. Hepatotoxicity varies from subclinical transient elevations of liver enzymes to manifestation of jaundice. The abnormal liver functions usually return to normal within two to four weeks. Patients with liver metastases and/or elevated liver enzymes prior to initiation of treatment, developed rapid deterioration of their liver functions during therapy. It was suggested that a decrease in the dosage of maytansine to .25 mg/m^2 daily times three in patients with abnormal liver function tests might be best (Cabanillas and co-workers, 1978).

Neurotoxicity occurred in 13% of 55 courses when maytansine was given at .6 mg/m^2 daily times three schedule (Eagan and co-workers, 1978). The dose-related neurotoxicity of maytansine was both central and peripheral (Blum and Kahlett, 1978; Eagan and co-workers, 1978). The central toxicity was characterized by lethargy, dysphoria, insomnia, agitated depression, and light-headedness. The peripheral neuropathy, which resembled the *Vinca* alkaloid neuropathy, was in the form of paresthesias, jaw pain, loss of deep tendon reflexes, muscle pain, and weakness. Greater neurotoxicity was noted in patients with previous *Vinca* alkaloid or carcinomatous neuropathy. Blum *et al.* observed a qualitative similarity between the neurotoxicity of maytansine and the cumulative *Vinca* alkaloid toxicity.

Hematologic toxicity was seen infrequently, and usually presented as transient thrombocytopenia which occurred between day 14 and 21, and recovered within 4-7

[2] Drs. Susan M. Sieber and Paul E. Gormley - personal communication.

TABLE 5 Phase I Studies with Maytansine

Dose Schedule	Dosage (mg/m^2)			
	Starting	Highest escalation	MTD *	
Daily x 3 every 2-3 wks	.01	0.9	.5 (.25)+	Cabanillas
Daily x 5 every 3 wks	.01	0.8	.4-.5	Blum
Every other day x 3 every 4 wks	.015	0.9	.75	Eagan
Single dose every 3 wks	.03	2	2	Chabner

* Maximum tolerated dose

+ For patients with liver disease

TABLE 6 Major Toxicities of Maytansine

(Phase I-II Trials)

Dose Schedule (Reference)	Range of Dosage (mg/m^2)	No. of Courses (patients)	Toxic Effects in %				
			Nausea & Vomiting	Diarrhea	LFT[a]	Neurotoxicity	Phlebitis
Daily x 5 (Cabanillas)	.15-5	30	16	33	-	-	14
	.6-.9	23	60	65	13	-	-
Daily x 5 (Blum)	.32-.8	51	+	+	47	+	12
Single dose (Chabner)	1.2-2	9	89	56	22	-	-
Every other day x 3 (Eagan)	.45-2.7[b]	40[c] + 21[d]	15	13	18	5	-
Daily x 3[e] (Eagan)	.6	73	22[f]	42[f]	-	-	-
		55	-	36[g]	11[g]	13[g]	-

a) Liver function tests
b) Total dosage of drugs administered per course
c) No. of initial treatment courses
d) No. of subsequent escalation courses
e) Phase II study
f) Immediate toxicity
g) Delayed toxicity

days (Chabner and co-workers, 1978; Eagan and co-workers, 1978). Cabanillas and co-workers observed myelosuppression in 9 of 88 courses, 8 of which occurred in patients with abnormal liver function tests prior to therapy. In general, myelosuppression was mild, reversible, and not dose-related. The overall incidence is 10%.

Other toxicities that occurred occasionally are summarized below. Local phlebitis has been observed (Cabanillas and co-workers, 1978; Blum and Kahlert, 1978) and noted to occur more frequently at the higher dose levels. The overall incidence was 12% of the courses given (Blum and Kahlert, 1978). This problem can be circumvented by rapid bolus injection and by diluting the drug in larger volumes (250-500 mg) of fluid. Constipation has been noted in 25% of the cases (Eagan and co-workers, 1978; Blum and Kahlert, 1978). It usually occurs during the second week of each course but recovers by day 21 (Blum, 1978). Some patients reported anorexia and a metallic taste post chemotherapy until the second week of each course (Blum and Kahlert, 1978). Stomatitis and alopecia have been infrequently reported toxicities of maytansine. No renal toxicity was encountered.

ANTITUMOR ACTIVITY

Antitumor activity was seen in the initial clinical trials in acute lymphocytic leukemia (ALL), lymphoma, ovarian cancer, breast cancer and melanoma (Chabner and co-workers, 1978; Cabanillas, 1978) (Table 7). A phase II study using maytansine in lung cancer has been completed and maytansine was shown to have minimal activity in this type of malignance (Eagan and co-workers, 1978). Responders in ALL and lymphoma had received prior vincristine (VCR), suggesting lack of absolute cross-resistance. The number of patients being studied in other tumors was too small in numbers for evaluation. Further trials to confirm this activity are indicated. Ongoing phase II trials in other solid tumors are shown in Table 8.

In contrast to the preclinical toxicologic findings in dogs and monkeys, myelosuppression was infrequent and mild; renal toxicity was not seen. Neurotoxicity, present in mice (Sieber and co-workers, 1976) but absent in animals, occurred in humans. However, the preclinical studies were able to predict gastrointestinal side effects. Although maytansine was inactive against the subline of the P388 leukemia resistant to vincristine in mice, it did not show any cross resistance with vincristine in clinical trials. It has an antitumor spectrum similar to vincristine in experimental animals and has a much greater ability to bind to tubulin. Because of its potent stathmokinetic effects and its lack of myelosuppression, it will warrant studies in combination with other drugs.

SUMMARY

Maytansine is a highly interesting antitumor agent which is undergoing extensive clinical trials under the auspices of NCI.

Other dosage schedules, together with drugs to reduce symptoms of gastrointestinal toxicity, will be studied. Possibly better analogs will be made available since the supply problem of maytansinoids may be alleviated by the discovery of a microbial fermentation to produce them.

TABLE 7 Clinical Responses in Maytansine
(Phase I-II Trials)

Tumor Type	No. of response/No. of patients	(%)	Investigator
ALL	2/4	(50)	Chabner
Lymphoma[+]	1/3	(33)	Chabner
Ovary	1/5	(20)	Chabner
Breast	1/7	(14)	Cabanillas
Melanoma	1/7	(14)	Cabanillas
Lung			
non-small cell	4/42	(10)	Eagan[*]
small cell	0/5	(0)	

+ Non-Hodgkins type

* Phase II trials

TABLE 8 Maytansine
Ongoing Phase II Studies

Tumor Type	Institution
Breast	M. D. Anderson, Mayo SWOG[1], Wayne State
Colon	Mayo
Head and Neck	H & N task force
Lung	BCRC[2]
Lymphoma	ECOG[3]
Melanoma	Mayo, M. D. Anderson, Wayne State
Mesothelioma	ECOG[3]
Ovary	Wayne State
Pelvic Malignancies	GOG[4]
Sarcoma	ECOG[3]
Solid Tumor	SEG[5]

1 SWOG - Southwest Oncology Group

2 BCRC - Baltimore Cancer Research Center

3 ECOG - Eastern Cooperative Oncology Group

4 GOG - Gynecologic Oncology Group

5 SEG - Southeast Oncology Group

BIBLIOGRAPHY

Bhattacharyya, B. and J. Wolff (1977). Maytansine binding to the Vinblastine sites of tubulin. FEBS Letters, 159-162.
Blum, R. H. and T. Kahlert (1978). Maytansine: A Phase I study of an ansa macrolide with antitumor activity. Cancer Treatment Rep., 62, 435-438.
Cabanillas, F., V. Rodriguez, S. W. Hall, M. A. Burgess, G. P. Bodey and E. J. Freireich (1978). Phase I study of maytansine using a 3-day schedule. Cancer Treatment Rep., 62, 425-428.
Chabner, B. A., A. S. Levine, B. L. Johnson and R. C. Young (1978). Initial clinical trials of maytansine, an antitumor plant alkaloid. Cancer Treatment Rep., 62, 429-434.
Corey, E. J., L. O. Weigel, D. Floyd and M. C. Bock (1978). Total synthesis of (±) N-methylmaysenine. J. Amer. Chem. Soc., 100, 2916-2918.
Eagan, R. T., J. N. Ingle, J. Rubin, S. Frytak and O. G. Moertel (1978). Early clinical study of intermittent schedule for maytansine (NSC 153858): Brief communication. J. Natl. Cancer Insts., 60(1), 93-96.
Eagan, R. T., E. T. Creagan, J. N. Ingle, S. Frytak and J. Rubin. Phase II evaluation of maytansine (NSC 153858) in patients with metastatic lung cancer. In press (Cancer Treatment Rep.).
Hanka, L. J. and M. S. Barnett (1974). Microbiological assays and bioautography of maytansine and its homologues. Antimicrob. Agents Chemother., 6, 651-652.
Higashide, E., M. Asai, K. Ootsu, S. Tanida, Y. Kozai, T. Hasegawa, T. Kishi, Y. Sugino and M. Yonedo (1977). Ansamitocin, a group of maytansinoid antibiotics with antitumor properties from Nocardia. Nature, 270, 721-722.
Johns, D. G., M. K. Wolpert-DeFilippes, F. Mandelbaum-Shavit, B. A. Chabner, V. H. Bono, Jr. and R. H. Adamson (1978). Pharmacological action of maytansine. Current Chemotherapy, 1179-1182.
Kupchan, S. M., Y. Komoda, W. A. Court, G. J. Thomas, R. M. Smith, A. Karim, C. J. Gilmore, R. C. Haltwanger and R. F. Bryan (1972). Maytansine, a novel antileukemic ansa macrolide from Maytenus ovatus. J. Amer. Chem. Soc., 94, 1354.
Kupchan, S. M., Y. Komoda, G. J. Thomas and H. P. Hintz (1972a). Maytanprine and maytanbutine, new antileukemic ansa macrolides from Maytenus buchananii. J. Chem. Soc., Chem. Comm., 1065.
Kupchan, S. M., A. R. Branfman, A. T. Sneden, A. K. Verma, R. G. Dailey, Jr. and Y. Komoda (1975). Novel maytansinoids. Naturally-occurring and synthetic antileukemic esters of maytansinol. J. Amer. Chem. Soc., 97, 5294.
Kupchan, S. M., Y. Komoda, A. R. Branfman, A. T. Sneden, W. A. Court, G. J. Thomas, H. P. J. Hintz, R. M. Smith, A. Karim, G. A. Howie, A. K. Verma, Y. Nagao, R. G. Dailey, Jr., V. A. Zimmerly and W. C. Sumner, Jr. (1977) The maytansinoids. Isolation, structural elucidation, and chemical interrelation of novel ansa macrolides. J. Org. Chem., 42, 2439.
Kupchan, S. M., A. T. Sneden, A. R. Branfman, G. A. Howie, L. I. Rebhum, R. W. Wang, W. E. McIvor and T. C. Schnaitman (1978). Structural requirements for antileukemic activity among naturally occurring and semisynthetic maytansinoids. J. Med. Chem., 21, 31-37.
Luduena, R. F. Maytansine interacts with microtubules by a novel mechanism of action (1978). Nature. In press.
Mandelbaum-Shavit, F. M., M. K. Wolpert-DeFilippes and D. G. Johns (1976). Binding of maytansine to rat brain tubulin. Biochem. Biophys. Res. Commun., 72, 47-54.
Sieber, S.M., M. K. Wolpert, R. H. Adamson, R. L. Cysyk, V. H. Bono and D. G. Johns (1976). Experimental studies with maytansine - A new antitumor agent. Bibliotheca Haematologica, 43, 495-500.
Sieber, S. M., J. Whang-Peng, C. Botkin and T. Knutsen (1978). Teratogenic and cytogenetic effects of some plant-derived antitumor agents (vincristine, colchicine, maytansine, VP-16-213 and VM-26) in mice. Teratology, 18(1), 31-47.

Wolpert-DeFilippes, M. K., R. H. Adamson, R. L. Cysyk and D. G. Johns (1975). Initial studies on the cytotoxic action of maytansine, a novel ansa macrolide. Biochem. Pharmacol., 24, 751-754.

Wolpert-DeFilippes, M. K., V. H. Bono, R. L. Dion and D. G. Johns (1975). Initial studies on maytansine induced metaphase arrest in L1210 murine leukemia cells. Biochem. Pharmacol., 24, 1735-1738.

The Development of Bruceantin as a Potential Chemotherapeutic Agent*

Albert T. Sneden

Department of Chemistry, Virginia Commonwealth University, Richmond, Virginia 23284, U.S.A.

ABSTRACT

Bruceantin is the primary antileukemic principle of Brucea antidysenterica, a tree used in the treatment of cancer in Ethiopia. Bruceantin and eight companion quassinoids were first isolated in 1972 by the late Dr. S. Morris Kupchan and co-workers at the University of Virginia. Bruceantin exhibited significant inhibitory activity against several tumor systems including the P388 lymphocytic leukemia, the L1210 lymphoid leukemia, the Lewis Lung carcinoma, and the B16 melanocarcinoma, all in mice. Bruceantin was also shown to irreversibly inhibit protein synthesis in HeLa cells, rabbit reticulocytes, and reticulocyte lysates and to partially inhibit DNA synthesis in HeLa cells. Structure-activity studies of bruceantin and its homologs showed that the C-15 side chain ester is required for optimal anti-leukemic activity and probably plays a role in transporting bruceantin across the cell wall. An intact diosphenol moiety in the A ring of bruceantin is also required for activity. Bruceantin has passed preclinical toxicologic and pharmacologic studies and is now undergoing Phase I clinical trials in the United States.

INTRODUCTION

Among the most promising of the new tumor-inhibitory compounds isolated from plants during the last ten years is the potent antileukemic quassinoid bruceantin (Fig. 1). Bruceantin was isolated in 1972 from the stem bark of the Ethiopian tree *Brucea antidysenterica* Mill. (Simaroubaceae) by the late Dr. S. Morris Kupchan and co-workers at the University of Virginia (Kupchan and co-workers, 1973b, 1975a). Fractionation of the ethanolic extract was guided by assays against the P388 lymphocytic leukemia *in vivo* and *in vitro* against a cell culture derived from a human epidermoid carcinoma (KB). In addition to bruceantin, eight companion quassinoids (Fig. 1) were also isolated from *B. antidysenterica*, and three of these, bruceantarin, bruceantinol, and isobruceine B exhibited significant activity against both the P388 and the KB systems. Bruceantin, bruceantarin, bruceantinol, bruceine B, and dehydrobruceantin were also isolated from the Ghanaian tree *Brucea guineensis* G. Don (Kupchan and co-workers, 1975a).

---

*Dedicated to the memory of Dr. S. Morris Kupchan.

Bruceantin, bruceantarin, bruceantinol, and the previously known (Polonsky and co-workers, 1967) bruceine B are all esters of bruceolide (Fig. 1). They differ from other antileukemic quassinoids such as holacanthone (Wall and Wani, 1970) and glaucarubinone (Gaudemer and Polonsky, 1965; Kupchan and co-workers, 1976a)(Fig.2) by having the C-8 hydroxymethyl moiety linked to C-13 as an ether bridge rather than to C-11 as a hemiketal bridge and a hydroxyl group at C-3 rather than C-1. The dehydro compounds (Fig. 1) have a 2-hydroxy-3-keto arrangement of the diosphenol A ring rather than the 3-hydroxy-2-keto arrangement of the parent compounds.

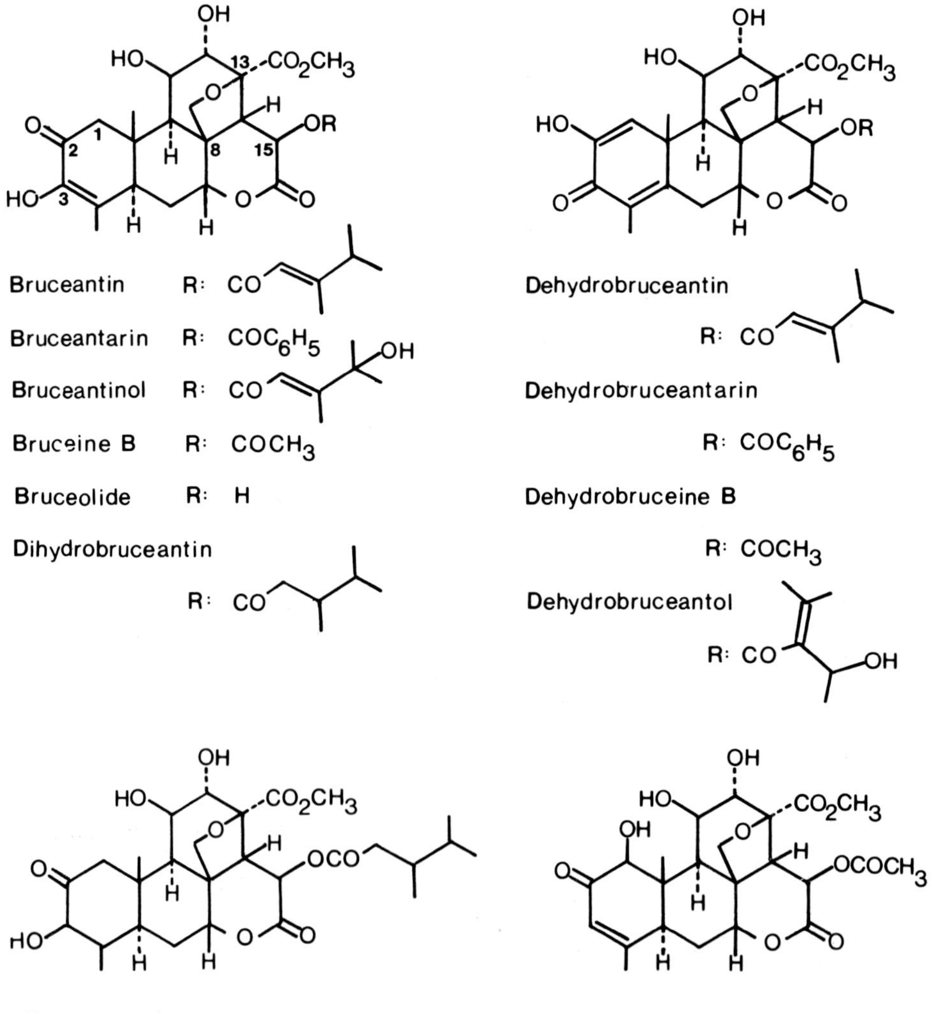

Fig. 1. Bruceantin and its congeners.

Holacanthone

R^1: H R^2, R^3: CH_3

Glaucarubinone

R^1: H R^2: CH_3 R^3: C(OH)(CH_3)C_2H_5

Dehydroailanthinone

R^1, R^2: $=CH_2$ R^3: $CH(CH_3)C_2H_5$

Quassimarin

Fig. 2. Other quassinoids showing antileukemic activity.

BIOLOGICAL ACTIVITY

Although a number of quassinoid principles were used as antiamoebic agents (Polonsky, 1973) and several genera of the Simaroubaceae family were used for anti-cancer treatments in folk medicine (Hartwell, 1971), bruceantin was one of the first quassinoids to exhibit significant tumor-inhibitory activity.[2] Bruceantin proved to be highly active against several tumor systems (Table 1) including the P388 lymphocytic leukemia (PS), the L1210 lymphoid leukemia (LE), an adriamycin resistant P388 leukemia (PA), a cytoxan resistant P388 leukemia (PO), the B16 malanocarcinoma (B1), and the Lewis Lung carcinoma (LL). Bruceantin was found (Liao, Kupchan, and Horwitz, 1976) to irreversibly inhibit protein synthesis in HeLa cells, rabbit reticulocytes, and reticulocyte lysates. Bruceantin also partially inhibited DNA synthesis in HeLa cells but had no effect on RNA synthesis.

The anti-tumor activity of bruceantin led to the screening of several other quassinoid compounds, including glaucarubinone, dehydroailanthinone (Kupchan and Lacadie, 1975b), and quassimarin (Kupchan and Streelman, 1976b) (Fig. 2). Among these were several congeners of bruceantin (Fig. 1). Of the naturally occurring congeners of bruceantin, bruceantarin and bruceantinol showed significant activity against the P388 leukemia (Table 2), while bruceine B and bruceolide showed only marginal activity. The dehydro compounds were inactive.

---

[2] Holacanthone (Wall and Wani, 1970) was evidently the first quassinoid to exhibit significant antitumor activity.

TABLE 1 Activity (T/C*) of Bruceantin Against In Vivo Tumor Systems

Dose (mg/kg)	Tumor Systems PS	LE	B1	LL	PA	PO
2.25					149	
1.50					145	
1.00	140	141	164	133	155	198
0.66					149	205
0.50	220	135	168	128		
0.44					125	209
0.25	195	128	160	128		
0.125	135	127	160			
0.062	165		145			
0.031	150					

*T/C=Test evaluation/Control evaluation X 100 = Percent evaluation. All results are evaluated on the basis of median survival time.

STRUCTURE-ACTIVITY RELATIONSHIPS

From the data in Table 2, it is apparent that there are two major structural features of bruceantin and its congeners which are involved in the activity of the compounds, the C-15 ester and the A ring diosphenol moiety. Variations in the structure of the C-15 ester greatly affect the antileukemic activity (Kupchan and

TABLE 2 Antileukemic Activity of Quassinoids from Brucea Antidysenterica Against the P-388 Lymphocytic Leukemia In Vivo (T/C)

Compound	Dose (mg/kg) 4.0	2.0	1.0	0.5	0.25	0.125	0.06	0.03
Bruceantin	Toxic	Toxic	140	220	195	135	165	150
Bruceantarin		150	145	135	125	135	140	100
Bruceantinol		194	238	211	200	183	183	133
Bruceine B			130	115	110	105	100	100
Bruceolide	147	110	137	117	99			
Dehydrobruceantin		110	120	105	100	135	115	120
Dehydrobruceantarin	100			133*				111**
Dihydrobruceantin	163	168	163	140	136	136	122	113
Tetrahydrobruceantin	126	108	103	103	106	115	105	

*Tested at 0.4 mg/kg
** Tested at 0.04 mg/kg

co-workers, 1976a). Bruceantin and bruceantinol, both of which have α,β-unsaturated side chain esters, are the most active compounds against the P388 leukemia system. Bruceantarin which has a C-15 benzoate ester and dihydrobruceantin, in which the unsaturated ester has been reduced to the saturated ester, both show diminished activity. The C-15 acetate, bruceine B, and the C-15 alcohol, bruceolide, both show only marginal activity at best.

The C-15 ester was proposed to be involved in the transport of the compound across the membrane of the cell (Kupchan and co-workers, 1975a). Horwitz and co-workers (Liao, Kupchan, and Horwitz, 1976) found that differences in the concentrations of bruceantin and its congeners required for inhibition of protein synthesis in HeLa cells and rabbit reticulocytes paralleled the relative order of antileukemic activity. Bruceantin and dihydrobruceantin were found to be most active in the inhibition of protein synthesis, requiring concentrations of 10^{-8} M for 50% inhibition. Bruceantarin and bruceine B were approximately ten times less active than bruceantin, and bruceolide was about 100 times less active than bruceantin. When the inhibition of protein synthesis in reticulocyte lysates was examined, however, all five compounds were approximately equal in activity. These results strongly supported the hypothesis that the main function of the ester is transport and suggested that longer, hydrophobic esters should be more effective and result in increased activity.

The A ring diosphenol moiety appears to be directly involved in the reactions which result in tumor-inhibitory activity and inhibition of protein synthesis, and this moiety must be intact for optimum activity. Dehydrobruceantin and dehydrobruceantarin, in which the diosphenol moiety has been altered, show no activity (Table 2) against the P388 leukemia. Tetrahydrobruceantin, in which the double bond of the A ring has been reduced also shows no activity against the P388 leukemia. Further evidence for the requirement of an intact A ring was found by Horwitz and co-workers (Liao, Kupchan, and Horwitz, 1976). Tetrahydrobruceantin was approximately 200 times less potent as an inhibitor of protein synthesis in HeLa cells and rabbit reticulocytes than bruceantin and dihydrobruceantin and about 100 times less potent in reticulocyte lysates. Thus, it is evident that an intact diosphenol A ring is required for significant biological activity.

The hypothesis that the tumor-inhibitory activity of many natural products may be due to their ability to selectively alkylate free sulfhydryl or amino groups in biological macromolecules has been discussed by several researchers (Kupchan, 1973a; Fujita and Nagao, 1977). The ability of bruceantin and its congeners to alkylate thiol moieties was investigated by Kupchan and Valdenegro (Valdenegro-Benitez, 1976) using the reaction of thioglycolic acid anilide with bruceantarin, dihydrobruceantin, and tetrahydrobruceantin (Fig. 3).[3] The reaction was carried out using boron trifluoride etherate as a catalyst, and the C-2 hemithioketal adducts of bruceantarin and dihydrobruceantin were isolated. However, tetrahydrobruceantin did not react in a similar fashion. Instead, a substitution product[4] was obtained. This difference in the chemical reactivity between bruceantarin and dihydrobruceantin, compounds with the diosphenol A ring intact, and tetrahydrobruceantin, which has a reduced A ring, toward thioglycolic acid anilide suggests that the difference in biological activity might result from a similar difference in reactivity toward a biological nucleophile such as a thiol group on a growth regulatory enzyme.

---

[3] Bruceantin was unstable under the reaction conditions and formed no stable adduct.

[4] The exact position of substitution (either C-2 of C-3) was undetermined.

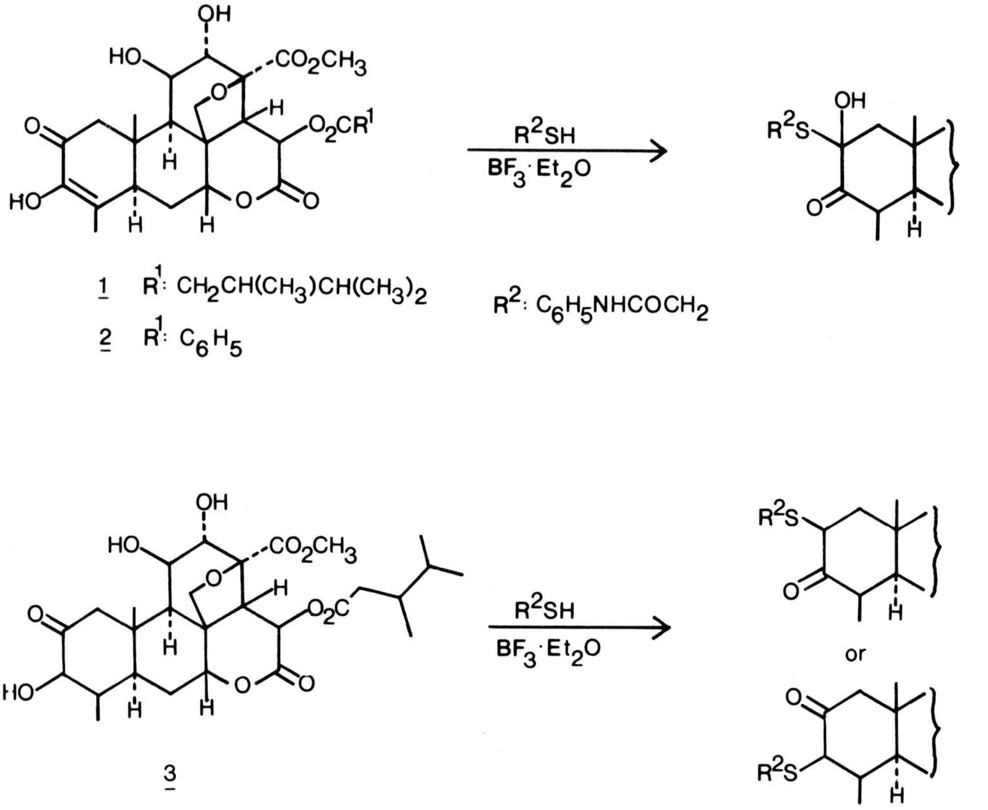

1 R^1: $CH_2CH(CH_3)CH(CH_3)_2$
2 R^1: C_6H_5
R^2: $C_6H_5NHCOCH_2$

Fig. 3. The reaction of dihydrobruceantin (1), bruceantarin (2), and tetrahydrobruceantin (3) with thioglycolic acid anilide.

TOXICOLOGIC STUDIES

Because of its high activity in several animal tumor systems, bruceantin was selected for clinical trial by the National Cancer Institute. Preclinical toxicologic studies using mice, dogs, and monkeys (Castles and co-workers, 1976, 1977a, 1977b; Hamlin and co-workers, 1977) indicated that the principal toxic effect was on the cardiovascular system, particularly an increased permeability of the peripheral vasculature. Bruceantin also had some effect on the hematopoietic system, liver and kidneys, and lymphoid tissues. Among the toxic symptoms noted were edema, diarrhea, emesis, and erythema, and hemorrhages were noted in some tissues upon necropsy.

In dogs, single doses of more than 10 mg/m^2 caused death within 12 to 16 hours. At single doses of 10 mg/m^2 or less toxicity appeared to be reversible for the most part.[5] Dogs given five daily doses of 5 mg/m^2/day showed adverse clinical signs and one dog died on day 3. However, dogs given five daily doses of 2.5 mg/ m^2/day showed milder toxic effects and the effects appeared to be reversible after treatment was completed. Monkeys given five daily doses of 5 mg/m^2/day had severe toxic responses, while doses of 2.5 mg/m^2/day produced somewhat milder effects (no deaths) and the toxicity again appeared to be reversible after completion of treatment. Dogs given three dose regimens of five daily doses of 2.5 mg/m^2/day followed by nine days rest exhibited toxic symptoms during treatment periods only. Dogs given six weekly doses of 5 mg/m^2/week showed toxic symptoms which lessened with time and disappeared upon completion of treatment. The latter two studies indicate that bruceantin's toxicity is not cumulative.

Bruceantin entered clinical trials in the fall of 1977 and is currently undergoing Phase I studies in three hospitals. Although it is too early to define bruceantin's clinical promise, bruceantin has certainly been shown to be very effective in animal tests to date.

REFERENCES

Castles, T. R., J. C. Bhandari, C. C. Lee, A. M. Guarino, and D. A. Cooney (1976). Preclinical toxicologic evaluation of bruceantin (NSC-165563) in mice, dogs, and monkeys. U.S. NTIS, PB Rep., PB-257175.

Castles, T. R., J. C. Bhandari, C. C. Lee, A. M. Guarino, and D. A. Cooney (1977a). Preclinical toxicologic evaluation of bruceantin (NSC-165563) in dogs. U.S. NTIS, PB Rep., PB-269584.

Castles, T. R., R. L. Bridges, J. C. Bhandari, and C. C. Lee (1977b). Preclinical evaluation of bruceantin (NSC-165563) in mice, dogs, and monkeys. Toxicology and Applied Pharmacology, 41, 192.

Fujita, E. and Y. Nagao (1976). Tumor inhibitors having potential for interaction with mercapto enzymes and/or coenzymes - A Review. Bioorg. Chem., 6, 287-309.

Gaudemer, A. and J. Polonsky (1965). Structure de la glaucarubinone, nouveau principe amer isole de Simaruba glauca. Phytochem., 4, 149-153.

Hamlin, R. L., F. S. Pipers, K. Nguyen, P. Mihalko, and R. M. Folk (1977). Acute cardiovascular effect of bruceantin (NSC-165563) following continuous intravenous infusion to anesthetized beaglehounds. U.S. NTIS, PB Rep., PB-264128.

Hartwell, J. L. (1971). Plants used against cancer. A survey. Lloydia, 34, 204-255.

Kupchan, S. M. (1973a). Selective alkylation: a mechanism of tumor inhibition. Intra-Science Chem. Rep., 8, 57-66.

Kupchan, S. M., R. W. Britton, M. F. Ziegler, and C. W. Sigel (1973b). Bruceantin, a new potent antileukemic simaroubolide from Brucea antidysenterica. J. Org. Chem., 38, 178-179.

Kupchan, S. M., R. W. Britton, J. A. Lacadie, M. F. Ziegler, and C. W. Sigel (1975a). The isolation and structural elucidation of bruceantin and bruceantinol, new potent antileukemic quassinoids from Brucea antidysenterica. J. Org. Chem., 40, 648-654.

Kupchan, S. M. and J. A. Lacadie (1975b). Dehydroailanthinone, a new antileukemic quassinoid from Pierreodendron kerstingii. J. Org. Chem., 40, 654-656.

Kupchan, S. M., J. A. Lacadie, G. A. Howie, and B. R. Sickles (1976a). Structural

[5]At a single dose of 10 mg/m^2, one dog died on day 2 and one dog exhibited severe toxic symptoms which gradually disappeared after day 4.

requirements for biological activity among antileukemic glaucarubolone ester quassinoids. J. Med. Chem., 19, 1130-1133.

Kupchan, S. M. and D. A. Streelman (1976b). Quassimarin, a new antileukemic quassinoid from Quassia amara. J. Org. Chem., 41, 3481-3482.

Liao, L.-L., S. M. Kupchan, and S. B. Horwitz (1976). Mode of action of the antitumor compound bruceantin, an inhibitor of protein synthesis. Molecular Pharmacology, 12, 167-176.

Polonsky, J. (1973). Quassinoid bitter principles. Fortschritte der Chemie Organischer Naturstoffe, 30, 101-150.

Polonsky, J., Z. Baskevitch, A. Gaudemer, and B. C. Das (1967). Constituants amers de Brucea amarissima, structures des bruceines A, B, et C. Experientia, 23, 424-430.

Valdenegro-Benitez, C. A. (1976). A chemical model for a potential mechanism of action of antileukemic quassinoids. M.S. Thesis, University of Virginia.

Wall, M. E. and M. C. Wani (1970). The isolation and structure of holacanthone, a potent experimental antitumor agent. 7th Int. Symp. Chem. Natl. Prod., IUPAC, Riga, 614 (Abstr.).

Mechanism of Action of ICRF 159

A. M. Creighton

Cellular Pharmacology and Antitumour Chemistry Laboratory,
Imperial Cancer Research Fund, London WC2A 3PX

ABSTRACT

The antitumour drug ICRF 159 [$(\pm)$1,2-Bis (3,5-dioxopiperazin-1-yl)] propane, Razoxane or NSC 12993 has been shown by several laboratories to interfere with cell division and to reduce the gross rate of DNA synthesis on a per culture basis. Sharpe and co-workers (1970) have shown that for PHA-stimulated lymphocytes, the sensitive part of the cell cycle for inducing a mitotic arrest is late G2 or early prophase. Subsequently, ICRF 159 has often been referred to as causing a "block" in or around the G2/M border of the cell cycle. In this paper, data is presented which suggests that this "block" is not a generality and does not occur with cultured cell lines at clinically significant dose levels and in the absence of colcemid etc. Cells move through "mitosis" a little slower than normal but the "daughter cells" have difficulty in separating and frequently rejoin to give a 4n cell which is almost certainly in a G1 biochemical state. Thus early accumulations of 4n cells demonstrated by FMF analyses do not necessarily indicate a G2/M block. This distinction might be important if a clinical combination regime is being constructed on the assumption that the drug causes an accumulation of cells in a particular phase of the cell cycle implying a particular biochemical status.

KEY WORDS: ICRF 159; $(\pm)$-1,2-Bis (3,5-dioxopiperazine-1-yl) propane; Razoxane; NSC 12993; G2/M block; cell cycle; cell synchrony; cytofluorographic analysis; mitosis.

INTRODUCTION

The bisdioxopiperazines are a class of antitumour agents which were first developed at the ICRF Laboratories some ten years ago. The best known of these is ICRF 159 (Fig. 1), now available in the U.K. under the name of Razoxane.

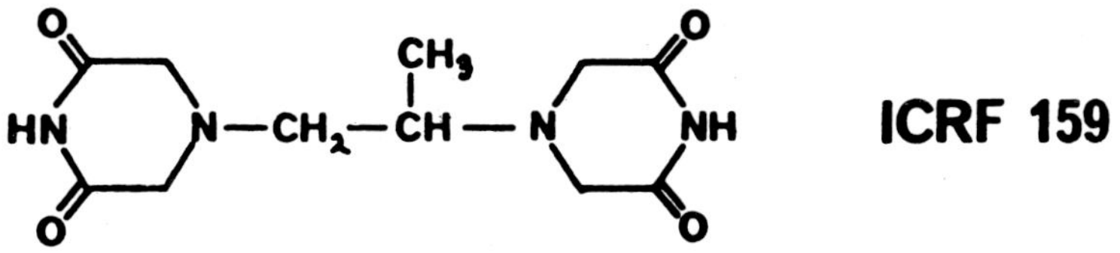

ICRF 159

Fig. 1.

The story really began following Furst's (1963) retrospective observation that most useful antitumour drugs were actual or potential chelating agents. Although not accepting the hypothesis that chelation played a major role in the mechanism of action of most of these drugs, it did stimulate my interest in chelation as a source of new cytotoxic agents. The potent chelating agent EDTA was known to be ineffective against experimental tumours (Leiter, Wodinsky and Bourke, 1959) presumably because of its highly polar structure which prevents adequate absorption. A few derivatives of EDTA with reduced polarity were synthesized in the hope that they would be able to penetrate cell membranes and subsequently break down by metabolism or hydrolysis to give cytotoxic, chelating species.

Two esters were inactive but the third compound synthesized, ICRF 154 (the bis cyclic imide from EDTA) was shown to be remarkably potent in the L1210 and Sarcoma 180 screening tests by Dr. Hellmann and his colleagues of the Cancer Chemotherapy Department at the I.C.R.F. (Creighton, Hellmann and Whitecross, 1969). This compound was rapidly superceded by the methyl analogue ICRF 159 which was slightly more potent and, being more soluble, significantly better absorbed. First evidence of its clinical utility was demonstrated by Hellmann and co-workers (1969) in some acute leukaemias and lymphosarcomas.

Animal studies and the current status and clinical prospects of the drug have been well covered by recent reviews by Bakowski (1976) and Bellet and co-workers (1977). Apart from the hoped-for cytotoxicity, ICRF 159 was found to possess a number of interesting, unexpected and potentially very useful properties which are briefly listed below with some lead references. In the Lewis lung system, the drug inhibits the formation of pulmonary metastases, apparently by stabilising the blood vessels of the primary tumour and preventing the release of viable tumour cells into the host's blood circulation (Hellmann and Burrage, 1969). ICRF 159 (and also EDTA in appropriate systems) affords some protection against the cardiotoxicity of daunomycin (Herman, Mhatre and Chadwick, 1974). It is synergistic with X-radiation in both mouse (Hellmann and Murkin, 1974) and man (Ryall and co-workers, 1974), a property that might be related to better vascularisation and hence oxygenation of the tumour. However, studies with cultured cells suggest that the drug is somehow able to reduce cells' ability to accumulate sub-lethal damage by X-radiation (Taylor and Bleehen, 1977a). It has also been shown to be synergistic with a wide range of drugs against experimental tumours (Kline, 1974; Wampler, Speckhart and Regelson, 1974; Woodman and co-workers, 1975).

Although it has been demonstrated that ^{14}C-labelled ICRF 159 penetrates cultured cells and then hydrolyses to liberate a chelating species which is too polar to diffuse in by itself (Dawson, 1975), the role of chelation, if any, in causing the principal cytotoxic lesion is still uncertain. There is some ground for the speculation that the reduction of cellular efficiency manifested by the reversible delays at different phases of the cell cycle (see below) could well be related to partial chelation of vital cations. The structurally-unrelated chelating agent 1,10-phenanthroline has been shown to have similar reversible effects on the cycle progression of lymphoblasts (Falchuk and Krishnan, 1977). The more critical derangement of the mitotic process which follows exposure around the G2/M border (see below) seems to involve a much more specific interaction. Structure/activity relationships, some of the chemical properties of the bisdioxopiperazines and the role of chelation have been discussed by Creighton, Jeffery and Long (1978).

EFFECTS OF ICRF 159 ON CULTURED CELLS

The first descriptions of the effects of ICRF 159 on cultured cells were by Carter (1968) who observed that although mouse L cells required a dose of at least 100μg/ml to arrest their division, lower doses allowed mitosis to proceed but abnormally, so that as daughter cells moved apart they often remained linked by strands of nuclear material. Eventually, the daughter cells either rejoined to form a tetraploid G1 cell or separated with uneven complements of genetic material to give potentially non-viable daughter cells. This effect can be demonstrated by detaching mitotic L cells from a culture exposed to the drug for an hour, replating and returning them to the incubator for a further 3 hours. Untreated cells separate evenly and completely while the drug-treated cells have the appearance shown (Fig. 2). If the mitotic cells are hypotonically swollen, fixed and stained directly following one hour's exposure to 12.5μg/ml of ICRF 159, they show the characteristic appearance of poorly condensed, "sticky" chromatin (Fig. 3).

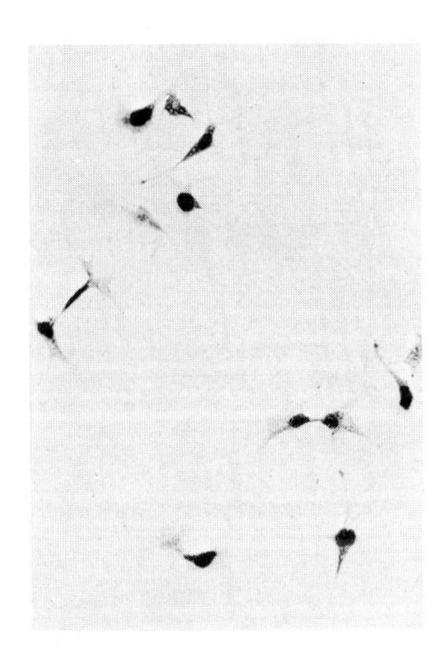

Fig. 2. Post-mitotic cells following one hour of drug treatment

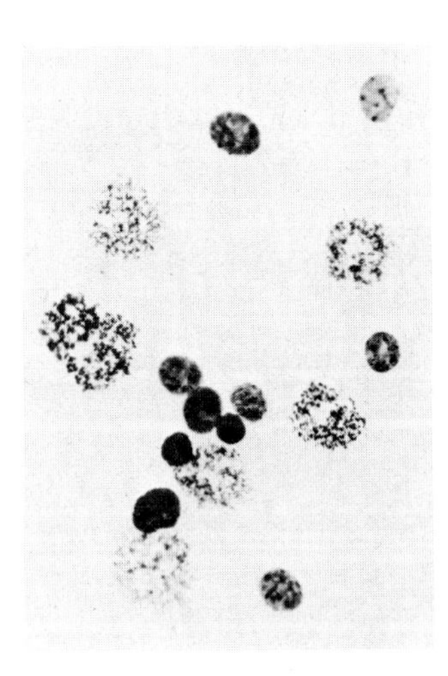

Fig. 3. Mitotic cells following one hour of drug treatment

The phenomenon is well illustrated in a different way by a cytofluorographic analysis of L cells treated for 24 hours with the drug (Fig. 4). In the control "spectrum" the 4n peak is composed of normal G2 and mitotic cells but, following treatment, it contains an increasing proportion of double-size G1 cells while at the higher doses a double-size G2/M peak (8n) appears. Similar responses have also been observed by Taylor and Bleehen (1977b) with EMT6 cells. The importance of not confusing the two types of 4n cell is discussed by Tobey, Deaven and Oka (1978). They also point out that drug-induced tetraploid cells are normally among the first to die-out in a population.

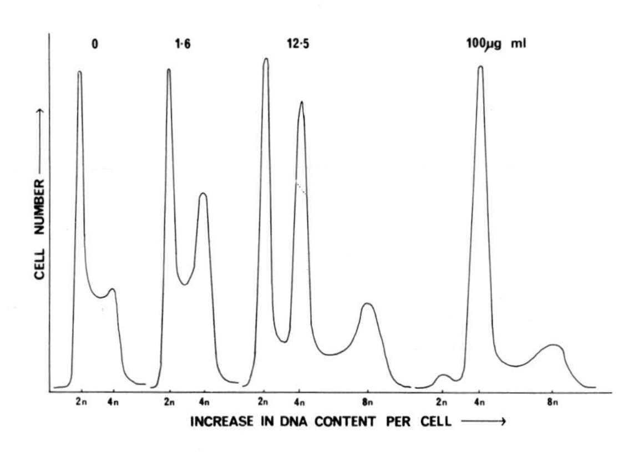

Fig. 4. DNA distribution of L cells 24 hr after addition of ICRF 159

Creighton and Birnie (1970) found that ICRF 159 substantially reduced the gross rate of DNA synthesis in mouse-embryo fibroblasts on a per culture basis with a smaller inhibition of RNA synthesis and hardly any effect on protein synthesis. Hellmann and Field (1970) showed that the drug had to be present only during a very brief (but unidentified) period of the cycle to affect the viability of HEp2 cells. Subsequently it was found that doses as low as 10µg/ml would prevent PHA-stimulated lymphocytes progressing to metaphase (in the presence of colcemid) providing the drug was present as the cells moved through the G2/M border (Sharpe, Field and Hellmann, 1970). Following these last observations, the drug has often been described as causing a G2 "block" but we have seen no evidence for it in our systems except at very high dose levels or in the presence of colcemid or related spindle poisons. It would seem to be important to clarify this point since clinical combination regimes might be constructed on the assumption that the drug causes an accumulation of cells in a particular phase of the cycle which implies a particular biochemical status.

In our experience with a variety of cultured cell lines (mainly mouse L cell and BHK 21S) there are two principal effects of the drug on cells. The first is a delay in progression which is reversible and the second the effect on viability if present at the sensitive part of the cycle (Creighton, 1974; Creighton and Long, in preparation). Using ^{3}H-dT autoradiography one can demonstrate with asynchronous L cells that there is an immediate rise in the proportion of cells in S-phase following treatment, which lasts for 8-10 hours (Fig. 5). This is interpreted as a delay in progression through S-phase with probably little or no effect on entry. FMF analyses also confirm an increase in the proportion of S-phase cells during the first 8 hours or so (Fig. 7).

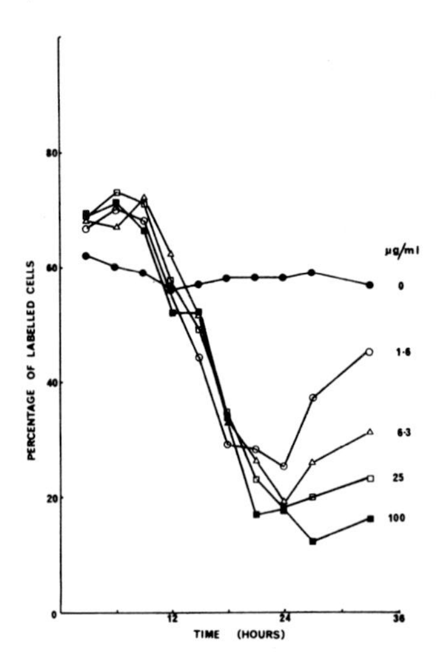

Fig. 5. Effect of ICRF 159 on the progression of asynchronous L cells

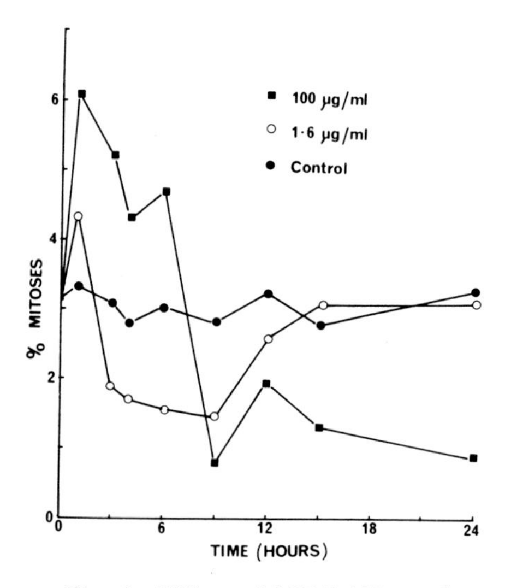

Fig. 6. Effect of ICRF 159 on the mitotic indices of asynchronous L cells

There is an immediate rise in the mitotic index of asynchronous L cells which lasts for several hours following the addition of ICRF 159 (Fig. 6). In conjunction with time-lapse film data, this is interpreted as meaning that cells close to mitosis are not significantly delayed in entry but all cells do take longer to pass through "mitosis" in the presence of the drug. Analysis of film data indicates that L cell mitoses may take up to double the normal time for completion. However in experiments measuring the rates of accumulation of mitoses in which vinblastine is present to prevent passage of cells beyond metaphase, ICRF 159 does cause an immediate delay of entry into mitosis (White and Creighton, 1977) (Fig. 8). This delay would seem to depend on the presence of the spindle poison.

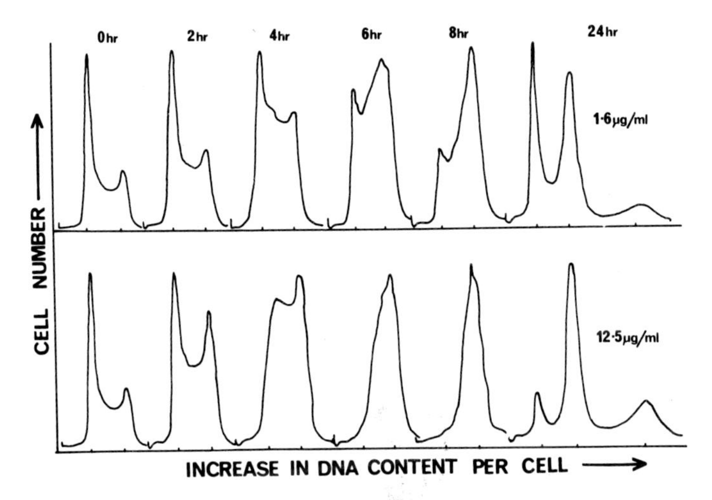

Fig. 7. DNA distribution of L cells following exposure to ICRF 159

Fractions of labelled mitoses (Fig. 9) suggest that L cells treated with 12.5µg/ml ICRF 159 are only delayed about 2 hours in transit from the end of S-phase to metaphase. In the context of the raised mitotic indices, this latter slow-down probably relates to early G2.

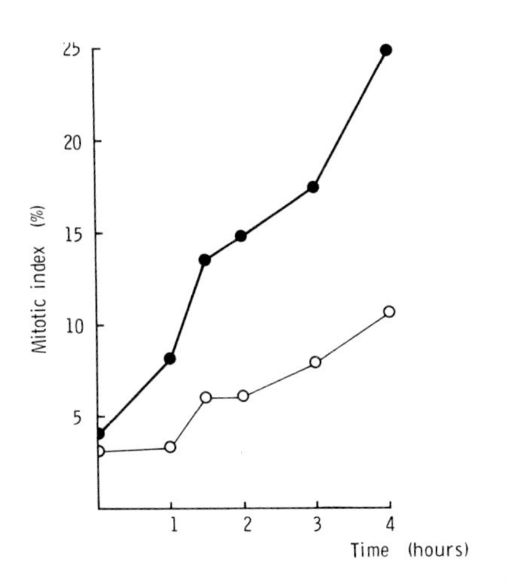

Fig. 8. Effect of ICRF 159 on rate of entry of BHK 21S cells into mitosis

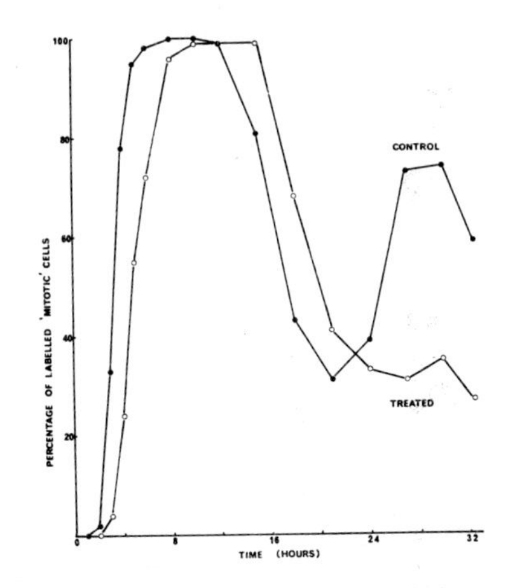

Fig. 9. Effect of ICRF 159 on fractions of labelled mitoses

Experiments with synchronised L cells pulsed with ^{3}H-dT at various times after mitotic collection and subsequent exposure to the drug, also show a delay in cycle progression and this is proportional to dose (Fig. 10). Pulsing synchronised cells for 3 hour periods at different times after mitotic collection always produced a subsequent delay in progression. A time-lapse cinemicrographic analysis showed a similar response proportional to dose (Fig. 12). The vast majority of cells which were not "lost" through moving out of the field, did complete a second "mitosis". At this stage the concentrations of active drug had fallen off considerably (ICRF 159 has a half life of about 10 hours at pH 7.2) and the times between the median first and second post-drug treatment mitoses were similar to the controls.

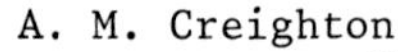

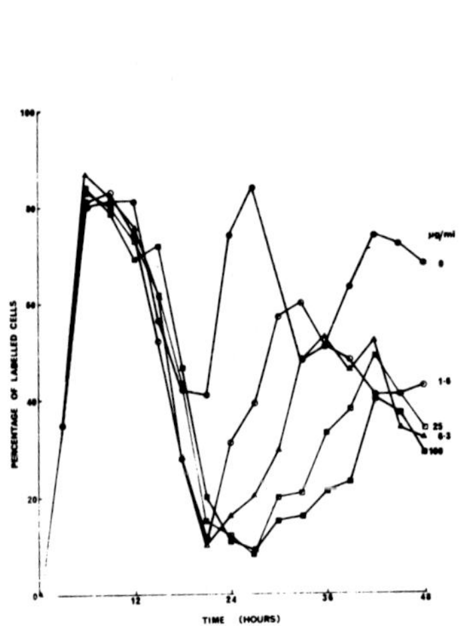

Fig. 10. Effect of different doses of ICRF 159 on the progression of synchronised L cells

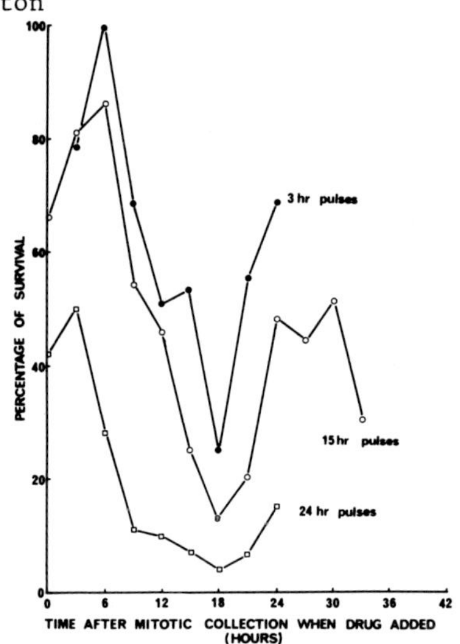

Fig. 11. Effect of pulses of ICRF 159 on the survival of synchronised L cells

Colony-forming assays with synchronised cells pulsed with the drug at different stages of th cell cycle confirmed that the G2/M border (about 18 hours post-collection) was the critica phase (Fig. 11). Even with the longer duration pulses, the drug had to be added at around o just before the "G2/M border" for maximum toxicity. The fact that a long pulse initiated onl 3 hours earlier is less effective reflects both the delay in progression to the sensitive stage an the gradual hydrolysis of the drug while the cells are reaching that stage. Pre-treatment wit the drug for as little as an hour is sufficient to reduce the viability of collected mitotic L cell by 60% (Fig. 13).

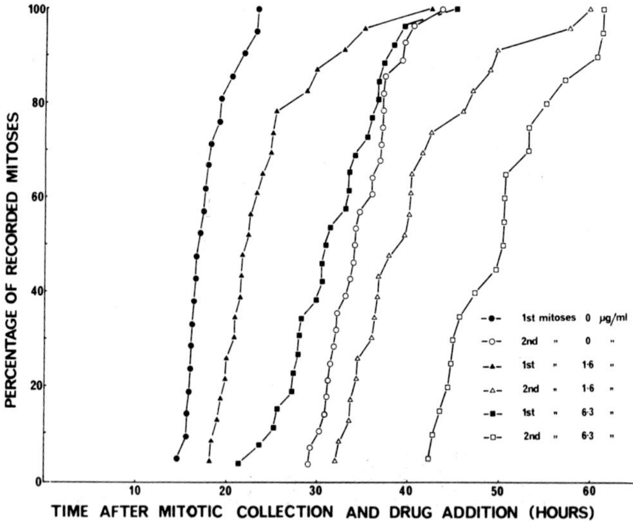

Fig. 12. Effect of ICRF 159 on intermitotic times of synchronised L cells

Many of the effects of ICRF 159 on the progression and survival of L cells described above have also been observed with BHK 21S cells (Stephens and Creighton, 1974; Stephens, 1975). The last slide describes the response of BHK 21S cells to ICRF 159 (Fig. 14). It confirms

Hellmann and Field's (1970) observation that the length of exposure to the drug is more important than the dose but it also illustrates that where a drug effects both progression and survival, the highest doses may not be the most toxic.

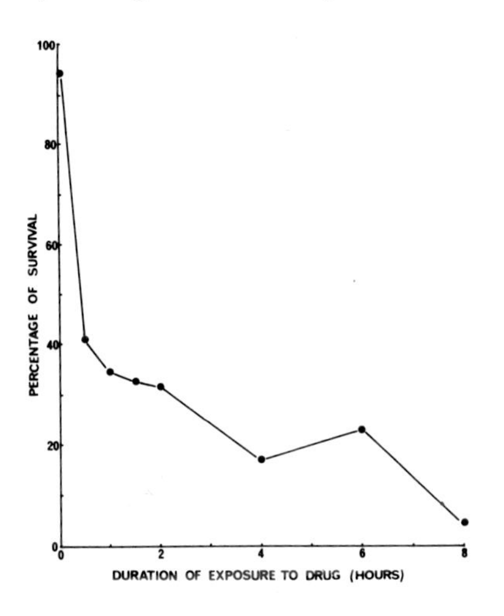

Fig. 13. Effect of pretreatment with ICRF 159 on survival of mitotic L cells

Fig. 14. Effect of length of exposure and dose of ICRF 159 on survival of BHK 21S cells

CONCLUSIONS

Evidence has been presented that ICRF 159 must be present just before mitosis to markedly reduce viability as first suggested by Sharpe and co-workers (1970). There is also an independent and reversible delay of progression if the drug is present at other stages of the cycle but no indication of a G2 block or arrest at clinically significant doses and in the absence of colcemid-like agents.

Grateful acknowledgment is made to Dr. Peter Riddle for the time-lapse facilities and to Domenico Delia for the FMF data.

REFERENCES

Bakowski, M.T. (1976). ICRF 159, (±)-1,2-di(3,5-dioxopiperazin-1-yl) propane, NSC-129,943; Razoxane. Cancer Treatment Rev., **3**, 95-107.

Bellet, R.E., M. Rozencweig, D.D. von Hoff, J.S. Penta, T.H. Wasserman, and F.M. Muggia (1977). ICRF 159: Current status and clinical prospects. Europ. J. Cancer, **13**, 1293-1298.

Carter, S.B. (1968). Unpublished results.

Creighton, A.M. (1974). The effects of ICRF 159 on synchronised L cells at various stages of the cell cycle. Abstracts XIth Intern. Cancer Congress, Florence, **3**, 423.

Creighton, A.M., and G.D. Birnie (1970). Biochemical studies on growth-inhibitory bisdioxopiperazines. I. Effect on DNA, RNA and protein synthesis in mouse-embryo fibroblasts. Int. J. Cancer, **5**, 47-54.

Creighton, A.M., K. Hellmann, and S. Whitecross (1969). Antitumour activity in a series o bisdiketopiperazines. Nature, **222**, 384-385.

Creighton, A.M., W.A. Jeffery, and J. Long (In the press). Bisdioxopiperazines. Proc. VIt Intern. Symp. Med. Chem., University of Sussex, 1978. Cotswold Press.

Dawson, K.M. (1975). Studies on the stability and cellular distribution of dioxopiperazines i cultured BHK 21S cells. Biochem. Pharmacol., **24**, 2249-2253.

Falchuk, K.H., and A. Krishan (1977). 1,10-Phenanthroline inhibition of lymphoblast cel cycle. Cancer Res., **37**, 2050-2056.

Furst, A. (1963). "Chemistry of Chelation in Cancer". Charles C. Thomas, Springfield Illinois.

Hellmann, K., and K. Burrage (1969). Control of malignant metastases by ICRF 159. Nature **224**, 273-275.

Hellmann, K., and E.O. Field (1970). Effect of ICRF 159 on the mammalian cell cycle significance for its use in cancer chemotherapy. J. Nat. Cancer Inst., **44**, 539-543.

Hellmann, K., and G.E. Murkin (1974). Synergism of ICRF 159 and radiotherapy in treatmen of experimental tumors. Cancer, **34**, 1033-1039.

Hellmann, K., K.A. Newton, D.N. Whitmore, I.W.F. Hanham, and J.V. Bond (1969). Preliminary clinical assessment of ICRF 159 in acute leukaemia and lymphosarcoma Brit. Med. J., **1**, 822-824.

Kline, I. (1974). Potentially useful combinations of chemotherapy detected in mouse tumou systems. Cancer Chemotherapy Reports, Part 2, **4** (1), 33-43.

Herman, E.H., R.M. Mhatre, and D.P. Chadwick (1974). Modification of some of the toxic effects of daunomycin (NSC-82,151) by pretreatment with the antineoplastic agent ICRF 159 (NSC-129,943). Toxicol. and Applied Pharmacol., **27**, 517-526.

Leiter, J., I. Wodinsky, and A.T. Bourke (1959). Screening data from the C.C.N.S.C. screenin laboratories II. Cancer Chemotherapy Reports, Part 2, **19** (6), 368.

Ryall, R.D.H., I.W.F. Hanham, K.A. Newton, K. Hellmann, D.M. Brinkley, and O.K. Hjertaa (1974). Combined treatment of soft tissue and osteosarcomas by radiation an ICRF 159. Cancer, **34**, 1040-1045.

Sharpe, H.B.A., E.O. Field, and K. Hellmann (1970). Mode of action of the cytostatic agen ICRF 159. Nature, **226**, 524-526.

Stephens, T.C. (1974). An investigation of the effects of the antitumour agent ICRF 159 or the growth of BHK-21S cells in culture. Ph.D. thesis, University of London.

Stephens, T.C., and A.M. Creighton (1974). Mechanism of action studies with ICRF 159 Effects on the growth and morphology of BHK-21S cells. Brit. J. Cancer, **29** 99-100.

Taylor, I.W., and N.M. Bleehen (1977a). Interaction of ICRF 159 with radiation, and its effect on sub-lethal and potentially lethal radiation damage in vitro. Brit. J. Cancer, **36**, 493-500.

Taylor, I.W., and N.M. Bleehen (1977b). Changes in sensitivity to radiation and to ICRF 159 occurring during the life history of monolayer cultures of the EMT6 tumour cell line. Brit. J. Cancer, **35**, 587-594.

Tobey, R.A., L.L. Deaven, and M.S. Oka (1978). Kinetic response of cultured chinese hamster cells to treatment with 4'-[(9-Acridinyl)-amino] methanesulphon-m-anisidide-HCl. J. Nat. Cancer Inst., **60**, 1147-1153.

Wampler, G.L., V.J. Speckhart and W. Regelson (1974). Phase I clinical study of adriamycin-ICRF 159 combination and other ICRF 159 drug combinations. Proc. Am. Soc. Clin. Oncol., **15**, 189.

White, K., and A.M. Creighton (1977). Studies of resistance to ICRF 159 in cell line BS/159-1. Brit. J. Cancer, **36**, 421.

Woodman, R.J., R.L. Cysyk, I. Kline, M. Gang, and J.M. Venditti (1975). Enhancement of the effectiveness of daunorubicin (NSC-82151) or adriamycin (NSC-123127) against early mouse L 1210 leukemia with ICRF 159 (NSC-129943). Cancer Chemotherapy Reports, Part 1, **59** (4), 689-695.

Development of Actinomycin Analogs

R. H. Adamson*, S. M. Sieber* and J. D. Douros**

**Laboratory of Chemical Pharmacology, National Cancer Institute, Bethesda, Maryland, U.S.A.*
***Natural Products Branch, National Cancer Institute, Silver Spring, Maryland, U.S.A.*

ABSTRACT

The Developmental Therapeutics Program, Division of Cancer Treatment, is interested in developing actinomycin D analogs because of the high degree of clinical activity of actinomycin D, albeit against a limited spectrum of tumors; the possibility that less toxic analogs may be developed; and because analogs can be produced biosynthetically. The introduction of new analogs into the clinic will be determined not only by experimental antitumor data, but also, and perhaps more importantly, by preclinical toxicology. Preliminary studies in animals indicate that the pipecolic acid analog, $Pip_{1\beta}$, is as active as actinomycin D and less toxic.

KEYWORDS

Actinomycins, experimental antitumor activity, actinomycin analogs, structure-activity relationships.

INTRODUCTION

The actinomycins were first isolated by Waksman and Woodruff (1940), and since that time a number of structural variants have been produced and studied. All the actinomycins contain the same 2-amino-4,6-dimethyl-phenoxaz-3-one-4,5-dicarboxylic acid chromophore with a variety of polypeptide side chains attached via the carboxyl functions. The structure of actinomycin D (also known as C_1 and actinomycin IV) is shown in Fig. 1.

Actinomycin D has good clinical activity against several tumors including gestational choriocarcinoma, gonadal choriocarcinoma and mixed embryonal carcinoma of the testis, unilateral nephroblastoma (Wilm's tumor) and childhood rhabdomyosarcoma. It is not well absorbed when given by the oral route (Friedman and Cerami, 1973), and is therefore administered by intravenous injection. Studies in animals have shown that the compound is rapidly cleared from the circulation after intravenous dosing (Schwartz, Sodergren and Ambaye, 1968), and that it does not penetrate the blood-brain barrier (Friedman and Cerami, 1973). Actinomycin D is a remarkably potent agent, inhibiting rapidly proliferating cells of both normal and neoplastic origin. Its toxicity is related primarily to hematopoietic and gastrointestinal systems. From a

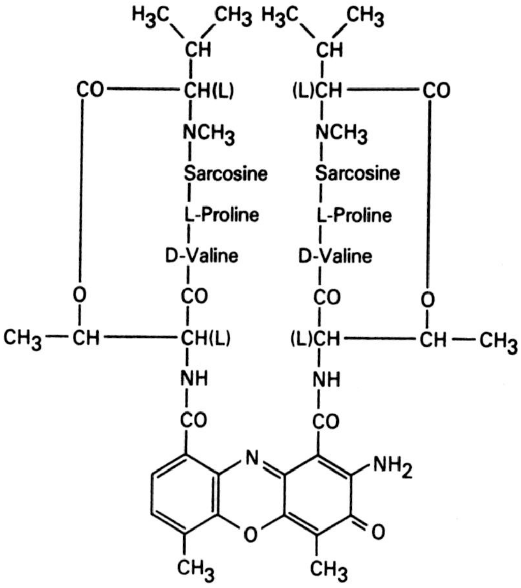

Fig. 1. Actinomycin D

clinical standpoint, the most hazardous manifestation of the toxicity of actinomycin D is myelosuppression, the first manifestation of which is often a decrease in the platelet count; pancytopenia may develop rapidly in affected patients. The gastrointestinal effects of actinomycin D are usually apparent a few hours after dosing and include anorexia, nausea, and vomiting; patients administered the drug may also experience abdominal pains and diarrhea. Ulcerations of the buccal mucosa and skin reactions in patients treated with actinomycin are also common. Radiosensitization has been shown to occur at sites of concomitant or prior radiotherapy. Since injury may also occur as a result of local toxic action, the drug should be administered by intravenous tubing (Calabresi and Parks, 1975).

There are several reasons why the development of actinomycin D analogs is an area of great interest and active research. First, although actinomycin D is highly effective in the treatment of malignancies, it is active against a very narrow spectrum of tumors; the discovery of actinomycin D derivatives with activity against a broader spectrum of tumors is therefore a primary goal in analog development. Second, it may be possible to develop analogs with equivalent activity, but with reduced toxicity. Third, analogs of actinomycin D could be produced with either increased or decreased radiosensitizing properties as desired. And fourth, and most importantly, by using amino acids or amino acid analogs that are chemically similar to the amino acid constituents in actinomycin, one can achieve controlled or directed biosynthesis of analogs.

An ideal analog of actinomycin D would therefore have a wide spectrum of antitumor activity, reduced toxicity, the ability to penetrate the blood brain barrier, and the cost of its production, separation and purification would be low.

CONTROLLED BIOSYNTHESIS OF ACTINOMYCINS

The technique of controlled biosynthesis is based on the ability of microorganisms to utilize certain exogenous nitrogen sources in synthesizing actinomycins. The nitrogen-containing compounds are amino acids or amino acid analogs which are chemically similar to endogenous substances. They can influence actinomycin formation in two ways: 1) by leading to an increase in the amount of a specific actinomycin formed, and 2) by leading to the formation of new actinomycins. For example, the yield of actinomycin D can be increased from 10% to 83% of total actinomycins produced by S. *chrysomallus* when DL-valine is added to the culture medium (Schmidt-Kastner, 1956). Modified congeners of actinomycin can be produced by supplying specific chemical precursors to the organism. Alterations in the peptide ring at the proline site were produced when either higher (pipecolic acid) or lower (azetidine-2-carboxylic acid) proline analogs were supplied as the nitrogen source during the synthetic phase of S. *antibioticus* (Katz, 1974). Thus proline is replaced in either chains A or B by pipecolic acid or azetidine-2-carboxylic acid, and the analogs formed are termed $Pip_{1\beta}$ or azetomycin-1, respectively. The antitumor effects of these 2 analogs will be discussed below.

ANTITUMOR ACTIVITY OF ACTINOMYCINS

Evaluation of the antineoplastic properties of numerous actinomycin analogs has been carried out by various research laboratories (Brockman and co-workers, 1956; Katz, 1974; Muller, 1962; Reich, 1963; and Stock, 1966), as well as by several industrial companies including Merck, Sharp & Dohme, Schering Corp. and Kyowa Hakko. In addition, over 100 analogs have been evaluated by the NCI for experimental antitumor effects.

Table 1 lists the rodent tumors used in the NCI studies to evaluate the antitumor activity of actinomycin analogs and to compare their chemotherapeutic effects with those of actinomycin D. Initially, these include leukemia P388 (highly sensitive to actinomycin D), B16 melanoma (moderately sensitive to actinomycin D), and leukemia L1210 (marginally responsive to actinomycin D). Subsequent evaluation is performed using colon 26, the $CD8F_1$ mammary carcinoma, Lewis lung carcinoma, and human xenografts CX_1 (colon), MX_1 (mammary), and LX_1 (lung).

A number of active actinomycin analogs have emerged from these studies. The two analogs in which we at the NCI are currently most interested are the pipecolic acid analog ($Pip_{1\beta}$) and the azetidine-2-carboxylic acid analog (Azet 1). Table 2 presents comparative data for activity of these compounds against leukemia P388. When given at optimal doses on a day 1-9 schedule, both $Pip_{1\beta}$ and Azet 1 have greater antitumor activity than does actinomycin D. Treatment with Azet 1 increased the lifespan of leukemic mice 522%, and produced a "cure" rate of 60%.

Table 3 compares the antitumor activity against B16 melanoma of actinomycin D, Azet 1 and $Pip_{1\beta}$ when the compounds were administered according to 3 different schedules. Actinomycin D and Azet 1 were only moderately active against this tumor regardless of dosing schedule. In contrast, the pipecolic acid analog had greater activity on all three schedules and led to 30% "cures" when given on days 1-9 after tumor implantation.

TABLE 1 Rodent Tumors Used for Assay of Activity of Actinomycin D and Analogs

Initial Evaluation	Sensitivity to Actinomycin D
Leukemia P388	high
B16 melanoma	moderate
Leukemia L1210	marginal
Subsequent Evaluation	
Panel Tumors	
Colon 26	moderate
$CD8F_1$ mammary carcinoma	-
Lewis lung carcinoma	-
Human Xenografts	
Colon (CX_1)	-
Mammary (MX_1)	-
Lung (LX_1)	-

TABLE 2 Comparative Activity of Actinomycin D and Two Analogs Against Leukemia P388*

Compound	Optimal Dose (mg/kg)	%ILS	%LTS
Act. D	0.1	212	12
$Pip_{1\beta}$	0.2	315	25
Azet 1	0.2	522	60

*CDF_1 males were inoculated ip with 10^6 tumor cells, and drug treatment begun 24 hours later. Drugs were injected ip daily for 9 days. ILS = increased life span; LTS = long term survivors (>60 days).

Leukemia L1210 is only marginally responsive to actinomycin D, and neither of the two analogs showed greater activity than did the parent compound in this system (Table 4). Similarly, (Table 5) no difference in activity could be seen among the compounds when tested against colon 26, one of the NCI panel tumors.

We are presently evaluating these two compounds in the other panel tumors and against the human xenografts. The toxicity of these analogs in the rat and dog is also being evaluated. Preliminary toxicity studies in the rat have shown that at optimal antitumor doses, the $Pip_{1\beta}$ analog had the lowest, and the Azet 1 analog the highest, gastrointestinal toxicity and irritant properties as compared to actinomycin D. However, these studies are preliminary and further toxicity tests in dogs will be carried out when additional quantities of compound become available.

TABLE 3 Comparative Activity of Actinomycin D and Two Analogs Against B16 Melanoma*

Compound	Schedule	Optimal Dose (mg/kg)	%ILS	%LTS
Act. D	Day 1-9	0.075	86	0
	Day 1,5,9	0.15	55	0
	Day 5,9,13	0.1	40	0
Azet 1	Day 1-9	0.75	52	20
	Day 1,5,9	0.45	68	0
	Day 5,9,13	0.30	33	0
$Pip_{1\beta}$	Day 1-9	0.8	125	30
	Day 1,5,9	1.6	173	10
	Day 5,9,13	1.6	136	0

*BDF_1 males were inoculated ip with 0.5 ml of a 1:10 tumor brei, and drugs administered ip according to the schedules indicated. ILS = increased life span; LTS = long term survivors (>60 days).

TABLE 4 Comparative Activity of Actinomycin D and Two Analogs Against Leukemia L1210*

Compound	Optimal Dose (mg/kg)	%ILS	%LTS
Act. D	0.1	51	0
$Pip_{1\beta}$	0.8	48	0
Azet 1	0.3	60	0

*CDF_1 males were inoculated ip with 10^5 tumor cells and drug treatment begun 24 hours later. Drugs were injected ip daily for 9 days. ILS = increased life span; LTS = long term survivors (>60 days).

TABLE 5 Comparative Activity of Actinomycin D and Two Analogs Against Colon 26*

Compound	Optimal Dose (mg/kg)	%ILS	%LTS
Act. D	0.5	53	0
$Pip_{1\beta}$	5.0	51	0
Azet 1	0.75	57	0

*CDF_1 males were inoculated ip with 0.5 ml of a 1:100 tumor brei, and drugs administered ip on days 2, 6 and 10. ILS = increased life span; LTS = long term survivors (>60 days).

Because a large number of actinomycin analogs of defined chemical structure have been tested for antitumor activity, a picture of those chemical features necessary for antitumor effects is beginning to emerge. Thus far, the conclusions one can make with regard to structure activity relationships are:

(1) the free chromophore is inactive
(2) the amino group at C2 is essential for optimal activity
(3) the carbonyl group at C3 is essential
(4) the lactose rings and polypeptide chains are essential but variable

Figure 2 illustrates why the amino moiety at C2 and the carbonyl functions at C3 are essential. These two groups may be responsible for hydrogen bonding to 2'-deoxyguanosine moieties of DNA. On the basis of X-ray diffraction studies, Reich and Goldberg (1964) proposed that the carbonyl group at C3 in actinomycin binds to the 2-amino group of guanine, and that the 2-amino moiety of actinomycin binds to the N3 of guanine and to the oxygen of the furanose ring. Others have presented evidence that there are two types of binding, one weak and one strong, and that the strong binding sites are perhaps a sequence of guanine-cytosine pairs (Gellert and co-workers, 1965). Regardless of details, actinomycin does bind tightly but reversibly to DNA. As a consequence, this DNA cannot serve as a template for RNA synthesis, and RNA and ultimately protein synthesis is arrested.

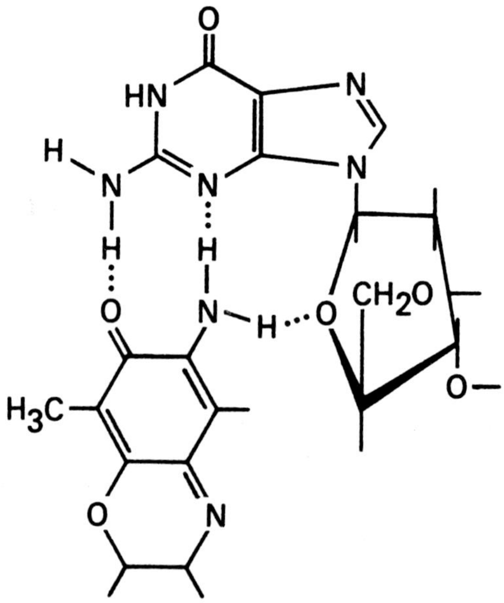

Fig. 2. Proposed binding site of actinomycin to 2'-deoxyguanylic acid residues of DNA (Reich and Goldberg, 1964)

CONCLUSIONS

In this manuscript, the ideal analog of actinomycin D was defined as one which would have a wide spectrum of antitumor activity, reduced toxicity, the ability to penetrate the blood brain barrier and a low cost of production, separation and purification. No analog tested thus far meets all of these specifications. However, results from recent studies on the pipecolic acid analog, $Pip_{1\beta}$, indicate that it does have a broader spectrum of antitumor activity and less host toxicity than actinomycin D. Further testing is needed before the introduction of $Pip_{1\beta}$ into clinical trials.

BIBLIOGRAPHY

Brockmann, H., G. Bohnsack, B. Franck, H. Grone, H. Muxfeldt, and C. Suling (1956). Zur konstitutions der actinomycine. Angew. Chem., 68, 70-71.

Calabresi, P., and R. E. Parks, Jr. (1975). Chemotherapy of neoplasic diseases. Natural products. In L. S. Goodman and A. Gilman (Eds.), The Pharmacological Basis of Therapeutics, 5th ed. Macmillan Publishing Co. Inc., New York. pp. 1287-1288.

Friedman, P. A., and A. Cerami (1973). Actinomycin. In J. F. Holland and E. Frei, III (Eds.), Cancer Medicine, Lea and Febiger, Philadelphia. pp. 835-839.

Gellert, M., C. E. Smith, D. Neville, and G. Felsenfeld (1965). Actinomycin binding to DNA: mechanism and specificity. J. Mol. Biol., 11, 445-457.

Katz, E. (1974). Controlled biosynthesis of actinomycins. Cancer Chemother. Rep., 58, 83-91.

Muller, W. (1962). Bindung von actinomycinen und actinomycin-derivaten an deoxyribonucleinsäure. Naturwissenschaften, 49, 156-157.

Reich, E. (1963). Biochemistry of actinomycins. Cancer Res., 23, 1428-1441.

Reich, E., and I. H. Goldberg (1964). Actinomycin and nucleic acid function. In J. N. Davidson and W. E. Cohn (Eds.), Progress in Nucleic Acid Res. and Mol. Biol., Vol. 3, Academic Press, New York. pp. 183-234.

Schmidt-Kastner, G. (1956). Uber neue biosynthetische actinomycine. In Medizin und Chemie. Abhandlungen aus dem Medizinisch-Chemischen Forschungsstatten der Farbenfabriken Bayer Aktiengesellschaft (Bayer-Leverkusen), Vol. 5, Verlag Chemie-GMGH, Weinheim/Bergstrasse. pp. 463-476.

Schwartz, H. S., J. E. Sodergren, and R. Y. Ambaye (1968). Actinomycin D: drug concentrations and actions in mouse tissues and tumors. Cancer Res. 28, 192-197.

Stock, J. A. (1966). Antitumor antibiotics. II. The actinomycins. In R. J. Schnitzer and F. Hawking (Eds.), Experimental Chemotherapy, Vol. 4, Academic Press, New York. pp. 243-267.

Waksman, S. A., and H. B. Woodruff (1940). Bacteriostatic and bactericidal substances produced by a soil actinomyces. Proc. Soc. Exp. Biol. Med., 45, 609-614.

Report on Symposium No. 25: Development of New Antitumour Agents

Norbert Brock

From the first empirical observation of an antitumour effect to the development of an effective drug, there is a great deal of chemical and pharmacological work. A. Di MARCO (Italy), as chairman, stresses the strong impact of the knowledge of the mechanism of action of the antitumour agents now used for the research of new and more selective drugs. Many presently used drugs may interact with the genomic material in an indiscriminate way: the more popular approach is oriented to the clarification of the interaction of these agents with DNA. It is now necessary to extend these studies to the interaction with native chromatine, and their interference with the mechanism of the regulation of gene expression. This may lead to a deeper understanding of the biochemical lesion of malignant cells and eventually help the research of agents selectively acting on cancer.

The first three lectures are concerned with the development of plant products for the treatment of cancer.

J.D. DOUROS (USA) reported on MAYTANSINE. This ansa macrolide was originally isolated by Kupchan from the East African plants *Maytenus serrata* and *Maytenus buchananii*. The mitosis-inhibiting compound induces metaphase arrest and is active against some experimental tumours at extremely low dose schedules. MAYTANSINE is a highly interesting antitumour agent, which is undergoing extensive clinical trials under the auspices of the NCI. Other dosage schedules together with drugs to reduce symptoms of gastrointestinal toxicity will be studied. Possible better analogues will be made available since the supply problem of maytansinoids may be alleviated by the discovery of a microbial fermentation to produce them.

BRUCEANTIN (A.T. SNEDEN, USA) is the primary antileukaemic principle of *Brucea antidysenterica*, a tree used in the treatment of cancer in Ethiopia. The active ingredient was, together with other quassinoids, isolated by Kupchan *et al.* (USA) in 1972. The substance is active in various mouse tumours and inhibits protein synthesis and, partially, DNA synthesis in Hela cells and other cell systems. Structure-activity studies led to important knowledge. BRUCEANTIN has passed preclinical toxicology and pharmacology and is now undergoing Phase I clinical trials in the United States.

TAXOL (M.W. WALL, USA) is a unique diterpene isolated from the stem-bark of the western yew *Taxus brevifolia*. Guided by bioassay, the active principle was isolated from a crude alcoholic extract. The chemical structure of TAXOL could be determined by spectroscopic techniques and X-ray analysis. By mild methanolysis, the substance disintegrates easily into two halves, which individually are biologically inactive. TAXOL itself and the combination of both of the split products are active in a broad spectrum of experimental leukaemia systems and some solid tumours. The mechanism of the antineoplastic activity may be due to its action as a mitotic spindle position.

The following lectures are concerned with new developments of the antitumour antibiotics. The introduction of new ACTINOMYCINE ANALOGUES (R.H. ADAMSON, USA) in the clinic will be determined not only by experimental antitumour data but also — and perhaps more important — by preclinical toxicology. The NCI is currently evaluating two promising analogues: the pipecolic acid analogue and the azetidine-2-carbonylic acid analogue in preclinical tests and in preclinical toxicology.

Mode of action and structure-activity relationship of antitumour ANTHRACYCLINES (doxorubicin and related compounds) were the main topic of F. ARGAMONE's (Italy) presentation. All available evidence could confirm that cell DNA is the main receptor for these drugs. Interaction with other cell constituents (proteins, phospholipids and mucopolysaccharides) however is possibly involved in the expression of antitumour activity of toxic side-effects. Structural and/or stereochemical modifications in the sugar moiety may induce substantial variation in the biological efficacy of the drugs. When the C-4' position is modified, as for instance in the new semi-synthetic analogues 4'-epiadramycin, 4'-deoxyadriamycin and 4'-O-methyladriamycin, these compounds are more effective and less toxic than the parent compound. Interesting analogues were seen by substitution of the aromatic ring, amongst which 4'-demethoxydaunomycin is an orally active compound with high efficacy in experimental leukaemia.

H. UMEZAWA (Japan) reported on extensive trials to increase the selectivity of BLEOMYCIN and to reduce its side-effects on the lung. The rationale for the development of new analogues of BLEOMYCIN has come from understanding its mechanism of action. BLEOMYCIN hydrolases are responsible for inactivating the drug before it can cause strand scission of DNA. More than 300 BLEOMYCINS have been synthesised by precursor-fed fermentation or chemical derivation. From the results of screening, Pepleomycin (PEP) was selected for clinical trial because of its favourable properties. The antitumour effect of PEP was stronger than that of BLEOMYCIN. Pulmonary toxicity of PEP was a quarter less (mouse, dog) than that of BLEOMYCIN. Clinically, PEP was more effective in smaller doses than BLEOMYCIN in the same anti-cancer spectrum.

Y.F. SHEALY (USA) reported on newer developments in the field of TRIAZENO IMIDAZOLES and related compounds. Based on DTIC, a variety of related triazenylimidazoles were synthesised including the analogous monomethyltriazenyl imidazole (MIC). Simultaneously, benzenoid triazenes and triazenyl derivatives of other heterocyclic rings were prepared. Amongst these were a number of chemotherapeutic active substances but the majority of them did not have the good efficacy of the parent compound DTIC. Other compounds were unstable so that DTIC is still the No. 1 compound clinically (especially in cases of malignant melanoma).

N. BROCK (Fed. Rep. of Germany): IFOSFAMIDE, like cyclophosphamide, belongs to the group of the N-2-chloroethyl-amino-oxazaphosphorines. It is inactive *in vitro* and becomes activated to cytotoxic metabolites in the liver of warm-blooded animals. Its cancerotoxic selectivity depends on the formation of the primary metabolites 4-hydroxy/aldoifosfamide. *In vivo*, ifosfamide shows a high cancerotoxic activity against various experimental tumours. Its margin of safety is better than that of cyclophosphamide due to its cumulative behaviour. The leukotoxicity of ifosfamide in man is less pronounced than that of other alkylating compounds. In clinical use, ifosfamide has markedly widened the spectrum of tumours amenable to chemotherapy. Even in cyclophosphamide-resistant patients and in cases which no longer respond to any other therapy, it is possible to obtain remissions. The problem of urotoxic side-effects was overcome by developing a specific antidote.

A.M. CREIGHTON (England) reported on ICRF-159 (1,2-(3,5-dioxopiperazine-1-yl)-propane). The substance has the following properties: antimetastatic activity, synergism with other cytostatic drugs (i.e. cyclophosphamide, adriamycin), effects as a radiosensitizer as well as cytostatic effects of its own. The latter was clinically seen in lymphomas mainly, but also in colorectal carcinomas. More recent studies on the mechanism of action seem to show that the substance induces DNA duplication while blocking nuclear division, thus leading to polyploidy. It is necessary to further test and differentiate the efficacy of ICRF-159 in controlled clinical studies.

In his closing remarks, N. BROCK (Fed. Rep. of Germany) stressed that the Symposium showed new progress and awakened hopes for the future. In all future research, the problem of the selectivity of the cancerotoxic activity of antitumour agents should be given our special attention. In order to progress in this field, an exact knowledge of the mechanism of action, as well as the pharmacokinetics of the various substances is necessary, not to mention a detailed insight into the metabolism of host and cancer cells. In the experimental stages, we must have common methodical procedures among the different work and study groups. This applied to the challenge of a correct dosage/action and time/action analysis for the individual active components and to the use of standard substances and reference tumours.

The progress which has been made should be used to the good of the patients more quickly than has been the case in the past and, in this way, an even closer link between experimental and clinical cancer research — again for the good of the patient — should be aimed at.

Studies on the Anticancer Action of 10-Hydroxycamptothecin

Institute of Materia Medica, Academia Sinica, Shanghai, The People's Republic of China*

ABSTRACT

10-Hydroxycamptothecin was isolated from the fruits of Camptotheca acuminata Decne. The sodium salt of this compound was found to have a broad antitumor spectrum in animal experiments. Its content in the tumor was comparatively high and persisted longer. It was excreted mainly in the feces. It was shown to have a stronger inhibitory action on DNA synthesis than on RNA of cancer cells. Clinical trials of the drug on 63 patients showed remarkable therapeutic effect on liver carcinoma and tumors of head and neck (chiefly carcinoma of salivery glands). Its toxicity was less marked than that of camptothecin, especially in the urinary tract.

Keywords: Hydroxycamptothecin, Anticancer action.

INTRODUCTION

Camptotheca acuminata Decne is a famous tree in which China has rich resources. Wall and co-workers (1966) isolated an active antitumor principle, camptothecin, from it, but clinical trials with this drug showed a rather high toxicity. They (1969) again succeeded in isolating 10-hydroxycamptothecin (10-OH-CN) and 10-methoxy-camptothecin from the tree. In animal tests these compounds were found to produce certain antitumor activity against murine leukemia P-388. Since 1968 we have carried out a systematic study on the antitumor action of various principles of Camptotheca acuminata Decne and found that comptothecin has certain therapeutic effect (Division of Tumor Pharmacology, 1975; Hsu, 1977). Later 10-OH-CN was also found to be a useful anticancer drug. This paper is to present the main results of our investigation on the antitumor action of this drug.

CHEMISTRY

The alkaloids from various parts of Camptotheca acuminata Decne were investigated. It was found that the content of camptothecin was

---

*Cooperating with Yangpu District Central Hospital of Shanghai and other institutions.

higher in the fruits (0.002%) than that in other parts of the tree. Later, 10-OH-CN, 10-methoxycamptothecin, 11-methoxycamptothecin, deoxycamptothecin and other principles were isolated from the fruits (Fig. 1). A dozen of analogs (mainly position 12-th substitutes) were synthesized too (Pan, 1975).

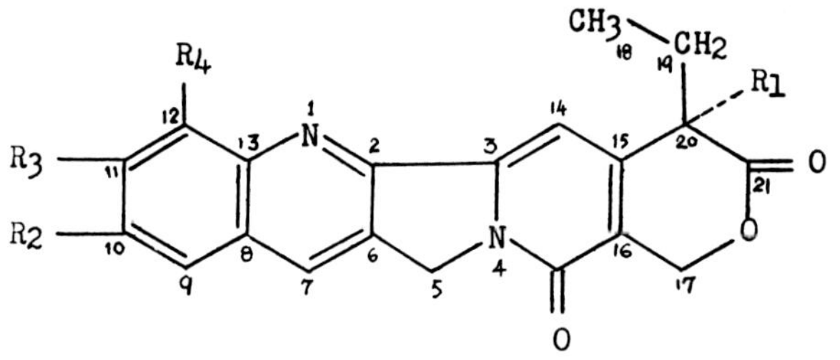

	R_1	R_2	R_3	R_4	
I	OH	H	H	H	Camptothecin
II	OH	OH	H	H	10-Hydroxycamptothecin
III	OH	OCH_3	H	H	10-Methoxycamptothecin
IV	OH	H	OCH_3	H	11-Methoxycamptothecin
V	H	H	H	H	Deoxycamptothecin
VI	OH	H	H	X	X=Cl, Br, OCH_3, OH, NO_2, NH_2, $NHCOCH_3$, SH, COOH, $COOCH_3$

Fig. 1. Chemical Structure of Camptothecin and Analogs

PHARMACOLOGIC ACTIONS

The action of the sodium salt of 10-OH-CN on a variety of animal tumors was determined. The methods for testing the antitumor effect have been previously described (Hsu, 1975). The experimental results are presented in Tables 1 and 2. 10-OH-CN exhibited an obvious inhibitory action on these tumors, especially ascites tumor. Oral administration of the compound also produced a significant inhibition (57.6%) on S-180 in mice.

TABLE 1 Therapeutic Effects of 10-OH-CN on Ascites Tumors

Tumor	Dosage (mg/kg)	No. of animals		Mean survival (days)		Prolongation (%)
		C	T	C	T	
EAC	1	40	20	10.7	36	236.4
	2	40	20	10.7	40.7	280.3
HepA	1	10	10	17	37.3	119.4
	2	21	17	14.1	32.7	131.9
ARS (L_2)	2	21	10	18	44	144.4
Yoshida tumor	1	6	6	11.7	39.8	240.2

TABLE 2 Effects of 10-OH-CN on Solid Tumors

Tumor	Dosage (mg/kg)	No. of animals		Mean tumor wt (gm)		Inhibition (%)
		C	T	C	T	
S-180	1	19	8	2.9	2.4	17.2
	2	36	17	2.8	1.8	35.4
	6*	27	16	2.6	1.1	57.6
S-37	1	10	10	2.8	2.2	21.4
	2	18	18	2.3	1.3	43.5
U-14	1	10	10	4.1	3.5	14.6
	2	10	10	4.1	1.9	53.7
Walker-256	1.5	24	19	8.3	7.1	14.4
	2	16	14	6.8	3.6	47.1

* Oral administration

For studying the distribution and excretion ^{3}H-10-hydroxycamptothecin was synthesized by us with a specific radioactivity of 56 µC/mg. Healthy mice or mice bearing ascites hepatoma were employed. The labelled compound was injected via the tail vein in a dose of 10 mg/kg. Following injection, various tissue samples were taken for radioactivity measurement (Chang, 1975). At 1 hour the radioactivity was highest in bile and intestinal contents, fairly low in carcinoma cells, lower in small intestine, liver, bone marrow, stomach and lung, and lowest in large intestine, heart, brain, muscle, thymus and spleen. Twenty-four hours later, the radioactivity level in bile was about the same as that at 1 hour, persistently high in intestinal content and cancer cells, and lower in all other tissues, being lowest in brain and spleen. The radioactivity in blood dropped rapidly. The first biological half-life was 4.5 minutes and the second 29 minutes. The labelled compound was excreted mainly in the feces. The radioactivity of excretion was 29.6% within 24 hours and 47.8% within 48 hours. The radioactivity excreted from the urine was less, being 8.8% of the total injected radioactivity in 24 hours and 12.8% in 48 hours. By means of the method determining the nucleic acid content and the incorporation of ^{3}H-methyl-thymidine into cancer cells (Shen, 1962), it was found that 10-OH-CN caused a definite inhibition on DNA synthesis (46-87%). It was also demonstrated that 10-OH-CN had certain depressive action on the concomitant immunity of tumor-bearing mice.

CLINICAL OBSERVATION

In Phase I and II clinical studies 253 cancer patients have been treated with the sodium salt of 10-OH-CN. Table 3 shows the results of Phase II clinical trial. The therapeutic dose of this drug was 4-8 mg, administered intravenously once daily or every other day. In some cases 2 or 10 mg/day was used. The total dose for one therapeutic course varied from 50 to 360 mg. Of the 19 cases of primary liver carcinoma, 8 responded effectively. The effective rate in cases of cancer of the head and neck was 39.8%.

TABLE 3 Therapeutic Effects on Cancer Patients

Tumor	No. of patients	Effective	Ineffective
Primary liver carcinoma	19	8(42%)	11
Cancer of head and neck*	28	11(39%)	17
AML	8	5	3
ALL	3	2	1
Ca. of the cardiac end of stomach	4	2	2
Urinary bladder carcinoma	1	1	
Total	63	29(46%)	34

* including 24 cases with carcinoma of salivery glands and 4 cases of malignant lymphoma (3/4 effective)

DISCUSSION

The fruits of Camptotheca acuminata Decne were proved to be a good source of 10-OH-CN. The main toxic actions of the drug are irritation of gastrointestinal tract and depression of bone marrow, but they are less severe than that of camptothecin. Clinical trials also demonstrated that this drug had therapeutic value for some cancerous diseases with mild untoward reactions, in particular the irritant action on urinary tract was much less than that caused by camptothecin. Therefore, it deserves further investigation. The content of the labelled drug is rather high and persists in the cancer cells. This probably plays an important role in the therapeutic action of the drug. While comptothecin is excreted chiefly in the urine, excretion of 10-OH-CN is mainly via feces. Clinically sodium camptothecin often causes hematuria and other urinary irritant symptoms, while 10-OH-CN rarely shows such side effects. This may be explained by the different route of their excretion. Similar to camptothecin, 10-OH-CN produces a stronger inhibition on DNA than on RNA. Further study of the mechanism of its action and clinical usefulness is in progress. Chemical synthesis of dl-10-OH-CN has already been completed in this Institute. Animal experiments showed that given in a dose of 2 or 4 mg/kg the drug prolonged the survival of mice bearing Ehrlich ascites carcinoma, ascitic reticulum cell sarcoma or lymphosarcoma. Its inhibitory action on S-180 and B-22 (brain tumor) was also noticeable. Its subacute LD_{50} in mice was 7.6 ± 1.8 mg/kg. In rabbits given this compound at 1 and 3 mg/kg/day intravenously for 10 days, no marked change was observed in white blood cell counts, Hgb level and ECG.

REFERENCES

Chang, S. Y., S. F. Yio, and B. Hsu (1975). Absorption, distribution and excretion of ^{14}C-AT-581 in animal. Chinese Med. J., 55, 121-125.
Division of Tumor Pharmacology, Depart. of Pharmacol., Shanghai Inst. of Materia Medica (1975). Experimental study of the anticancer action of camptothecin. Chinese Med. J., 55, 274-278.
Hsu, B., S. Y. Chang, J. T. Cheng, X. F. Le, J. L. Yang, F. G. Wu, Z. Hua, L. J. Wang, J. X. Han, F. L. Lu, and Z. W. Wang (1975). Studies on the sensitivity of several tumour models in screen of anticancer agents. Kexue Tongbao, No. 5, 242-246.
Hsu, J. S., T. Y. Chao, L. T. Lin, and C. F. Hsu (1977). Chemical constituents of the anticancer plant Camptotheca acuminata Decne II. Chemical constituents of the fruits of Camptotheca acuminata Decne. Acta Chimica Sinica, 35, 193-200.
Pan, P. C., S. Y. Pan, Y. H. Tu, S. Y. Wang, and T. Y. Owen (1975). Studies on the derivatives of camptothecin. Acta Chimica Sinica 33, 71-74.
Shen, M. L., J. T. Chen, and B. Hsu (1962). Studies on antitumor drugs. XII. Effect of actinomycin K on the amount of and on the incorporation of ^{32}P into nucleic acids of Ehrlich carcinoma cells. Acta Biochemica et Biophysica Sinica, 2, 218-225.
Wall, M. E., M. C. Wani, C. E. Cook, and K. H. Palmer (1966). Plant antitumor agent. I. The isolation and structure of camptothecin, a novel alkaloidal leukemia and tumor inhibitor from Camptotheca acuminata, J. Am. Chem. Soc., 88, 3888-3889.
Wani, M. C., and M. E. Wall (1969). Plant antitumor agents. II. The structure of two new alkaloids from Camptotheca acuminata. J. Org. Chem., 34, 1364-1367.

Biological Basis for Cancer Chemotherapy

New Animal Models in Cancer Chemotherapy

A. Goldin, J. M. Venditti, F. M. Muggia, M. Rozencweig and V. T. DeVita

Division of Cancer Treatment, National Cancer Institute, National Institutes of Health, Bethesda, Maryland 20014

ABSTRACT

In the Division of Cancer Treatment at the National Cancer Institute, USA, a new prospective screening program has been instituted. It includes human tumor xenografts growing in athymic mice and corresponding murine tumor models. The new screening program has the primary objective of discovering new and more effective antitumor agents. The flow of drugs through the new screen is presented as well as a listing of questions that are being addressed to the new screen. Examples are cited in which there is increased interest in specific drugs for the clinic based on activity in the new screening panel. Data todate indicate that drugs with high broad-spectrum activity in the new screen may have a good likelihood of evidencing definitive activity in the clinic. The usefulness of human tumor xenografts for not only screening, but also drug development and fundamental investigations is outlined.

KEYWORDS

Animal models, screening, drug evaluation, drug development, preclinical-clinical correlations, human tumor xenografts, athymic mice.

INTRODUCTION

The utilization of new animal models encompasses two primary interests:
a) Screening for the identification of new and more effective antitumor agents.
b) Preclinical investigation of the newly discovered antitumor agents with a view to the most efficacious utilization in the clinic. This entails preclinical characterization of the candidate antitumor drugs and toxicologic, pharmacologic and biochemical investigation of their mode of action in relation to clinical application. In the current report attention is focused on the recently introduced new prospective screen in the Division of Cancer Treatment (DCT) at the National Cancer Institute (NCI) which has incorporated model systems that may indeed be useful in the discovery of new and more effective antitumor agents and their further investigation for the clinic.

It has been well recognized that despite the advances in the chemotherapy of clinical neoplasia, resulting in large measure from the use of drugs in combination, there has been successful therapy in only a small fraction of the various types of cancer (Zubrod, 1972, DeVita and Schein, 1973). There has been therefore, an important concern as to whether a screening program which relied primarily on the murine leukemias L1210 and P388, plus additional test systems such as B16 melanoma and Lewis lung carcinoma for special testing, provided the greatest likelihood of the identification of the most highly effective drugs against clinical solid tumors. With the availability of athymic mice, in which it is possible to successfully transplant human tumors, along with the development of new murine tumor models, including mouse colon and mammary tumor models, an opportunity was provided for a reevaluation of the ongoing antitumor screening program. In 1976, the DCT, NCI revised the screening program, incorporating in addition to well established experimental tumor systems more recently developed animal solid tumor models and human tumors growing as xenografts in athymic mice. The newly established screening program provides the opportunity for the conduct of a coordinated systematic screening experiment which is designed to provide a basis for a prospective analysis of the validity of the various screening systems that comprise a screening program.

THE NEW ANTICANCER SCREENING PROGRAM; TUMOR MODELS AND PERSPECTIVES

In the development of the new DCT screen, a number of possibilities had been considered (Goldin and others 1978):

a) Establishment of a large battery of experimental tumor systems. Adoption of an extensive spectrum of experimental tumor systems would tend to maximize the discovery of clinically active compounds by identification of compounds that would not be identified (false negatives) by a limited screening system. It could also conceivably result in the availability of compounds with the greatest activity in the clinic. With such an approach a large number of false positives might be identified (compounds active in one or more of the experimental tumor systems but inactive in the clinic), and some ranking order would then have to be established to avoid overloading of the clinical capacity. It was considered that this approach was not logistically feasible since it was projected that it would require the testing of a large number of materials (approximately 5000) in the total tumor spectrum in order to obtain a significant number of materials that would be active in one or more of the tumor systems of an extensive tumor spectrum.

b) Utilization of a primary and secondary spectrum. This would include: 1) a limited primary spectrum of test systems including leukemias L1210, P388, B16 melanoma, Lewis lung carcinoma, plus representative murine colon, lung and breast tumor models and corresponding human tumor xenografts in athymic mice. 2) The establishment of a large secondary spectrum of tumor systems representative of those employed in the various countries of the world. With this approach too, the level of effort for each test compound was considered to be too extensive for an organized large scale screening program.

c) The new screen. It was with the above considerations in mind that the schema of the new screen was developed since it was considered to be feasible logistically and could have the potentiality of improving drug selectivity for the clinic. The flow of drugs through the new screen is presented in the schema of Fig. 1. Leukemia P388 is being employed as a "prescreen" and was selected for the following reasons: a) it responds to clinically active drugs of various classes in a manner qualitatively similar to leukemia L1210.

Fig. 1

Flow of drugs through DCT screens

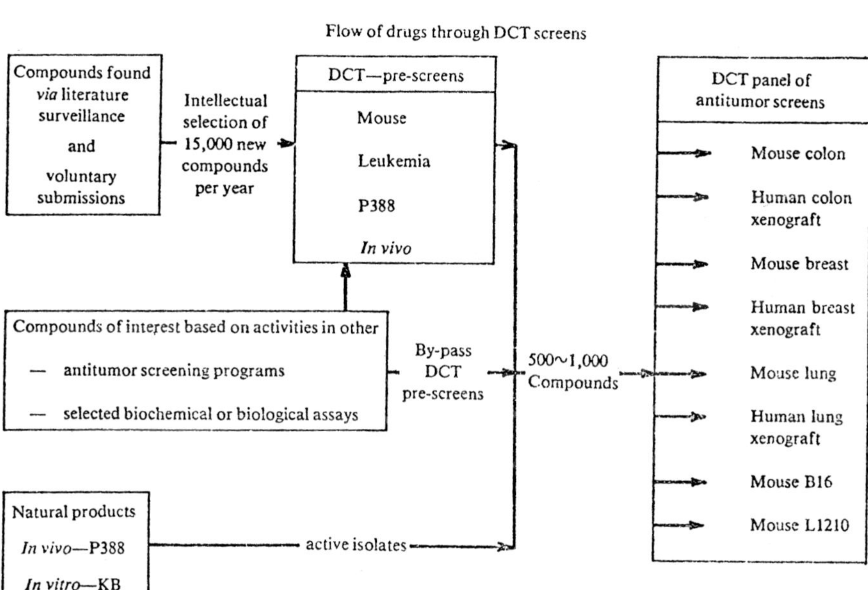

b) it is quantitatively more sensitive than leukemia L1210 and can therefore meet a primary objective of the prescreen which is designed to select a population of materials that will yield a high percentage of actives in one or more of the tumors in the panel.

The schema includes a panel of antitumor screens which currently include: mouse colon tumors 26 and 38, human colon tumor xenografts CX-1 and CX-2; mouse mammary tumor CD8F1, human mammary tumor xenograft MX-1; mouse Lewis lung carcinoma, human lung tumor xenograft LX-1; and mouse B16 melanoma and leukemia L1210. The specific protocols for screening, utilizing leukemias L1210 and P388, B16 melanoma and Lewis lung carcinoma have been reported previously (Geran and others 1972). Corbett and others (1977) have reported on the origin and experimental methodology employed in the screening with the carcinogen induced transplantable colon tumors 26 and 38. The spontaneous mammary carcinoma system in CD8F1 mice, where the first generation transplant is employed for drug evaluation, has been described by Martin and others (1975). The human tumor xenografts are carried in serial transplantation in athymic mice. Colon tumor CX-2, mammary tumor MX-1 and lung tumor LX-1 were developed by Giovanella and others (1977) (Venditti and others 1978), and the colon CX-1 tumor was developed by A. Bogden, Mason Research Institute.

The level of screening (40,000 new materials per year) was reduced to 15,000 per year in order to accommodate for the more extensive screening effort of the new screen. Instead of screening compounds selected entirely at random an effort is being made to select as many compounds as possible based on review of the world's literature and through voluntary submissions. These compounds are tested in the prescreen in vivo against leukemia P388, and all of the compounds that are active are moved forward for testing in the entire panel of antitumor screens. In addition, since initial testing in a single prescreen such as leukemia P388 might be too restrictive, compounds that are considered to be of interest based on activity in other antitumor screening programs, or compounds that are selected on the basis of special biochemical or biological assays, may bypass the P388 prescreen and go directly to testing in the screening panel. They are also tested in the P388 prescreen but activity there is not mandatory for testing in the entire screening panel.

Natural products including fermentation broths and plant and animal extracts are tested in vivo in leukemia P388 and also in vitro in the KB system and those which demonstrate activity are then tested in the entire antitumor screening panel. The P388 system has proven to be more sensitive to such materials than leukemia L1210 and is sufficiently sensitive to identify activity of low concentration of active materials. The KB system has also proven useful particularly for bioassay of natural product fractions during the process of concentration and purification.

Employing the "prescreen" approach approximately 500-1000 compounds per year are becoming eligible for testing against the DCT experimental screening panel.

A number of questions are being addressed prospectively to the new NCI screen (Goldin and others, 1978). 1) Does the new screen increase the number of true positive compounds (active in the screen and active in the clinic)? 2) Does it reduce the number of false positives (active in the screen but inactive in the clinic)? 3) Does it reduce the number of false negatives (inactive in the screen but active in the clinic)? 4) To what extent do human tumor xenografts and animal tumor screens select the same or different drugs as active? 5) Are the human tumor xenografts more effective than the animal tumors in predicting for clinical antitumor activity? 6) Are the xenograft positives more active in the clinic than those selected by animal screens? 7) Is there a correspondence of activity against animal tumors and/or human tumor xenografts with activity against clinical tumors for specific organ systems or specific histologic types. 8) Are compounds that bypass the prescreen because of activity in other screening programs or in selected biochemical or biological assays more effective in the screening panel and in the clinic than compounds initially selected for further testing by the prescreen? 9) What contribution will the data of the new screening panel make to prediction of clinical effectiveness of new drugs, utilizing mathematical procedures such as multivariate discriminant analysis (Venditti and others 1978).

As part of the totality of effort of the DCT program the necessary clinical data are being obtained in a panel of human tumor types for compounds emerging as active in the new screen, in order to obtain answers to these prospectively addressed questions.

Although the revised screening program is relatively new, examples may already be cited in which there is increased interest in compounds for the clinic stemming from the demonstration of activity in the new screening panel (Table 1). The new acridine derivative methanesulfon-m-anisidide, 4'-(9-acridinyl) amino-

-(AMSA) (Cain and Atwell, 1974) has been observed to be active in seven of the animal test systems in the new screen including leukemias L1210 and P388, Lewis lung carcinoma, B16 melanoma, colon carcinoma 26, colon carcinoma 38, and the mammary tumor in CD8F1 mice. It showed borderline activity against colon xenograft CX-2 and was inactive against breast xenograft MX-1. Preliminary clinical studies suggest possible activity in leukemia, and breast, ovarian and renal cell tumors.

TABLE 1 ACTIVITY OF DRUGS IN TUMOR PANEL SYSTEMS

	L1210	P388	B 16 Melanoma	Lewis Lung	Colon 26	Colon 38	Colon Xeno-graft CX-1	Colon Xeno-graft CX-2	Mammary Xeno-graft MX-1	$CD8F_1$ Mam-mary	Lung Xeno-graft LX-1
AMSA	+	+	+	+	+	+	NT	±	-	+	NT
PALA	-	±	+	+	+	±	-	-	+	+	-
Maytansine	+	+	+	-	-	-	NT	NT	NT	-	NT
DON	±	±	-	-	+	±	-	+	+	+	+
AT 125	+	+	-	-	-	-	-	-	+	-	±
Anguidine	+	+	+	-	-	+	NT	-	-	+	NT
Cis-pt II	+	+	+	+	+	+	NT	NT	+	+	-
Hexamethyl-melamine	±	-	±	+	±	±	NT	-	+	±	-
Bleomycin	±	+	+	-	-	-	+	NT	-	+	NT
Chlorozotocin	+	+	+	±	+	+	NT	NT	±	+	NT

NT = Not treated

Phosphonacetyl-L-aspartic acid (PALA), an inhibitor of aspartate transcarbamylase involved in the conversion of carbamylphosphate plus L-aspartate to carbamyl-L-aspartate (Collins and Stark, 1971) although inactive in the treatment of leukemia L1210 and only marginally effective in the treatment of leukemia P388 was nevertheless active in the treatment of Lewis lung carcinoma (Johnson and others, 1976; Schabel and others 1978). In addition, in the new screen it was observed to be active against B16 melanoma, colon 26, mammary CD8F1 and mammary xenograft MX-1, and marginally effective against colon 38. PALA has broad spectrum activity against a variety of experimental solid tumors (Venditti and others 1978, Johnson and others 1978). PALA is currently in Phase I clinical trial.

The new plant product maytansine has been observed to be active against three animal tumors in the new screen, namely leukemias L1210 and P388, and B16 melanoma. It was inactive against Lewis lung carcinoma, colon 26, colon 38 and CD8F1 mammary tumor. In view of an apparently more limited preclinical spectrum of maytansine activity, the extent of activity in the clinic against various tumor categories will be of interest to determine.

There is renewed clinical interest in the glutamine antagonist 6-diazo-5-oxo-L-norleucine (DON) (Livingston and others, 1970) a drug that has been available for some years, on the basis of its activity against tumors of the new spectrum. In the initial report of Goldin and others (1966) DON was listed as demonstrating activity against Walker adenocarcinoma 256 (IM) and carcinoma 755. It had borderline activity against leukemia L1210. It has now also been observed to be active against the human tumor xenografts colon CX-2, breast MX-1 and lung

LX-1. It was inactive against colon CX-1, but active against colon 26.

The antimetabolite antibiotic (αS, 5S)- α -amino-3-chloro-4,5-dihydro-5-iso-oxazoleacetic acid (NSC 163501) (AT-125) (Hanka and others 1973) is an inhibitor of various bacterial and mammalian enzymes including L-asparaginase which catalyze the transfer of the amino group of glutamine. It has demonstrated activity in the new screen against leukemias L1210 and P388, and breast xenograft MX-1. It had no effect against colon xenografts CX-1 and CX-2 (Houchens and others 1978). This drug is not yet in clinical trial.

Anguidine (NSC 141537), a novel structure produced by several soil fungi has activity against leukemias L1210 and P388, B16 melanoma, mammary CD8F1 and colon carcinoma 38. There is no definitive data in the clinic as yet for this drug.

The platinum derivative cis-dichlorodiammine platinum (II) (cis-platinum II) has broad spectrum activity in the new screen including activity against leukemias L1210 and P388, Lewis lung carcinoma, B16 melanoma, CD8F1 mammary carcinoma, colon 26, colon 38 and mammary xenograft MX-1. This drug already has demonstrated evidence of clinical activity against testicular tumors, ovarian carcinoma, head and neck carcinoma, bladder carcinoma, carcinoma of penis, cervical carcinoma and osteosarcoma.

Hexamethylmelamine although not definitely active in the treatment of leukemia L1210 or in a number of other murine tumor screens was active in the treatment of human colon, bronchus, ovary and kidney tumors in mice thymectomized at birth and reconstituted with syngeneic bone marrow (Detre and others 1975; Mitchley and others 1975). These observations led to renewed interest in hexamethylmalamine as well as in the water soluble analogs of hexamethylmelamine such as pentamethylmelamine (Connors and others 1977). In the new screen hexamethylmelamine has shown activity against Lewis lung and the mammary xenograft MX-1. Borderline activity was observed with a number of the tumors. Hexamethylmelamine has shown activity in the clinic against lung carcinoma, ovarian carcinoma and bilharzial bladder cancer.

Bleomycin was active against leukemia P388, B16 melanoma, colon xenograft CX-1 and mammary CD8F1 tumor of the new screen. Clinically, it has been observed to be active against Hodgkin's disease, non-Hodgkin's lymphoma, head and neck carcinoma, testicular tumors and superficial squamous cell tumors including penile cancer.

Chlorozotocin has broad-spectrum activity in the new screen and in initial clinical studies has evidenced activity against non-Hodgkin's lymphoma.

In regard to the questions that are being asked prospectively, the observations todate indicate that the new screen has the capability of increasing the number of true positive compounds for the clinic. The generalization appears to be emerging that high broad-spectrum activity in a variety of experimental test systems prospectively increases the prediction for clinical activity for at least one and possibly more, or even a broad spectrum of human tumors. If this is correct, broad-spectrum activity would tend to diminish the incidence of false positives. By the same token, since there appears to be detection of compounds that would normally be missed by the earlier conventional screen it would tend to diminish the incidence of false negatives. A good example is provided by cis-platinum II which was found to be highly active in eight of the nine tumor panel systems in which it was tested. This compound, as indicated above, appears to have broad-spectrum clinical activity. The screen suggests that AMSA, which has been demonstrated to be active in seven (possibly eight) of the nine tumor systems in which it was tested, will have a high likelihood of clinical activity.

This generalization is consistent with the results of an earlier retrospective analysis of the relationship of experimental screening procedures and clinical predictability value in which it was indicated that "The current analysis highlights a somewhat fortunate circumstance, namely, that active compounds are, in general, active not only in one screening system, but also in a variety of screening systems" (Goldin and others 1966). This was also observed in the extensive analysis of Hirschberg (1963). It was further noted that "Also, such compounds, when clinically active are in general active against more than one type of clinical tumor" (Goldin and others 1966). It should be mentioned in this regard that most of the drugs that have been predicted by experimental antitumor systems but failed to become clinically useful agents were dropped from clinical trial due to severe, unpredictable, or irreversible toxicity (Johnson and Goldin 1975). In this sense they may not be designated as clearly "false positives."

The answers to the various questions being directed to the new screen will surely be resolved as additional active drugs are discovered and the clinical data forthcoming.

The new screen is subject to modification with respect to the tumor models employed and the way in which they are utilized, in accordance with ongoing experience. Bogden and others (1978) have demonstrated that human tumor xenografts may be grown under the renal capsule of immunocompetent mice and that it is possible to measure the response of these tumors to chemotherapeutic agents. A number of drugs were evaluated in the subrenal capsule system in normal BDF1 mice and also in the subrenal capsule of athymic mice, for comparison with activity against subcutaneously growing tumors in athymic mice. For example, with the MX-1 human breast tumor, CX-2 human colon tumor and LX-1 human lung tumor, there was good correspondence of activity for standard drugs such as L-PAM, cyclophosphamide, methotrexate, 5-fluorouracil, adriamycin, and methyl CCNU in the two subrenal capsule systems and against subcutaneously inoculated tumor in athymic mice. In the subrenal capsule assay system, employing athymic animals, hexamethylmelamine was observed to be active in the colon CX-1 and mammary MX-1 xenografts (Cobb 1978). This system, from a logistic point of view, may prove to be an important type of prescreen.

Salmon and others (1978) applied a soft agar assay to primary bioassay of ovarian carcinoma stem cells. *In vitro* sensitivity to adriamycin, melphalan, bleomycin, vinblastine and other drugs was investigated employing this system and differential *in vitro* sensitivity patterns were observed between patients, for specific drugs. Both retrospective and prospective studies suggested a correlation between *in vitro* sensitivity to drugs and clinical response. It is possible that this type of methodology may provide an *in vitro* approach for identification of new drugs active against specific types of neoplastic disease.

In another example, Ozols and others (1978) characterized a murine transplantable ovarian cancer and observed that it has many similarities to human ovarian cancer. Adriamycin given ip 48 hours after tumor transplantation, at an LD10 dose, produced prolongation of survival, whereas cyclophosphamide, 5-fluorouracil and cis-platinum II did not. However, when adriamycin was administered four and seven days after tumor transplantation it did not prolong survival. It was suggested that this murine ovarian tumor could provide a suitable model for studying combined modality therapy of ovarian cancer.

A variety of new animal models of potential interest for the chemotherapy of human solid tumors was reviewed at an International Union Against Cancer (UICC) workshop (Mihich and others 1974).

UTILIZATION OF HUMAN TUMOR XENOGRAFTS IN ATHYMIC MICE FOR DRUG EVALUATION AND DEVELOPMENT AND FUNDAMENTAL INVESTIGATIONS

Human tumor xenografts of various histologic types may serve as useful models for the preclinical characterization of candidate antitumor drugs and toxicologic, pharmacologic and biochemical investigations. In essence, a whole gamut of established procedures may be applied to investigation of drug activity against human tumor xenografts in athymic mice, and comparisons can then be made with activity in the standard murine models and activity in the clinic (Goldin and others 1975). New antitumor agents may be characterized for drug toxicity in normal mice and drug toxicity in conventional and athymic tumorous mice and contribute to the baseline of information for the introduction of new drugs into the clinic. The schedule dependency characteristics of new antitumor agents may be determined in athymic mice in accordance with standardized and novel protocols. Drug effects on the course of tumor growth may be determined. In this regard, the studies with the Ridgway osteogenic sarcoma model, which has been utilized for therapeutic staging (Schabel 1974; Goldin and others 1975) may be adapted to the investigation of drug activity in human tumor xenograft systems in athymic animals. Parameters of response that may be measured include: the influence of the drug on extent of tumor regression (complete and partial), duration of regression, median lifespan, range of deaths in the animals that relapse, and increase in lifespan of the animals. In addition, the tumor cell kinetics, tumor cell kill and "cure response" can be investigated.

With tumors such as the Ridgway osteogenic sarcoma it is possible to follow the response to therapy of the individual mouse (Schabel 1974). The possibility of following the growth of human tumors in individual athymic mice has been illustrated by Osieka and others (1977).

Studies may be conducted in athymic mice on the origin and treatment of tumors that have become resistant to therapy and the nature of cross-resistance to various antitumor agents, as well as studies of the treatment of naturally resistant human tumor xenografts. In correspondence with the clinic, human tumors of the same histologic type will not necessarily respond in the same manner to therapy. Osieka and others (1977) observed that two colon tumor lines CA and HT were highly resistant to therapy with methyl CCNU, while a third line, BE was markedly sensitive. The basis of resistance and sensitivity may be investigated in athymic mice, and drugs which are successful in the treatment of naturally resistant tumors may be of considerable interest for potential clinical testing. The baseline of drug evaluation in the athymic mice is being broadened by having available a number of tumors of similar histologic type with varying responsiveness to standard chemotherapeutic agents.

In general, the human tumors growing in athymic animals have not metastasized too readily, although invasiveness and metastasis has been demonstrated. The process of tumor cell infiltration and metastasis is being investigated in athymic mice as well as the influence of immunosuppressant agents and antitumor agents. Antitumor agents may be investigated with respect to their effectiveness in the treatment of naturally occurring or artificially produced metastatic tumor.

Human tumors growing in athymic animals may provide suitable models for investigation of combination chemotherapy and combined modalities including surgery plus chemotherapy, immunity plus chemotherapy and radiation plus chemotherapy.

Clearly, the models may be of considerable value in investigation of the basic mechanism of action of antitumor agents with respect to antitumor inhibitory properties in the athymic animals, and an attempt can be made to relate such findings to clinical utilization of the drugs in terms of elicitation of maximum activity in the clinic.

REFERENCES

Bogden, A.E., Esber, H.J., Haskell, P.M, and LePage, D.J. Growth and chemotherapy response of human tumor xenografts under the renal capsule of immunocompetent mice. Proc. American Assoc. Cancer Res. 19, 105, 1978

Cain, B.F., and Atwell, G.J. The experimental antitumor properties of three congeners of the acridylmethanesulphonanilide (AMSA) series. Eur. J. Cancer 10, 539-549, 1974.

Cobb, W.R. Response of human tumor xenografts to hexamethylmelamine and its derivatives in a rapid sub-renal capsule assay. Proc. American Assoc. Cancer Res. 19, 41, 1978.

Collins, K.D. and Stark, G.R. Aspartate transcarbamylase interaction with the transition state analog, N-(Phosphonacetyl)-L-aspartate. J. Biol. Chem. 246, 6599-6605, 1971.

Connors, T.A., Cumber, A.J., Ross, W.C.J., Clarke, S.A. and Mitchley, B.C.V. Regression of human lung xenografts induced by water-soluble analogs of hexamethylmelamine. Cancer Treat. Rep. 61, (5), 927-928, 1977.

Corbett, T.H., Griswold, D.P., Jr., Roberts, B.J., Peckham, J.C., and Schabel, F. M., Jr. Evaluation of single agents and combinations of chemotherapeutic agents in mouse colon carcinomas. Cancer 40, No. 5, 2660-2680, 1977.

Detre, S.I., Davies, A.J.S., and Connors, T.A. New models for cancer chemotherapy. Cancer Chemother. Rep. Part 2, 5, 133-143, 1975.

DeVita, V.T., and Schein, P.S. The use of drugs in combination for the therapy of cancer. New Engl. J. Med. 288, 998-1006, 1973.

Geran, R.I., Greenberg, N.H., Macdonald, M.M., Schumacher, A.M., and Abbott, B.J. Protocols for screening chemical agents and natural products against animal tumors and other biological systems (Third Edition) Cancer Chemother. Rep. Part 3, Vol. 3, No. 2, 1-5, 1972.

Giovanella, B.C., Stehlin, J.S. and Shepard, R.C. Experimental Chemotherapy of human breast carcinoma heteroptransplanted into nude mice. Proc. Second Internatl. Workshop on Nude Mice, Tokyo, 1977, pp 475-481.

Goldin, A., Johnson, R.K., and Venditti, J.M. Preclinical characterization of candidate antitumor drugs. Cancer Chemother. Rep. Part 2, Vol. 5, No. 1, 21-81, 1975.

Goldin, A., Serpick, A.A., and Mantel, N. A Commentary: Experimental screening procedures and clinical predictability value. Cancer Chemother. Rep. 50, 173-218, 1966.

Goldin, A., Schepartz, S.A., Venditti, J.M., and DeVita, V.T. Jr. Historical development and current strategy of the National Cancer Institute drug development program. In: Methods in Cancer Research. H. Busch and V. T. DeVita, Jr., (Eds.) Academic Press, Inc. 1978.

Hanka, L.J., Martin, D.G., and Neil, G.L. A new antitumor antibiotic, (S,α5S) -α-amino-3-chloro-4,5-dihydro-5-isoxazoleacetic acid (NSC 163501): microbial reversal studies and preliminary evaluation against L1210 mouse leukemia in vivo. Cancer Chemother. Rep. 57, 141-148, 1973.
Hirschberg, E. Patterns of response of animal tumors to anticancer agents. Cancer Res. (Supp.) 23 (No. 5, part 2), 521-980, 1963.
Houchens, D., Ovejera, A., Johnson, R. Bogden, A., and Neil, G. Therapy of mouse tumors and human tumor xenografts by the antitumor antibiotic AT-125 (NSC 163501). Proc. American Assoc. Cancer Res. 19, 40, 1978.
Johnson, R.K. and Goldin, A. The clinical impact of screening and other experimental tumor studies. Cancer Treat. Rev. 2, 1-31, 1975.
Johnson, R.K., Inouye, T., Goldin, A. and Stark, G.R. Antitumor activity of N-(phosphonacetyl)-L-aspartic acid; a transition-state inhibitor of aspartate transcarbamylase. Cancer Res. 36, 2720-2725, 1976.
Johnson, R.K., Swyryd, E.A., and Stark, G.R. Effects of N-(phosphonacetyl) -L-aspartate on murine tumors and normal tissues in vivo and in vitro and the relationship of sensitivity to rate of proliferation and level of aspartate transcarbamylase. Cancer Res. 38, 371-378, 1978.
Livingston, R.B., Venditti, J.M., Cooney, D.A., Carter, S.K. Glutamine antagonist in cancer chemotherapy. In: Advances in Pharmacology and Chemotherapy. Vol. 8, S. Garratini, A. Goldin, F. Hawking, and I.J. Kopin (eds.) New York Academic Press, 57-120, 1970.
Martin, D.S., Fugmann, R.A., Stolfi, R.L., Hayworth, P.E. Solid tumor animal model therapeutically predictive for human breast cancer. Cancer Chemother. Rep. Part 2, 5 No. 1, 89-109, 1975.
Mihich, E., Laurence, D.J.R., Laurence, D.M., and Eckhardt, S. UICC Workshop on new animal models for chemotherapy of human solid tumors. UICC Technical Rep. Series, Vol. 15, 1-50, 1974.
Mitchley, B.C.V., Clarke, S.A., and Connors, T.A. Hexamethylmelamine-induced regression of human lung tumors growing in immune deprived mice. Cancer Res. 35, 1099-1102, 1975.
Osieka, R., Houchens, D.P., Goldin, A. and Johnson, R.K. Chemotherapy of human colon cancer xenografts in athymic nude mice. Cancer 40, 2640-2650, 1977.
Ozols, R.F., Grotzinger, K.R., and Young, R.C. Murine ovarian cancer: A model for human disease. Proc. American Assoc. Cancer Res. 19, 234-1978.
Salmon, S.E., Hamburger, A.W., Soehnlen, B.A. Schmidt, H.J., and Alberts, D.S. Primary bioassay of ovarian carcinoma stem cells. Proc. American Assoc. Cancer Res 19, 231, 1978.
Schabel, F.M., Jr. New experimental drug combinations with potential clinical utility. Biochem. Pharmacol. 23 (Supp. 2), 163-176, 1974.
Schabel, F.M., Jr., Laster, W.R., Jr. and Rose, W.C. Experimental chemotherapy and tumor cell kinetics. Overview of experimental tumor systems. In: Treatment of lung cancer, F.M. Muggia and M. Rozencweig (eds), New York, Raven Press. (In press 1978)
Venditti, J.M., Goldin, A. Miller, I., and Rozencweig, M. Experimental models for antitumor testing in current use by the National Cancer Institute, USA. Statistical analysis and methods for selecting agents for clinical trials. Proc. of Takamatsu Symp. Tokyo. (In press 1978).
Zubrod, C.G. Chemical control of cancer. Proc. Nat. Acad. Sci. USA, 69, 1042-1047, 1972.

Biological Aspects of Drug Resistance

V. Ujházy

Cancer Research Institute of the Slovak Academy of Sciences, Bratislava, Czechoslovakia

ABSTRACT

Biological aspects of drug resistance development are briefly reviewed and documented by results achieved in the field of a, tumor cell population heterogeneity, b, cell-to-cell interaction, and c, karyology of drug resistance.
Key words: drug resistance, heterogeneity of cell population, chromosomal markers.

INTRODUCTION

Final cancer chemotherapy results are far behind the effectivity of initial therapy courses during the first contact of tumor cell population with cytostatics. In the majority of human tumors the degree of drug sensitivity is during successive therapy courses gradually decreasing and finally the tumor becomes resistant. So, the drug resistance development is one of the basic problems of the present cancer chemotherapy.

Biochemical mechanisms of drug resistance development have been extensively studied and a great number of molecular events related to drug resistance have been described / for. rev. see Bertino and Skell, 1973; Brockman, 1974; Goldin and Johnson, 1977; Mihich, 1973 /.

Two fundamental biological mechanisms of drug resistance development have been well recognized from the very beginning of cancer chemotherapy: 1. mutation, and 2. selection of resistant clones. The role of the mutation is accentuated by the fact, that almost all cytostatics possess certain degree of mutagenic activity. Resistant cells a priori present, or appearing following mutation in tumour cell population, are during cell proliferation selected by the specific pressure of cytostatics. It has been also found, in experimental models, that transmission of resistance - apparently by mechanisms different from those known in the bacterial drug resistance transfer - might be considered as a potential factor in mechanisms of drug resistance development /Szende, Fox and Fox, 1973; Ujházy, 1969/.

In the present paper the biological aspects of drug resistance will be discussed from the point of view of a, tumor cell population heterogeneity, b, cell-to-cell interaction within tumor cell population, and c, karyology of drug resistance.

a, TUMOR CELL POPULATION HETEROGENEITY AND DRUG RESISTANCE

It has been shown that cell populations of tumors are heterogenous as related to differences in generation time, portion of cycling and resting cells, number of chromosomes and sensitivity against cytostatic drugs /Hakansson and Tropé, 1974; Mittelman, 1971;

TABLE 1 Differences in Drug Sensitivity of Specimens from Different Areas of the Same Tumor
Dg: colon cancer
/expressed as % of ^{3}H-thymidine uptake in vitro/

Specimen	Control	5-FU	MTX	HN3	VINCRISTINE
1.	100.0	93.5	29.2	24.9	68.6
2.	100.0	66.6	20.8	11.4	13.6
3.	100.0	74.1	231.0	22.6	1.6

Pályi, Oláh and Sugár, 1977; Schabel and Simpson-Herren, 1978; Schiffer, Braunschweiger and Stragand, 1978; Siracký, Maťoška and Siracká, 1977 /. These differences inside a tumor cell population are consequences of mutation, stable phenotypic changes, environmental conditions, etc. Differences in drug sensitivity form one of the primary prerequisites for the drug resistance development by the "drug pressure" mechanism. Table 1.

Using the in vitro ^{3}H-thymidine uptake test for determination of drug sensitivity, considerable differences have been found also in human tumors between a, solid tumor cells and related ascitic cells, b, between samples from different metastatic nodes, and even c, between different samples from the same solid tumor. Table 1. During the contact of cytostatic drugs with heterogenous tumor cell population a strict selection may be monitored by prolongation of doubling time of the surviving resistant cell population to the values found in normal tissue cells. Table 2.

TABLE 2 Doubling Time of Metastatic Ovary Cell Population

		doubling time in hrs
before chemotherapy:	Pat. No. 1	187.5
	No. 2	177.7
	No. 3	172.5
after chemotherapy:	Pat. No. 1	463.6
	No. 2	364.2
	No. 3	491.3

/Siracký and co-workers, 1977/

b, CELL-TO-CELL INTERACTION AND DRUG RESISTANCE

Beside the mutation and selection, less attention has been devoted to the participation of the cell-to-cell interaction in drug resistance development. In 1969 Subak-Sharpe and co-workers showed that the lack of an enzyme / hypoxanthine-guanine phosphoribosyltransferase / could be reversed by the presence of cells, possessing the particular enzyme. There is not sufficient evidence, whether the modification of drug sensitivity by "metabolic co--operation" of two cells is a result of a specific enzyme, messenger RNA, or episomal DNA transfer.

The biochemical background for drug resistance may be represented both by deficiency of an enzyme / hypoxanthine-guanine phosphoribosyltransferase in 8-Azaguanine-resistant cells /, or increased activity of an enzyme / asparagine synthetase in L-Asparaginase-resistant cells /. A decrease in drug sensitivity could be observed in Gardner lymphoma cells following incubation with cytoplasmic fractions isolated from L-Asparaginase resistant cells / Ujházy and co-workers, 1971 /. Table 3.

TABLE 3 Modification of Drug Sensitivity by Cytoplasmic Fractions Isolated from Drug Resistant Cells

sensitive tumor cells	added cytoplasmic fractions from cells resistant to:	decrease of sensitivity
Yoshida sarcoma	alkylating agents	-
Ehrlich carcinoma	5-Fluorouracil	±
Gardner lymphoma	L-Asparaginase	+

In a similar in vitro study drug resistance could be transfered by culture medium from the resistant subline of cells / Szende, Fox and Fox, 1973 /. In an other model system a decrease in nitrogen mustard-sensitivity was observed in sensitive Walker tumor cells following 3 transplant generations with resistant Yoshida sarcoma cells / Ujházy, 1969/.

The knowledge of mechanisms by which drug resistance can be transfered in mammalian cells is still lacking the extensiveness of experimental and clinical data obtained from the investigation of resistance transfer in bacterial systems. Nevertheless, the possibility of influencing the sensitivity of a tumor cell which is in close contact with a resistant cell, can not be excluded from the variety of mechanisms, by which drug resistance develops in mammalian cells.

c, KARYOLOGY OF DRUG RESISTANCE

Various changes have been observed in chromosomal composition during development of drug resistance. Some of these findings were represented by changes of chromosome modal number, in the others there was loss of a particular chromosome / Biedler, Schrecker and Hutchison, 1963; Ujházy, 1968; Yosida, 1966; Yosida, Ohara and Law, 1967 /. The aim of these investigations was to find specific markers closely related to the observed resistance.
The classical example of disappearance of subtelocentric chromosome during Methotrexate-resistance development in L1210 leukemia was soon explained as more related to the increase of dihydrofolate reductase, moreover the chromosomal change persisted even after the reappearance of sensitivity.

As it was shown in a karyological investigation, following resistance induction against alkylating agents in rat Yoshida sarcoma, no difference was found in the number of chromosomes / Ujházy, 1968 /. Both the resistant and sensitive line of the tumor exhibited a modal number of 39 chromosomes. The only difference was in the group of subtelocentric chromosomes, where the largest chromosome disappeared. Using banding technique it was found, that the change was consequence of a deletion on the short chromatids. Identical change has been found in all 3 sublines in which the resistance was separately induced / Ujházy, 1974 /. In spite of the similar change this chromosome marker is not considered to be resistance-specific. In conclusion it may be stated, that a, resistant tumors may differ from their original sensitive cell populations in chromosomal composition, and b, changes observed are rather resistant subline-specific and not specific markers of resistance.

REFERENCES

Bertino, J. R., and R. T. Skell /1973/. Resistance to chemotherapeutic agents. Proc. 5th Int. Congress Pharmacology San Francisco 1972, Vol. 3, ed. Karger, Basel. pp. 376-392.

Biedler, J. L., A. W. Schrecker, and D. J. Hutchison /1963/. Selection of chromosomal variants in amethopterin-resistant sublines of leukemia L1210 with increased levels of dyhydrofolate reductase. J. Natl. Cancer Inst., 31, 357-601.

Brockman, R. W. /1974/. Mechanisms of resistance. In A. C. Sartorelli and D. G. Johns /Ed./, Antineoplastic and Immunosuppressive Agents I, ed. Springer Verlag New York.

Broome, J. D., and J. H. Schwartz /1967/. Differences in the production of L-asparagine in asparaginase-sensitive and resistant cells. Biochim. biophys. Acta, 138, 637-639.

Goldin, A., and R. K. Johnson /1977/. Resistance to antitumor agents. In H. J. Tagnon, and M. J. Staquet /Ed/, Recent Advances in Cancer Treatment, ed Raven Press, New York. pp. 155-161.

Hakansson, L., and C. Tropé /1974/. Cell clones with different sensitivity to cytostatic drugs in methylcholanthrene-induced mouse sarcomas. Acta Path. Microbiol. Scand. Sect. A., 82, 41-47.

Mihich, E. /Ed./, Drug Resistance and Selectivity, ed. Academic Press, New York, 1973.

Mittelman, F. /1971/. Chromosomes of fifty primary Rous rat sarcomas. Hereditas, 69, 155-186.

Pályi, I., E. Oláh, and J. Sugár /1977/. Drug sensitivity studies on clonal cell lines isolated from heteroploid tumor cell populations. I. Dose response of clones growing in monolayer cultures. Int. J. Cancer, 19, 859-865.

Schabel, F. M.Jr., and L. Simpson-Herren /1978/. Some variables in experimental tumor systems which complicate interpretation of data from in vivo kinetic and pharmacologic studies with anticancer drugs. In F. M. Schabel Jr. /Ed./, Antibiotics and Chemotherapy, Vol. 23, ed. S. Karger, Basel. pp. 113-127.

Schiffer, L. M., P. G. Braunschweiger, and J. J. Stragand /1978/. Tumor cell population kinetics following noncurative treatment.

In F. M. Schabel Jr. /Ed./, Antibiotics and Chemotherapy, Vol. 23, ed. S. Karger, Basel. pp. 148-156.

Siracký, J., J. Maťoška, and E. Siracká /1977/. Cell proliferation kinetics in gynecologic cancer. Neoplasma, 24, 327-332.

Subak-Sharpe, H., R. R. Bürk, and J. D. Pitts /1969/. Metabolic co-operation between biochemically marked mammalian cells in tissue culture. J. Cell. Sci., 4, 353-367.

Szende, B., M. Fox, and B. W. Fox /1973/. A transferable "resistance factor" from in vitro cultured MDMS-resistant Yoshida sarcoma cells. Brit. J. Cancer, 27, 245-252.

Ujházy, V. /1968/. Chromosomal studies with the nitrogen mustard sensitive and resistant Yoshida tumour. Neoplasma, 15, 657-661.

Ujházy, V. /1969/. Drug resistance study in a mixed nitrogen mustard-sensitive/resistant tumour. Neoplasma, 16, 467-468.

Ujházy, V. /1974/. Identical karyotipe in acquired drug-resistant sublines of Yoshida sarcoma. Neoplasma, 21, 665-669.

Ujházy, V., Š. Kužela, V. Krempaský, and E. Bohunická /1971/. Uptake of drug-resistant tumour cell fractions by drug-sensitive tumour cells, Neoplasma, 18, 627-630.

Yoshida, T. H. /1966/. Chromosomal alteration and the development of tumours XV. Change of chromosome pattern in 8-azaguanine and amethopterine-resistant sublines of the mouse lymphocytic neoplasm P388 cultured in vitro. Japan. J. Genetics, 41, 59-74.

Yoshida, T. H., K. Ohara, and L. W. Law /1967/. Chromosomal alteration and the development of tumours XVI. Karyological studies on sensitive and resistant sublines of the mouse lymphocytic leukemia L1210 to several antitumor agents. Japan. J. Genetics. 42, 339-348.

Selectivity of Antitumor Agents on Immunity*

E. Mihich and M. J. Ehrke

Grace Cancer Drug Center, Roswell Park Memorial Institute, New York State Department of Health, Buffalo, NY 14263, U.S.A.

Cancer chemotherapy has achieved notable successes in the sense that a certain number of patients with some forms of cancer can be brought into complete remission by the use of drugs and are free of detectable disease five years or longer after diagnosis[9]. Nevertheless, substantial limitations need to be overcome before chemotherapy can be generally used in the definitive management of cancer. These limitations are essentially all related to the fact that the available drugs do not exert uniquely selective effects on tumor cells as compared to normal cells. This lack of selectivity of action represents a major obstacle in overcoming resistance to treatment; indeed under these circumstances drug dosage cannot be increased beyond a critical level without incurring unacceptable limiting toxicity. Consequently, resistance at the clinical level may be actually the result of a relatively minor degree of resistance at the target tumor level.

In attempts to overcome these difficulties several approaches are being pursued in cancer therapeutics. These are: 1) the development of new and more effective anticancer drugs, 2) the clarification of the biochemical and pharmacological determinants of the selectivity of drug action against tumors, aiming at improving the utilization of the drugs studied, 3) the design of new treatments utilizing drugs in combination, 4) the development of more effective modes of drug administration, and 5) the design of combination treatments involving chemotherapy and other modalities such as surgery or radiotherapy. In recent years emphasis has also been placed on combining chemotherapy with immunotherapy.

Based on the suggestive evidence that tumor associated antigens may elicit specific host defense responses in humans, the possibility is being considered that cooperative interactions between antitumor effects of drugs and antitumor host defenses may be exploited in clinical cancer therapeutics. Indeed the cytoreductive antitumor effects of the potent but relatively non-specific antiproliferative and cytotoxic agents available to date may be augmented by the defenses of the host directed against the tumor, which are likely to be specific but relatively inefficient *per se*. Such cooperative interactions seem feasible despite the well known immunosuppressive potential of many anticancer drugs. In fact, contrary to earlier belief, it is now becoming increasingly evident that anticancer drugs may exert both immunosuppressive and immunoaugmenting effects through selective actions on specific cells and cell interactions at different stages of development of the

*The results discussed in this presentation were obtained in studies supported in part from Grant CA-15142 and Contract CM-57039 from the National Cancer Institute, USPHS.

immune response. These effects may indeed be due to the cytocydal or antiproliferative action of these agents causing imbalances among cell subsets in the immune system but this need not be, in each case, the mechanism responsible for the immunomodulations observed[10].

The Effects of Adriamycin (AM), Daunorubicin (DR) and other anticancer agents on some components of the immune system are under study in this laboratory and may serve to illustrate the potential for selectivity of action of certain anticancer agents on the immune system. The effects of AM and DR on the effector cells involved in the responses of mouse spleen cells against sheep erythrocytes (SRBC) as compared to those of other agents provide one example[3]. The inhibitory or augmenting effects of AM or DR on the development of a primary cell mediated cytolytic response of mouse spleen cells to allogeneic tumor cells represent another set of examples[4,12].

Effects of anticancer agents on mouse spleen effector functions against sheep erythrocytes

The recognition of the selectivity of action of anticancer agents on fully differentiated immune effectors may be of significance in cancer therapeutics. In fact, in humans, it is likely that the initial response to tumor-associated antigens has already occurred at the time of initial diagnosis and that immune effectors are already formed at the time chemotherapy is instituted[11].

The effects of AM and DR on spleen effector functions were compared with those of 22 other agents, most of them with recognized antitumor activity[3]. Briefly, the experimental design consisted of immunizing C3Hf/He female mice, 6 to 12 weeks old, with a single intraperitoneal (i.p.) administration of 5×10^8 SRBC and using spleen cell suspensions, obtained at different times after immunization, as effectors in the various assays where the function of immune cells was tested. Spleen cell suspensions were prepared according to standard procedures reported previously[2]. The assays were as follows: the complement dependent (CDCC) and complement independent (CICC) cellular cytotoxicity assays[8]; the antibody dependent cellular cytotoxicity (ADCC) assay where spleen cells from non-immunized mice were used as killer cells and the target cells were pre-coated with day 11 anti-SRBC antibody[2]; the phagocytosis of target SRBC coated by "endogenous" antibody[1], namely antibody secreted by antibody forming cells (AFC) from immunized mice during the CICC incubation period (CICC-PHg); the phagocytosis of target SRBC coated by "exogenous" antibody[2], namely antibody added to target cells before the addition of spleen cells from non-sensitized donors (ADCC-Phg). The phagocytic functions were measured as reported previously[2,5] based on the protection of SRBC engulfed by phagocytic cells from hypotonic lysis. Each of these five assays is based on measurements of ^{51}Cr-release from labeled target cells. The conditions used in these assays have been reported in detail recently[3].

The CDCC response, which peaks on day 4 or 5, is similar in its functional characteristics to the direct anti-SRBC plague-forming assay[7]. The CDCC assay involves primarily secretion of preformed IgM during the 90 minutes incubation, binding of the antibody to target cells, and complement functions leading to cell lysis and ^{51}Cr release. The CICC response peaks on day 10-12, and the assay is dependent on the IgG secreted by AFC during the 4 hours incubation; the killer cells bear Fc receptors and bind to the Fc portion of the IgG which has complexed with the target cells[3]. Therefore the lytic process of CICC is generally similar to that of ADCC; however, the CICC involves synthesis and secretion of antibody by AFC, binding of antibody to target, as well as the function of killer cells in spleen of immunized mice, whereas the ADCC assay uses target cells pre-coated with antibody and measures only the function of killer cells in spleen of non-immunized mice. Because the drugs studied were added to the assay incubation mixture, it derives that the CICC assay was sensitive to drug effects on the synthesis, secretion and binding of endogenous antibody to target cells as well as on killer cell functions whereas in the ADCC assay the exogenous antibody was already bound when the drug was added. Drug effects on phagocytosis as measured with the CICC-Phg and ADCC-Phg assays are to be considered keeping in mind the differences between the basic CICC and ADCC assays mentioned above. Each of the drugs studied was added to the assay incubation mixtures at least at three concentrations.

When AM and DR were tested in the five assay systems outlined above they inhibited ADCC-Phg and CICC-Phg and, to a much lesser degree, CICC. No difference in potency between the two drugs was noted. No effect was seen on ADCC and CDCC. The effects on CICC were prevented by the addition of day 11, heat-inactivated, anti-SRBC antiserum. This prevention and the lack of effect on ADCC suggest that these agents affect antibody production or secretion from AFC and not antibody binding to target cells or the function of killer cells. Moreover, they seem to affect the phagocytic function directly, because the ADCC-Phg was inhibited. The fact that the inhibition of CICC-Phg was somewhat greater than that of the ADCC-Phg is consistent with the notion that in the CICC-Phg the combined effects on AFC and phagocytic functions would become additive.

While other agents had the same spectrum of activities as that of the two anthracyclines (e.g. Bleomycin and Actinomycin D), quite different profiles were observed with other drugs[3]: thus some were found to inhibit only CICC-Phg (e.g. Maytansine), or to inhibit all the tests except CDCC (e.g. Vincristine), or to inhibit all the tests (e.g. prednisolone), or not to inhibit any of them (e.g. Arabinosylcytosine). This clearly indicates that certain anticancer agents may exert rather specific inhibitory effects on preformed immune effector functions, and that these effects can occur under conditions where the antiproliferative actions of the drug are irrelevant.

<u>Effects of AM or DR on the development of primary cell-mediated immune (CM) response of C57Bl/6 spleen against allogeneic P815 mastocytoma cells</u>.

The primary CMI response of C57Bl/6 spleen cells against P815 cells was measured utilizing immune effector cells obtained after immunization of the mouse or sensitization in culture. Under the *in vivo* conditions, mice were immunized on day 0 by the i.p. injection of 3×10^7 washed P815 ascites tumor cells and the CMI was assayed on day 10[6]. Under the culture conditions, non-sensitized spleen responder cells (R) were stimulated on day 0 with x-irradiated P815 stimulator cells (S) at different R:S ratios, and the CMI was measured on day 4[4,12]. In both cases the response was quantitated as the percent specific ^{51}Cr release from labeled P815 target cells incubated with effector cells at different effector to target cell ratios, for 1 to 4 hours[4,12,6]. In mice, either AM or DR was given at doses ranging from 1.25 to 8.0 mg/km; in culture either AM or DR was added at concentrations ranging from 1×10^{-6} to 1×10^{-7}M. Three basic experimental designs were followed, namely: A) the drug was added to the primary immunization culture at different times after the initiation of the cultures; B) the drug was given i.p. to non-sensitized mice on different days (-5 to -1) before their spleen cells were used as R in the primary immunization culture and no more drug was added in culture; C) the drug was given i.p. to sensitized mice on different days after *in vivo* primary immunization and the CMI assay was performed with the immune spleen effectors which had developed in the mouse.

Under experimental design A, AM was found to have inhibiting or augmenting effects depending on the R:S ratio and the culture conditions used. Under conditions which give near maximal lytic responses (45 to 55% specific ^{51}Cr release), the drug was inhibitory if added on say 0, less so if added on day 1, and not inhibitory if added on day 2. Pulses of drug were maximally inhibitory only if given for a minimum of 24 hours and within the first 36 hours of culture[4]. Under conditions which give submaximal lytic responses (5-15% specific ^{51}Cr release) AM augmented the response if added on day 0, 1 or 2, this suggesting the existence of a difference in time dependence between inhibitory and augmenting effects[4]. DR has only been tested under conditions of near maximal lytic responses and its effect was very similar to that of AM.

Under experimental design B, either AM or DR caused an augmentation of the response up to maximal levels obtainable in culture (see above) when it was given 5 or 3 days but not when it was given 1 day, before the spleen cells from the tested mice were used as R in the culture incubation mixture[12]. This augmentation was confirmed with AM using the various culture conditions used under design A, and it was greater using conditions in which the response of R cells from non-treated mice was submaximal[4]. The augmented response was also observed if nylon wool adherent cells were used as responders[12]. Moreover, the augmentation was in parallel with time-dependent enrichment of the spleen with cells of the monocyte-phagocytic type[12].

Under experimental design C, neither AM nor DR caused any augmentation of the CMI response, even if the drug was given on different days before immunization. When either drug was given after immunization, however, inhibition of the response was observed and the effect with DR appeared to be quantitatively greater. The inhibition was maximal when DR was given on day 2 and relatively moderate when it was given on day 4. In contrast, under the same conditions, procarbazine caused maximal inhibition of the CMI when given on day 4 or 6, and no inhibition when given on earlier days[6]. This difference suggests that there might be a difference in what target cells are most sensitive to these two drugs.

Concluding Remarks

As mentioned initially, it is expected that major therapeutic advantages would be achieved in the treatment of cancer if the toxicity of the potent agents available were reduced without reduction of antitumor effectiveness. The immune system includes tissues which are frequently the site of toxicity of anticancer agents and thus the function of this system may be altered by these drugs. If mechanisms of host defenses operate against human tumor, whether augmented by immune manipulations or not, and if they participate in determining the therapeutic consequences of certain chemotherapeutic treatments, the therapy-limiting effects of immunosuppression by anticancer drugs might be significant.

The possibility that anticancer agents exert specific effects on the development and/or functions of the different components of the immune system would be expected because the multiple biochemical and pharmacological determinants of drug action are different in different cell types and in the same cell type at different functional stages[13,14]. Indeed several examples of selectivity of drug action within the immune system have been reported in recent years.

The examples discussed herein indicate that anticancer drugs may have selective effects on preformed effector functions and that these effects are independent from their antiproliferative action. Moreover, the results obtained with AM and DR indicate that a drug may exhibit specificities of inhibition or of augmentation of selective immune functions which may be put in evidence under appropriate experimental conditions. Thus, whereas immunosuppression by anticancer drugs must be acknowledged as a definite potential action of this type of agent it is also likely that for most of these drugs, treatment conditions could be determined where immunosuppression may be minimized or, in fact, immuno-augmentation may be obtained. It seems appropriate at this time to infer that anticancer and other agents are capable of causing an imbalance of the immune response, and imbalance which may be expressed in inhibition or augmentation of the overall response depending on a number of factors.

In the light of the preceding considerations, it becomes important to identify the basis for the selective immune imbalances caused by drugs and to relate it to the mechanisms of regulation of the immune responses. Indeed a drug may cause specific effects on the immune system dependent both on pharmacological and biochemical parameters of action and on the functional status of the frequently opposing mechanisms of regulation of the target systems. Through the clarification of these relationships it may become ultimately possible to apply deliberate pharmacological intervention leading to therapeutically advantageous imbalances of the immune response.

REFERENCES

1. Bennett, J., Ehrke, J., Dave, C. and Mihich, E. (1977) Selective effects of methylglyoxal-bis(guanylhydrazone) on the development of antibody-forming cells in mice. *Biochem. Pharm., 26,* 723-728.

2. Cohen, S.A., Ehrke, M.J. and Mihich, E. (1975). Mouse effector functions involved in the antibody dependent cellular cytotoxicity to xenogeneic erythrocytes. *J. Immunology, 115,* 1007-2662.

3. Ehrke, M.J., Cohen, S.A. and Mihich, E. (1978). Selectivity of inhibition by anticancer agents of mouse spleen immune effector functions involved in responses to sheep erythrocytes. *Cancer Research, 38,* 521-530.

4. Ehrke, J., Tomazic, V., Eppolito, C. and Mihich, E. (1978). Effects of Adriamycin (AM) in mice or in culture on the primary immune response of C57Bl/6 spleen cells to allogeneic P815 mastocytoma cells. *Fed. Proc., 37,* 1652.

5. Hersey, P. (1973). Macrophage effector function. An *in vitro* system of assessment. *Transplantation, 15,* 282-290.

6. Hoffmann, C.C., Ehrke, M.J. and Mihich, E. (1978). Selective imbalances of mouse immune responses to allogeneic tumor cells by procarbazine. In: W. Siegenthaler and R. Luthy (eds.), *Current Chemotherapy,* Proceedings of the 10th Intl. Congress of Chemotherapy, Vol. II, Am. Soc. for Microbiology Publishers, Washington, D.C., pp. 1097-1099.

7. Jerne, N.K. and Nordin, A.A. (1963). Plaque formation in agar by single antibody producing cells. *Science, 140,* 405.

8. Mawas, C., Carey, T. and Mihich, E. (1973). Study of the immune responses to nucleated cells, I: *In vitro* functional evaluation of the immune effectors responsible for complement dependent cellular cytotoxicity. *Cellular Immunology, 6,* 243-260.

9. Mihich, E. (1978). Chemotherapy and immunotherapy as a combined modality of cancer treatment. In: Excerpta Medica Intl. Congress Series No. 420, *Advances in Tumour Prevention, Detection and Characterization,* Vol. 4, Excerpta Medica, Amsterdam, pp. 113-121.

10. Mihich, E. (1978). Drug selectivity in the suppression of the immune response. In: J.L. Turk (ed.), *Drugs and Immune Responsiveness,* Macmillan Press, London (in press).

11. Mihich, E. (1971). Preclinical evaluation of the interrelationships between cancer chemotherapy and immunity. In: T.C. Hall (ed.), *Prediction of*

Response in Cancer Chemotherapy and Immunity (National Cancer Institute Monograph #34), pp. 90-102.

12. Orsini, F., Pavelic, Z. and Mihich, E. (1977). Increased primary cell-mediated immunity in culture subsequent to Adriamycin or Daunorubicin treatment of spleen donor mice. *Cancer Research, 37,* 1719-1726.

13. Rustum, Y.M., Grindey, G.B., Hakala, M.T. and Mihich, E. (1976). Multifactorial cellular determinants of the action of antimetabolites. In: G. Weber (ed.) *Advances in Enzyme Regulation*, Vol. 14, Pergamon Press, New York and Oxford, pp. 281-295.

14. Rustum, Y., Cheng, Y-C., Pavelic, Z., Creaven, P. and Mihich, E. (1978). Design of adjuvant chemotherapy based on target cell determinants of drug action: possibilities and limitations. In: *Recent Results in Cancer Research*, Springer-Verlag, Berlin (in press).

The Duration of Myeloma Remissions and Survival Cannot be Explained by the Myeloma Cell-kill Achieved

Daniel E. Bergsagel

Ontario Cancer Institute, Toronto, Canada

ABSTRACT

The cell-kill hypothesis does not explain the effect of treatment on human myeloma. Effective chemotherapy induces stable remissions with a myeloma cell-kill of about 10^{-1}. The duration of the remission and survival do not correlate with the cell-kill achieved. Continuing treatment during a stable remission does not increase the cell-kill, or prolong the remission. The rate at which the myeloma growth rate increases is an important determinant of survival. A model is proposed to explain the effect of treatment on myeloma.

MYELOMA TUMOR CELL-KILL

One of the basic tenets which has guided the development of cancer chemotherapy during the past thirty years is the hypothesis that chemotherapeutic agents cause tumor regression, prolong survival and cure some patients by selectively killing tumor cells. In most patients the tumor cell population of about 10^{12} cells must be eliminated if the patient is to be cured. Chemotherapeutic agents must be selectively toxic for tumor cells, because the stem cell population of normal tissues, such as the bone marrow, cannot be decreased by more than 10^{-2} is the patient is to survive.

This cell-kill hypothesis has been tested with many transplanted mouse tumors, and provides an adequate explanation of the improvement which follows treatment with a chemotherapeutic agent. In Fig. 1 are illustrated the survival of normal mouse hemopoietic precursor cells which form colonies in culture (CFU-C), and stem cells which form colonies in the spleen (CFU-S), following increasing doses of melphalan.

These normal cells are much less sensitive to melphalan than the myeloma colony-forming cells of three mouse plasmacytomas, Adj. PC-5, MOPC 46B and MOPC 460D. The colony-forming cells of the Adj. PC-5 myeloma are the most sensitive cells. Mice bearing this tumor can be cured with a single dose of melphalan, even when treatment is delayed for 12-14 days. Treatment of mice bearing MOPC 46B prolongs survival, but does not cure any of the mice, whereas treatment of mice bearing MOPC 460D has very little effect on survival (Bergsagel, Ogawa and Librach, 1975; Ogawa, Bergsagel and McCulloch, 1973).

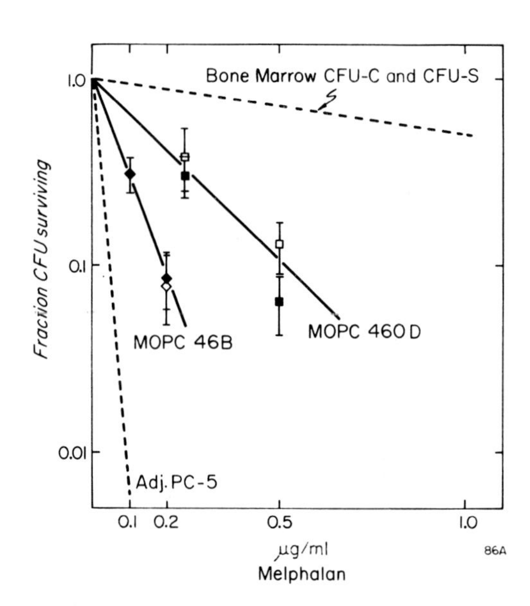

Fig. 1. The effect of increasing doses of melphalan on the survival of mouse hemopoietic precursor cells and the colony-forming cells of three mouse myelomas.

It is possible to estimate the total myeloma cell mass in humans on the basis of the amount of M-protein produced per day divided by the M-protein produced by single myeloma cells in vitro (Hamburger and Salmon, 1977). Estimates of the changes in the myeloma cell mass during the treatment of a patient are shown in Fig. 2.

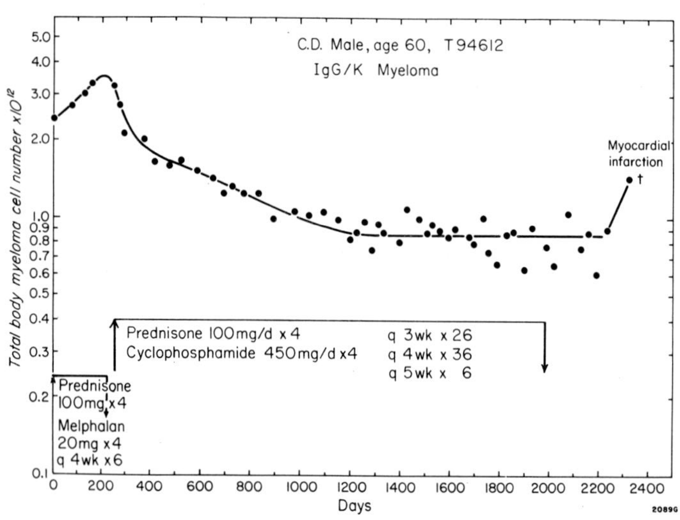

Fig. 2. Changes in the myeloma cell mass of a patient during treatment.

The amount of M-protein synthesized per day was calculated (Salmon and Wampler, 1977) and the myeloma cell mass estimated by assuming that each myeloma cell produced 8 picograms of M-protein per day (Hamburger and Salmon, 1977). It will be noted that the myeloma cell mass increased during treatment with melphalan, but decreased progressively for 2½ years after cyclophosphamide was started. A plateau was reached after 1200 days, despite continued treatment with cyclophosphamide and prednisone. We discontinued treatment because the remission was stable and we were unable to further decrease the myeloma cell mass. Thereafter the myeloma cell mass did not change significantly for almost a year. The last serum M-protein was elevated, suggesting an increasing myeloma cell mass. The patient died suddenly of an apparent myocardial infarction before cyclophosphamide could be restarted. This patient illustrates several unusual features of plasma cell myeloma: 1) some myeloma patients are resistant to melphalan but sensitive to cyclophosphamide; 2) after cyclophosphamide was started the myeloma cell mass decreased rapidly initially, then more slowly until the plateau phase was reached, suggesting that the fractional cell-kill of myeloma cells became progressively smaller as treatment was continued; 3) during the plateau phase continued treatment does not result in further improvement; 4) the plateau phase persists for prolonged periods after treatment is stopped; and 5) treatment which reduced the myeloma cell number by less than 10^{-1} resulted in a prolonged remission and survival for 6½ years.

The failure of continued maintenance therapy to prolong the duration of myeloma remissions or survival over that achieved by discontinuing treatment in those who achieve a stable remission, has been demonstrated by the Southwest Oncology Study Group (Southwest Oncology Group, 1975).

We have estimated myeloma cell-kill by the method illustrated in Fig. 3.

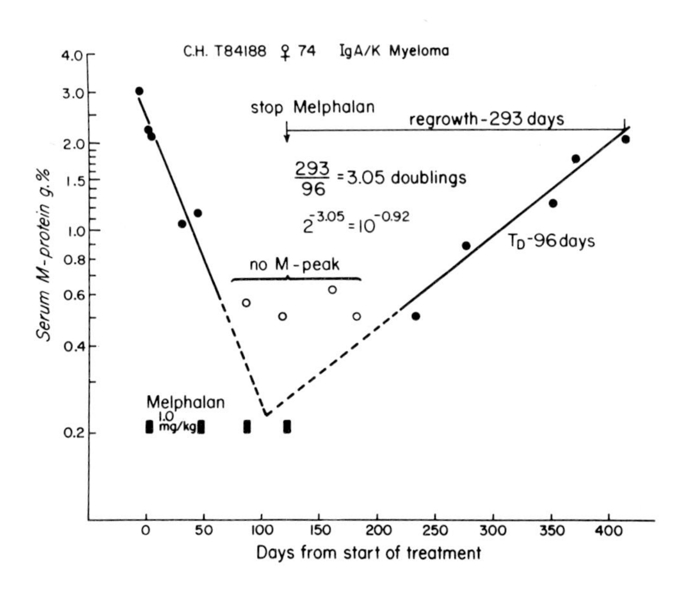

Fig. 3. A method for estimating myeloma cell-kill.

We found no correlation of myeloma cell-kill with survival (Fig. 4), but there was a good correlation of survival with the growth rate of the myeloma, as measured by the M-protein doubling time, during relapse (Fig. 5). We also found that there was a progressive shortening of the M-protein doubling time during the course of the disease (Bergsagel, 1977).

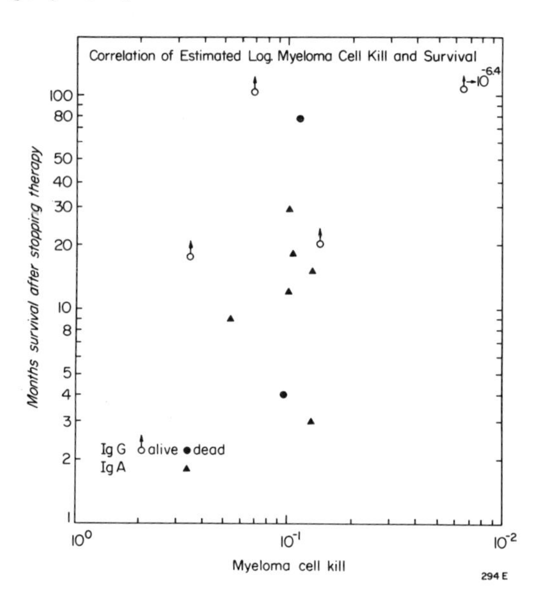

Fig. 4. Correlation of myeloma cell-kill and survival.

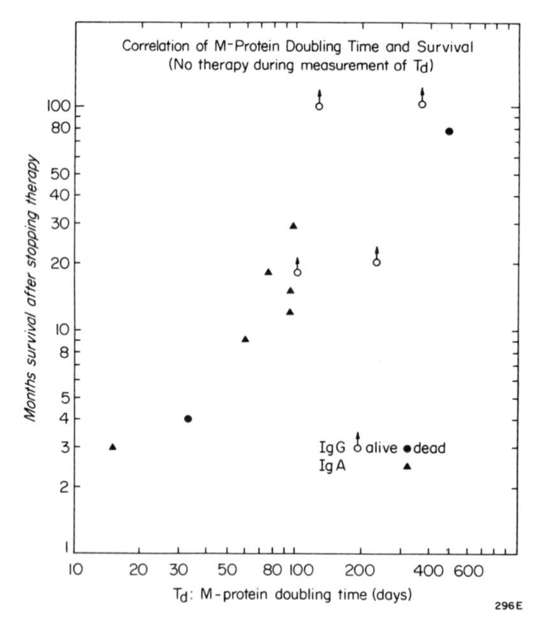

Fig. 5. Correlation of growth rate during relapse and survival.

Thus, the cell-kill hypothesis does not explain the effect of treatment on human plasma cell myeloma. Effective chemotherapy induces a good stable remission with a myeloma cell-kill of about 10^{-1}. The duration of the remission or survival does not correlate with the cell-kill achieved. Continuing treatment after a stable remission has been achieved does not reduce the myeloma cell number further, and does not prolong the duration of remissions or survival. Furthermore, there is a progressive increase in the growth rate of plasma cell myeloma during the course of the disease, and the rate at which this occurs is a very important determinant of survival.

Recent studies of the growth of human myeloma stem cells in culture (Hamburger and Salmon, 1977), and the appearance of T-lymphocytes in the marrow of relapsing myeloma patients which are capable of suppressing the mixed leukocyte response of peripheral blood lymphocytes to autologous marrow myeloma cells (Paglieroni, MacKenzie and Caggiano, 1978), suggest a hypothesis to explain the course of plasma cell myeloma and the effect of chemotherapy.

The facts on which the hypothesis is based are shown in Tables 1 and 2. Myeloma marrow contains a population of T-cells which are capable of suppressing the response of peripheral lymphocytes to autologous myeloma cells (Table 1).

TABLE 1

MLR RESPONSE OF PERIPHERAL BLOOD LYMPHOCYTES TO AUTOLOGOUS MITOMYCIN C-TREATED STIMULATOR CELLS

Stimulator Cells	Response
1. Unseparated myeloma marrow	-
2. Myeloma marrow plasma cell fraction	+
3. Myeloma marrow lymphocyte fraction	-
4. Combine 2 + 3	-
5. Combine 2 + E-rosette depleted 3	+

Paglieroni, MacKenzie and Caggiano, JNCI, Oct. 1978

These "suppressor" T-cells are found only in the marrows of myeloma patients in relapse; they are not present in the marrows of normals, benign monoclonal immuno-globulinopathy (BMI), or myelomas in remission (Table 2).

TABLE 2

DEMONSTRATION OF SUPPRESSOR T-CELLS AND MYELOMA STEM CELLS IN VARIOUS MARROWS

Marrow	Suppressor T-Cells*	Myeloma CFU**
Normal	absent	absent
Benign monoclonal immunoglobulinopathy	absent	present
Myeloma in relapse	present	present
Myeloma in remission	absent	present

* Paglieroni, MacKenzie and Caggiano, JNCI, Oct. 1978
** Hamburger and Salmon, JCI 60:846, 1977

2' 47G

On the other hand, myeloma colony-forming cells have been demonstrated in the marrows of patients with BMI, myelomas in relapse and remission (Hamburger and Salmon, 1977). Thus, the population of myeloma cells appears to be stable in the marrows of patients when "suppressor" T-cells cannot be demonstrated (i.e. BMI and myeloma in remission), and expands when "suppressor" T-cells are present.

A hypothesis which makes use of these observations is as follows. A malignant transformation affects a single committed plasma cell precursor which proliferates to form a monoclonal population of from 10^{10} to 10^{12} cells. Growth stops when this population is achieved, and we recognize patients with a stable monoclonal population of plasma cells as BMI. Progressive growth, resulting in the disease we recognize as plasma cell myeloma, does not occur unless there is also an increase in a population of "suppressor" T-cells which promotes the growth of the myeloma cells. Treatment reduces the population of "suppressor" T-cells, releases the myeloma cells from their growth-promoting effect and allows them to return to their former stable state. The duration of the stable remission is determined by the time required for the "suppressor" T-cells to reappear and reach an effective number. The growth rate during relapse is determined by the number of "suppressor" T-cells and the effectiveness of their growth-promoting effect. The number, and/or the effectiveness of the "suppressor" T-cell population would tend to increase with each relapse during the course of the disease. It is important to recognize that this hypothesis shifts the target for control of the disease from the myeloma cell to a possible cell which regulates its growth.

This hypothesis predicts that the M-protein doubling time should become progressively shorter as the population of marrow "suppressor" T-cells increase. The testing of this hypothesis awaits the development of a quantitative assay for the "suppressor" T-cells.

REFERENCES

Bergsagel, D.E. (1977). Assessment of the response of mouse and human myeloma to chemotherapy and radiotherapy. In B. Drewinko and R.M. Humphrey (Eds.), Growth Kinetics and Biochemical Regulation of Normal and Malignant Cells, The Williams and Wilkins Co., Baltimore, pp. 705-717.

Bergsagel, D.E., M. Ogawa, and S.L. Librach (1975). Mouse myeloma: A model for cell kinetics. Arch. Intern. Med., 135, 109-113.

Hamburger, A.W., and S.E. Salmon (1977). Primary bioassay of human myeloma stem cells. J. Clin. Invest., 60, 846-854.

Ogawa, M., D.E. Bergsagel, and E.A. McCulloch (1973). Chemotherapy of mouse myeloma: Quantitative cell cultures predictive of response in vivo. Blood, 41, 7-15.

Paglieroni, T., M.R. MacKenzie, and V. Caggiano (1978). Multiple myeloma: An immunologic profile. II Bone marrow studies. J. Nat. Cancer Inst., in press.

Salmon, S.E., and S.E. Wampler (1977). Multiple myeloma: Quantitative staging and assessment of response with a programmable pocket calculator. Blood, 49, 379-389.

Southwest Oncology Group (1975). Remission maintenance therapy for multiple myeloma. Arch. Intern. Med., 135, 147-152.

A 10-Year Follow-up of Combination Chemotherapy of Hodgkin's Disease by Cancer and Acute Leukemia Group B

L. Stutzman* and T. Pajak**

**Roswell Park Memorial Institute, New York State Department of Health, Buffalo, New York, U.S.A.*
***CALGB Statistics Office, Scarsdale, New York, U.S.A.*

ABSTRACT

Long-term results of three combination agent chemotherapy routines were compared in 247 patients. MOPP produced superior duration of complete remission compared to combination or sequential 5-drug routines, but overall 10 yr. survival has been nearly identical for those in the 3 routines. Factors of sex, former therapy, age and unusual sites of disease were found to be predictive of duration of response and survival. When these were combined by multivariant analysis, the prognostic implications of these factors was strengthened.

KEYWORDS

Hodgkin's disease, chemotherapy, prognostic factors, survival, multivariant analyses

INTRODUCTION

The advantages of combination chemotherapy for disseminated Hodgkin's disease are well accepted. While response rates have improved by utilization of such combinations, the real advantages are shown in the quality of response, with a greater proportion reaching complete remission, and the durability of such remissions. These routines now have been utilized for over 10 years and long-range analyses of survival characteristics and prognostic factors can be made. Certain characteristics of the patient population are important factors in predicting duration of both remission and survival.

MATERIALS AND METHODS

In 1967 a multicenter chemotherapy group protocol trial of 5-drug regimens, vincristine, procarbazine, vinblastine, chlorambucil and prednisone, used in combination or sequentially and compared to MOPP (mechlorethamine, vincristine, procarbazine and prednisone) was initiated. 247 patients with Stage III or IV disease were entered and continuously followed (Stutzman and Glidewell, 1975).

RESULTS

Although there were no differences in either partial (PR) or complete (CR) response rates among the three regimens, MOPP treated patients who achieved CR had significantly longer unmaintained remissions after 6 cycles of therapy as compared to the other routines (p.018) (Fig. 1). There have been no relapses in this CR group after 60 months of follow-up with 40% of MOPP patients remaining in CR and 20% of those in the other routines maintaining this status.

In contrast to the remission duration, the survival of all patients entering the study are similar regardless of the therapeutic routine with a 10-year survival of 27% (Fig. 2).

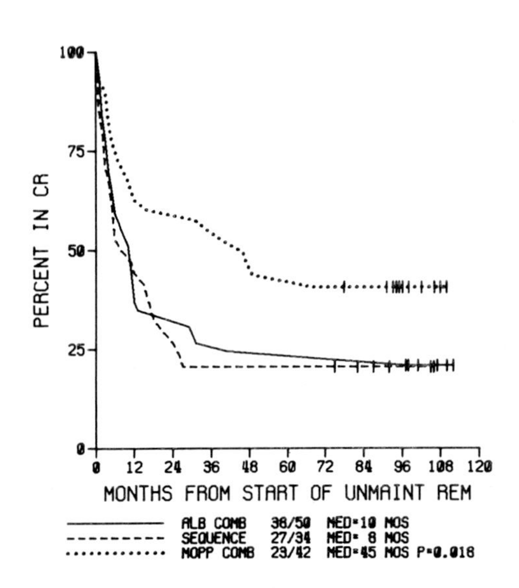

Fig. 1. Duration of complete response (CR) by type of combination chemotherapy regimen.

Fig. 2. Survival of all patients by type of combination chemotherapy regimen.

An analysis of factors important to favorable survival revealed significant differences in acturial curves (Breslow, 1970) by age; under 40>40-50>over 60 years (p.04) (Fig. 3); sex, female>male (p.04) (Fig. 4) and by prior treatment, none or local radiation ports>chemotherapy and/or extensive chemotherapy (p.0005).

Individual factors important for unmaintained complete response were sex, female> male, (p.02) (Fig. 6); no prior therapy>prior chemotherapy, (p.008) (Fig. 7), and usual sites of involvement, e.g. lung and lymph nodes>unusual sites, e.g. multiple skin lesions, CNS or bone, (p.009) (Fig. 8).

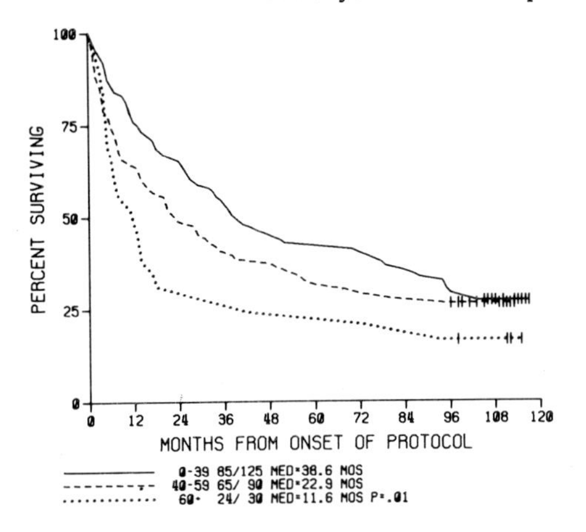

Fig. 3. Survival of all patients by age.

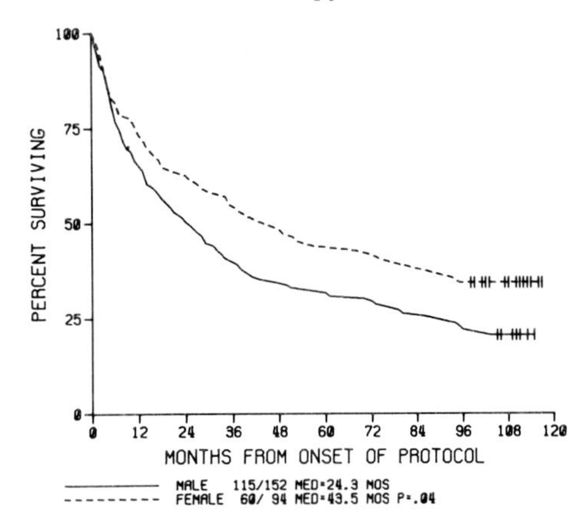

Fig. 4. Survival of all patients by sex.

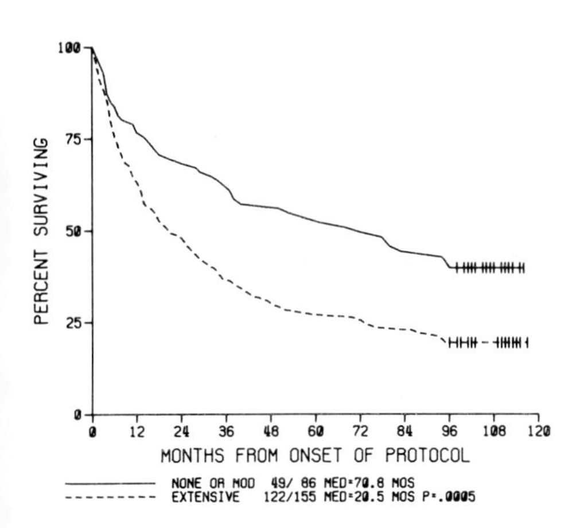

Fig. 5. Survival of all patients by prior therapy.

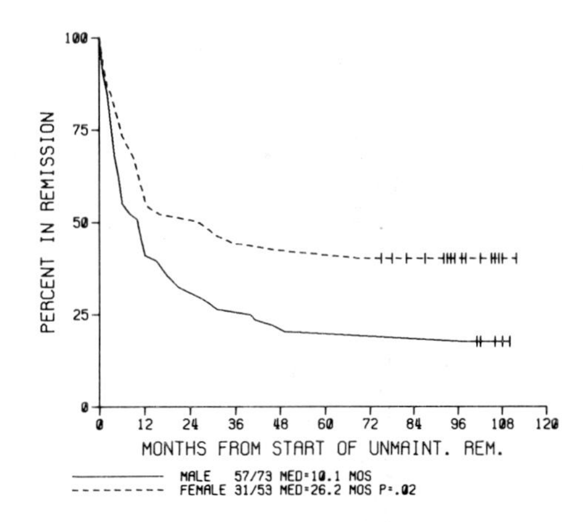

Fig. 6. Duration of complete response by sex.

In further exploration of prognostic factors, multivariant analyses of pretreatment characteristics were performed with the Cox regression model in stepwise fashion (Cox, 1972). Four favorable factors for unmaintained remission were found to be active by this analysis- MOPP therapy, female sex, no prior chemotherapy and no unusual sites of disease. The most favorable group, UNR-1, included patients with 3-4 of these factors; an intermediate group, UNR-2, included patients with any 2 of the factors; and the least favorable group, UNR-3, contained patients with only one or none of the favorable factors. These curves are shown in Figure 9 with median duration of CR of 43 months for UNR-1; 11 mo. for UNR-2; and 4 mo. for UNR-3.

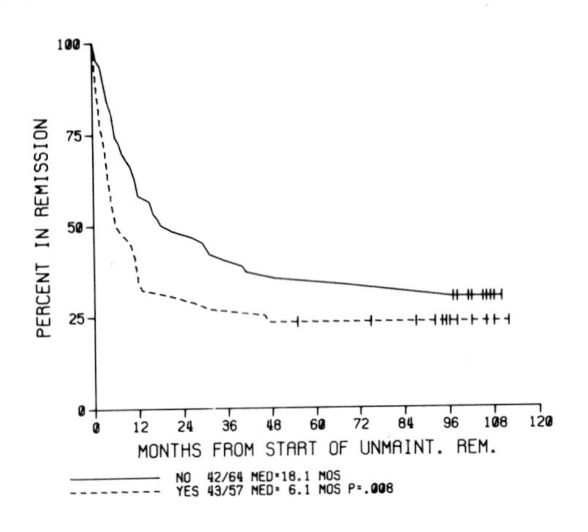

Fig. 7. Duration of complete response by prior chemotherapy.

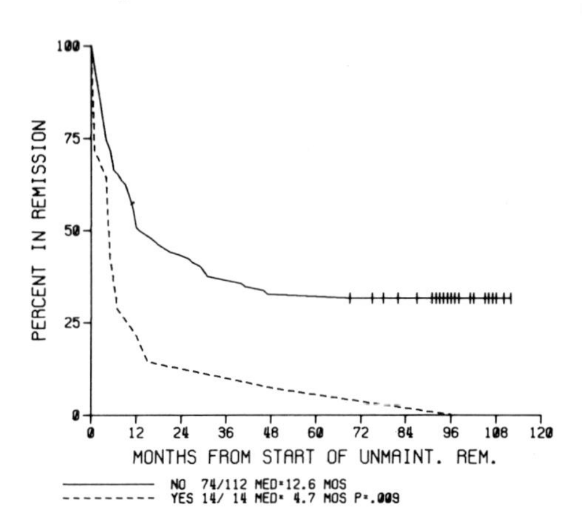

Fig. 8. Duration of complete response by unusual sites of disease.

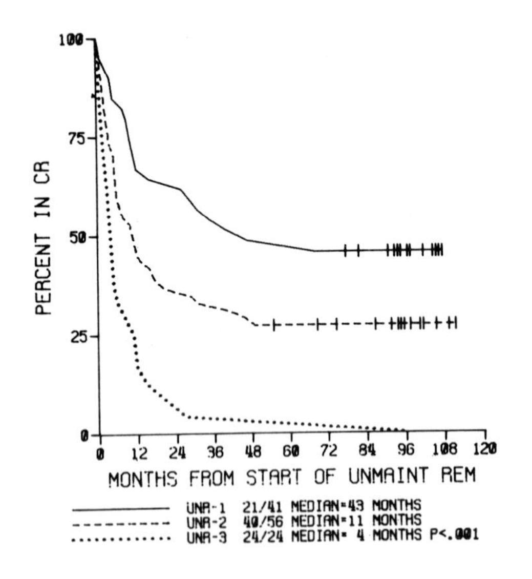

Fig. 9. Duration of complete response by risk factors of MOPP or other therapy, sex, prior chemotherapy and unusual site.

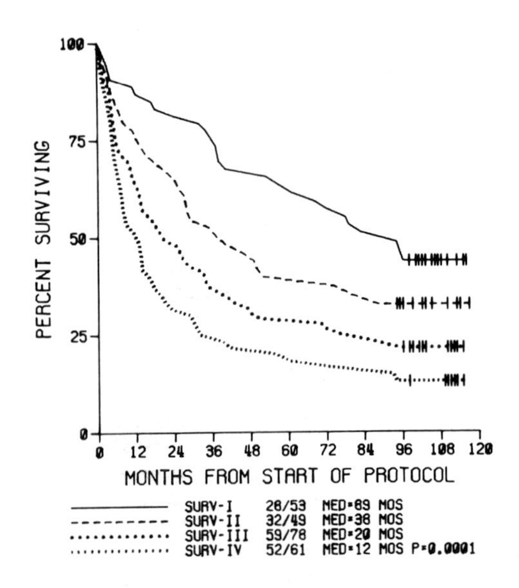

Fig. 10. Survival by risk factors of prior therapy, age & sex (See Table 1).

Similar multivariant analyses were performed for factors strongly influencing survival and three were significant- extent of prior therapy, age, and sex. No advantage of the MOPP treated group appeared in this phase of the analysis. Four risk groups were found (Table 1):

TABLE 1 Prognostic Survival Groups

Group	Prior Rx	Sex & Age
SURV-1	0	M<40;F<60
SURV-2	0	M 40-59
	+	F<40
SURV-3	+	M<40;F40-59
SURV-4	+	M>40;F>60

The curves by this analysis are found in Fig. 10. The median survivals were; SURV-1 89 mo, SURV-2 38 mo, SURV-3 20 mo and SURV-4 12 mo (p.0001).

Factors without prognostic significance were: fever, weight loss, Stage III vs IV, peripheral lymph node site, spleen involvement, pruritus and limited prior radiotherapy.

In a former analysis of this group of patients a median survival benefit was found for mediastinal (70 mo) or lung involvement (50 mo) as compared to liver disease (16 mo) (Stutzman & Kutcher, 1974). While these sites remain significantly different for survival benefit, but not for duration of remission, they do not retain importance as independent factors of prognosis on multivariant analyses.

DISCUSSION

These analyses demonstrate an appreciable portion of sustained disease-free patients who remain in CR after multiple agent therapy. There is also a high rate of responsiveness of the patients to salvage therapy following relapse after these multiple agent routines, usually by the utilization of repeat combination therapy with routines that contain at least some different cross-over agents, resulting in prolonged survival of a fourth of all the patients studied. Thus, 19 of 29 patients treated with either of the 5-drug routines were able to achieve complete remissions upon retreatment with a variety of combination chemotherapy routines.

DeVita and co-workers (1976), reported 58% 10-year survivors from a series of 194 patients treated with MOPP, and a 72% 10-year survival in those with CR status at the end of the therapy. Our results as reported here are less favorable, even among those treated with MOPP. Factors of age, prior therapy and sex in the two studies appear to explain the differences. Strong prognostic factors of duration of both remission and survival have been found, and these acquire much higher predictive value when present concurrently as demonstrated by multivariant analyses.

REFERENCES

Breslow, N. (1970). A generalized Krushal-Wallis test for comparing K samples subject to unequal patterns of censorship. Biometrika 57, 579-594.

Cox, D.R. (1972). Regression models and life tables. J. Roy Statistical Soc B 34, 187-202.

DeVita, V., Canellos, G., Hubbard, S., Chabner, G., Young, R. (1976). Chemotherapy of Hodgkin's disease (HD) with MOPP: A 10-year progress report. (Abstr. 531). 67th Ann. Mtg. of the Am. Assoc. for Cancer Res., May 4-8.

Stutzman, L. and Glidewell, O. (1973). Multiple chemotherapeutic agents for Hodgkin's disease. Comparison of three routines: A cooperative study by Acute Leukemia Group B. JAMA 225, 1202-1211.

Stutzman, L. and Kutcher, J. (1974). Patterns of response and relapse during chemotherapy of disseminated Hodgkin's disease. Proc. 65th Ann. Mtg. of Am. A. Cancer Res. and 10th Ann. Mtg. of Am. Soc. Clin. Oncol. 15, 181 (Abstr. 788).

Enzyme-Pattern-Targeted Chemotherapy and Mechanism of Pyrazofurin Action

George Weber, Edith Olah, May S. Lui and Diana Tzeng

Laboratory for Experimental Oncology, Indiana University School of Medicine, Indianapolis, Indiana 46223. U.S.A.

ABSTRACT

The purpose of this presentation is to outline the conceptual and experimental approaches of the design of enzyme-pattern-targeted chemotherapy and to document experimental progress in elucidating the cytochemical mechanism of action of pyrazofurin, a C-nucleoside. Hepatomas were grown in tissue culture and as solid tumors in rats. The effect of pyrazofurin was investigated by determining tumor phase and cell cycle specificity and biochemical action on the concentrations of nucleotides in tumor cells and in host liver. The results indicated that pyrazofurin had an LD_{50} of 0.4 μM in hepatoma 3924A in tissue culture and that the drug inhibited only logarithmically growing cells and failed to affect the plateau phase ones. In synchronized hepatoma cultures it was demonstrated that pyrazofurin was cell cycle phase specific because it was toxic particularly to early G_1 and early S phase cells. Uridine and cytidine, but not uracil or hypoxanthine, were able to protect hepatoma 3924A cells in tissue culture from the cytotoxic action of pyrazofurin. In Novikoff hepatoma cells pyrazofurin (1 μM) decreased concentrations of UTP (22%), CTP (9%), dTTP (50%) and dCTP (51%), but did not affect ATP, dATP, GTP or dGTP pools. Combination chemotherapy with pyrazofurin and galactosamine caused a synergistic depression of UTP and CTP pools to a significantly lower level than with either of the drugs alone. Combination chemotherapy also provided a synergistic cytotoxic action in these tumor cells. In animals carrying transplantable hepatoma 3924A, pyrazofurin and galactosamine combination treatment *in vivo* achieved synergistic depression of CTP concentrations in the tumor without affecting CTP pool in host liver. In conclusion, new evidence was provided for the feasibility of enzyme-pattern-targeted chemotherapy in tissue culture and in animals, and progress has been made in elucidating the cytochemical mechanism of action of pyrazofurin in tumor cells.

Key Words: enzyme-pattern-targeted chemotherapy; pyrazofurin; hepatoma; cell cycle phase specificity; ribonucleotide concentrations; deoxyribonucleotide concentrations, galactosamine; combination chemotherapy; molecular correlation concept; OMP decarboxylase.

INTRODUCTION

Clinical anti-cancer chemotherapy has been successful in the treatment and palliation of a number of neoplastic diseases, with particularly good survival rates in the more rapidly growing tumors (Zubrod, 1972). Important progress has been made also in the improved treatment scheduling and combination modalities in the therapy of such otherwise rapidly fatal diseases as osteogenic sarcoma (Frei and others, 1978). Advances in the area of human tumors also point out the difficulties and the poor therapeutic results obtained so far in some of the most common human cancers, including colo-rectal, lung and mammary carcinomas, and sarcomas of the soft tissues. Recent major reviews in this area, those of Holland and Frei (1973), and Sartorelli and Johns (1974), on anti-tumor agents, and considerations of the growth kinetics of tumors (Steel, 1977), indicate significant progress in our knowledge of certain aspects of pharmacology and cell biology. Both the success and the failure emphasize the acute need to gain a deeper insight into the strategic enzymatic and metabolic differences between normal and cancer cells and to improve the design of biochemical and pharmacological treatment methods and schedules.

In this Laboratory an ordered pattern of enzymic and biochemical imbalance was discovered in a spectrum of chemically-induced, transplantable, solid hepatocellular carcinomas (Weber, 1977a, 1977b). Subsequent studies showed that much of this pattern of enzymic imbalance was applicable to chemically-induced, transplantable renal tumors in rat, lymphomas in mice (Weber and others, 1977), transplantable colon tumors in mouse and rat (Weber and others, 1978b), and in primary liver (Weber and others, 1976) and renal cell carcinomas (Weber and others, 1978a) in human. It became feasible to attempt to design chemotherapy by making use of our knowledge of the enzymology of the hepatomas and of the specific liver-type functions that may be retained in some of the hepatoma cell lines. The approach introduced in this Laboratory was termed enzyme-pattern-targeted chemotherapy and defined as follows: "The specialized enzymic pattern characteristic of the tissue of origin of the tumor is exploited to target and enhance drug selectivity" (Jackson and Weber, 1976; Jackson, Williams, and Weber, 1976).

Recent studies in our Laboratory elucidated aspects of the mechanism of action of pyrazofurin (PF), and combination chemotherapy was designed in utilizing the liver-specific metabolic action of galactosamine in tissue culture systems and in tumor-bearing animals (Weber and others, 1979). The purpose of this presentation is to report our current conceptual and experimental approaches in the application of enzyme-pattern-targeted chemotherapy using primarily pyrazofurin and galactosamine in demonstrating the selectivity of the approach that can be achieved.

MATERIALS AND METHODS

Biological Systems

In vitro models. Hepatoma 3924A and Novikoff tumor were grown as monolayer cultures in McCoy's 5A medium (Grand Island Biological Co., Grand Island, NY) with 10% fetal calf serum in flasks and incubated in 5% CO_2, 95% air atmosphere. To keep the lines continually in the logarithmic growth phase, the cells were subcultured 3 times a week. The experimental details of induction of the plateau phase cultures, the techniques to achieve cell synchrony, the measurement of cell cycle time and the length of the different phases were cited elsewhere (Olah and others, 1979; Weber and others, 1979).

In vivo models. Hepatocellular carcinomas of different growth rates were studied. The emphasis was on hepatoma 3924A because earlier studies demonstrated the particular suitability of this tumor for freeze-clamp studies of the biochemistry (Weber, Stubbs, and Morris, 1971) and biochemical pharmacology (Lui, Jackson, and Weber, 1979) of tumor treatment. The biochemical and biological properties of this tumor spectrum and the approaches of the molecular correlation concept were outlined previously (Weber, 1977a, 1977b). The growth rates of the different solid hepatomas ranged from 2 weeks to about 1 year for the most rapidly to the slowest growing neoplasms.

Biochemical Studies

Preparation of extracts and enzymic assays. The tissue extracts were prepared following the carefully standardized procedures employed in this Laboratory (Weber and others, 1977). Determination of the activity of OMP decarboxylase (orotidine 5'-phosphate decarboxylase, EC 4.1.1.23) in the 100,000 X *g* supernatant fluid and assay of the protein concentration were cited elsewhere (Weber and others, 1978c).

Determination of concentrations of ribonucleotides and deoxyribonucleotides and utilization of the freeze-clamp method. The animals were anesthetized with ether and rapid tissue sampling was achieved by the freeze-clamp technic as employed in this Laboratory (Weber, Stubbs, and Morris, 1971). The methods utilized to measure the ribonucleotides and deoxyribonucleotides in the extracts prepared by the freeze-clamp method from host liver and hepatoma 3924A were reported elsewhere (Jackson and others, 1976; Weber and others, 1979).

Expression and evaluation of results. Enzyme activities were calculated in nmol of substrate metabolized per hr per mg protein, as specific activity. The tissue concentrations of ribonucleotides and deoxyribonucleotides were calculated as nmol per g wet weight of tissue. Results were subjected to statistical evaluation by the *t* test for small samples. Differences between means yielding a probability of less than 5% were considered significant.

RESULTS AND DISCUSSION

In this investigation we examined the approaches of enzyme-pattern-targeted chemotherapy that are relevant to test the postulated synergistic and selective action of the combination of the two compounds, galactosamine and PF. Earlier work showed that galactosamine produced a marked depression of UTP concentration in the liver (Decker, Keppler, and Pausch, 1973). Recently by use of the freeze-clamp method we demonstrated that among the various organs of the rat only in the liver was a decrease observed in UTP concentration (Weber and others, 1979). To provide a synergistic action with galactosamine in earlier studies we employed drugs that inhibited the utilization of UTP by using an inhibitor of CTP synthetase, 6-azauridine, and an inhibitor of DNA polymerase, cytosine arabinoside (Jackson, Williams, and Weber, 1976). In the present work we tested the potential usefulness of an inhibitor, pyrazofurin, that competitively blocks the activity of OMP decarboxylase, the terminal enzyme in the *de novo* synthesis of UMP.

Cytochemical Action of Pyrazofurin in Tumor Cells

PF is a C-nucleoside antibiotic which has an effect against certain viruses and inhibits growth of some transplantable tumors in rodents (Gutowski and others, 1975; Sweeney and others, 1973). The potential chemotherapeutic action in human tumors was also explored and the drug is currently under investigation (Cadman and others, 1978a, 1978b). The structure of PF, shown in Fig. 1, indicates that it can act as an analog of orotidine. Analysis of the biochemical pharmacology of PF suggested that the drug was activated by adenosine kinase to pyrazofurin

5'-phosphate and the phosphorylated drug exerted a competitive inhibition on the activity of OMP decarboxylase (Gutowski and others, 1975; Streightoff and others, 1969). The proposed site of the mechanism of action is shown in Fig. 2.

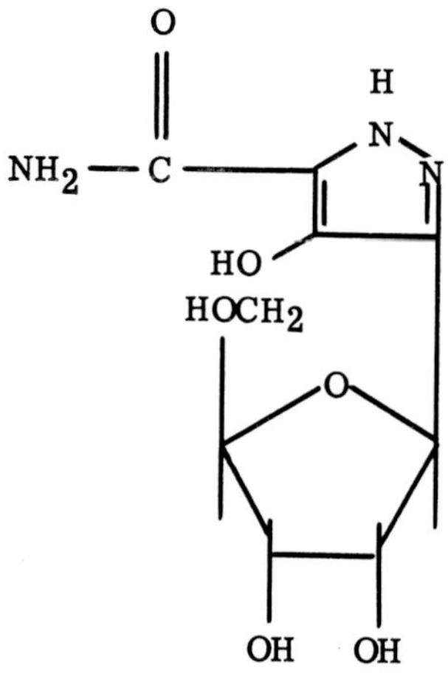

Fig. 1. Structure of pyrazofurin (4-hydroxy-5-beta-D-ribofuranosylpyrazole-3-carboxamide).

Increased OMP decarboxylase activity in hepatomas of different growth rates. Sweeney, Parton, and Hoffman (1974) reported an increase in OMP decarboxylase activity in 5 hepatomas, but they did not identify a linking of the behavior of this enzyme activity with transformation or progression. In order to elucidate the possible linkage of OMP decarboxylase activity with tumor malignancy we carried out a systematic study in a series of 13 hepatomas representing a wide spectrum of growth rate where there was a 26-fold difference in the replication rate of the tumors. The OMP decarboxylase specific activity was significantly increased in every neoplasm and it was particularly high in some of the rapidly growing tumors (Table 1). Since the enzymic activity was elevated in all the neoplasms, including the slowest growing hepatomas that took a year to attain the same size the rapidly growing tumors reached in 2 weeks, we identified the behavior of the activity of this enzyme as transformation-linked, as classified in Group 2 of the molecular correlation concept (Weber, 1977a, 1977b). We also noted that the enzymic activity was not progression linked, since calculation of the correlation coefficient indicated that the activity did not significantly correlate with tumor growth rate. Since the activity of OMP decarboxylase was increased in all the hepatomas examined, it was of interest to elucidate the cytochemical mechanism of action of PF which exerts a competitive inhibition of this enzyme.

Inhibition of growth of hepatoma 3924A cells in tissue culture; dose response studies. Exponentially growing cells were treated with different concentrations of PF. The cultures were exposed to PF for 1 hr, then the drug was removed and flasks were rinsed 3 times with phosphate buffered saline. The cells were trypsinized and seeded into culture flasks with 500 cells per 25 cm^2 flask. The pH of the medium was adjusted with 5% CO_2 in air. After an incubation period of 7 days, the medium was removed and the cells were stained with crystal violet.

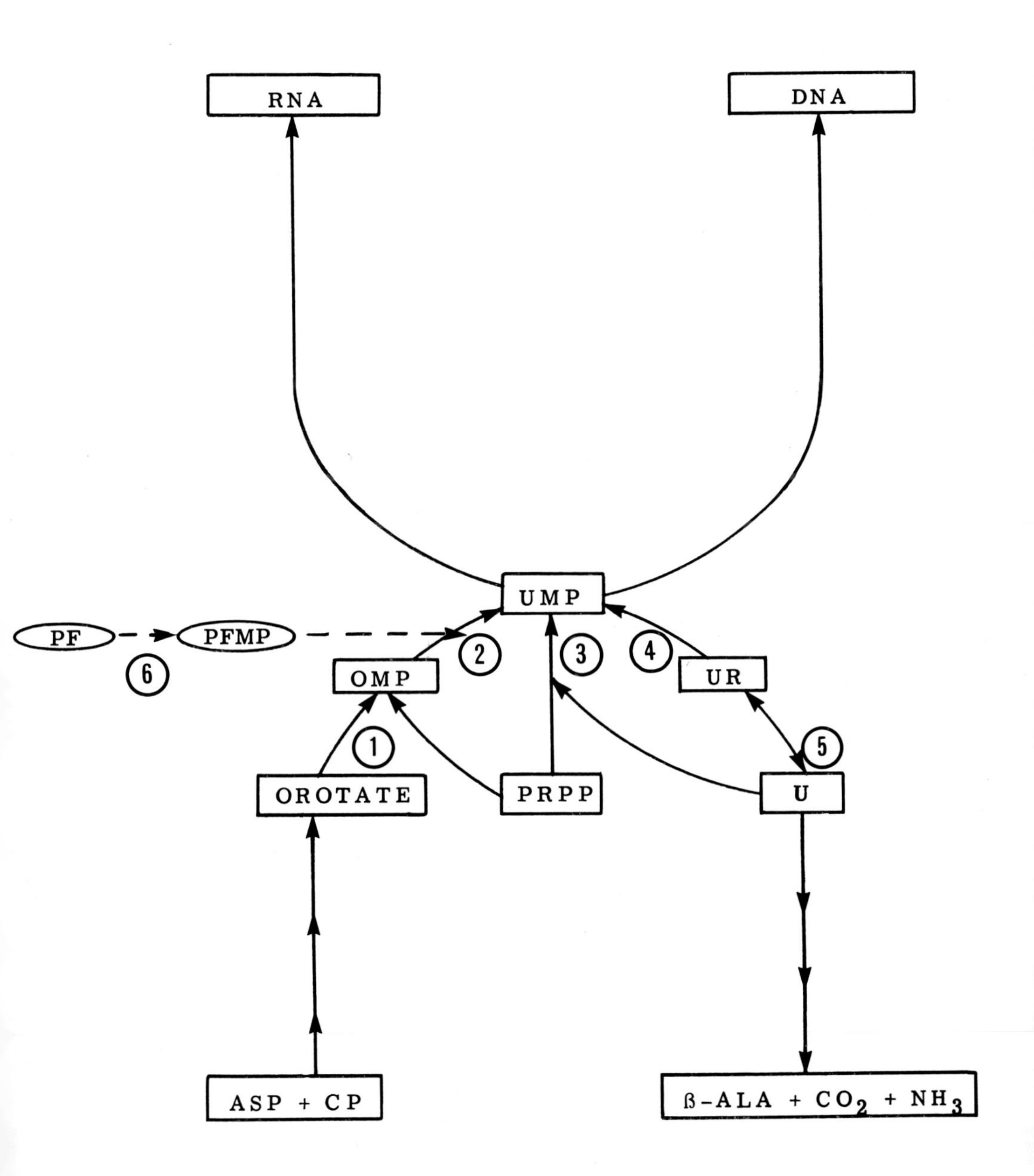

Fig. 2. Proposed site of action of pyrazofurin in pyrimidine metabolism.

PF, pyrazofurin; PFMP, pyrazofurin 5'-phosphate; 1, orotate phosphoribosyltransferase; 2, OMP decarboxylase; 3, uracil phosphoribosyltransferase; 4, uridine kinase; 5, uridine phosphorylase; 6, adenosine kinase.

TABLE 1 OMP Decarboxylase Specific Activity in Hepatomas of Different Growth Rates

Tissues	Growth rate (months)	OMP decarboxylase activity % of control
Control, normal liver		34.2 ± 0.5 100
Hepatomas		
20	15.4	243
9618A	13.0	361
16	9.3	341
21	7.0	288
47C	4.8	376
8999	4.0	393
7794A	3.0	447
9618B	2.1	725
7800	1.2	862
3924A	0.8	307
7288C	0.6	511
9618A2	0.4	353
3683F	0.3	740

The data were means ± S.E. of 4 to 10 rats in each group. Enzyme activities were calculated as nmol/hr/mg protein and expressed as % of respective control livers. All tumor enzyme activities were significantly higher than those in control livers ($p < 0.05$). Tumor growth rate was expressed as the mean transplantation time given in months between inoculation and growth to a diameter of approximately 1.5 cm.

Colony counts were made and the surviving fraction was calculated as percent of untreated cells. The survival curve of logarithmically growing cells treated with different doses of PF was of a sigmoid type and cell survival dropped sharply at PF concentrations higher than 0.5 μM. The LD_{50} was at a PF concentration of 0.4 μM (Olah and others, 1979). It appears relevant that the I_{50} for OMP decarboxylase activity of hepatoma 3924A extracts was observed with a pyrazofurin 5'-phosphate concentration of 0.5 μM.

Action of pyrazofurin in logarithmically growing and plateau phase cells. The growth curve of hepatoma 3924A cells and the time periods for sampling for logarithmic and plateau phase cells are shown in Fig. 3. One-hr treatment with PF of hepatoma 3924A cells in tissue culture in the logarithmic phase resulted in a biphasic dose-response curve, yielding an LD_{50} of 20 μM and an LD_{90} of 300 μM. The same drug concentrations in the plateau phase culture had no inhibitory effect on the growth of tumor cells. Thus, PF inhibited only growing cells and not resting ones (Olah and others, 1979).

Cell-cycle phase specificity of pyrazofurin in tumor cells. To elucidate the sensitivity to the drug at different phases of the tumor cell cycle, synchronized hepatoma 3924A cultures were treated for 1 hr with 10, 25 and 300 μM PF in every hr of the cell cycle. Similar treatment killed 30, 50 and 90% of the cells in the asynchronous tumor cell population. The experiments demonstrated a clear-cut,

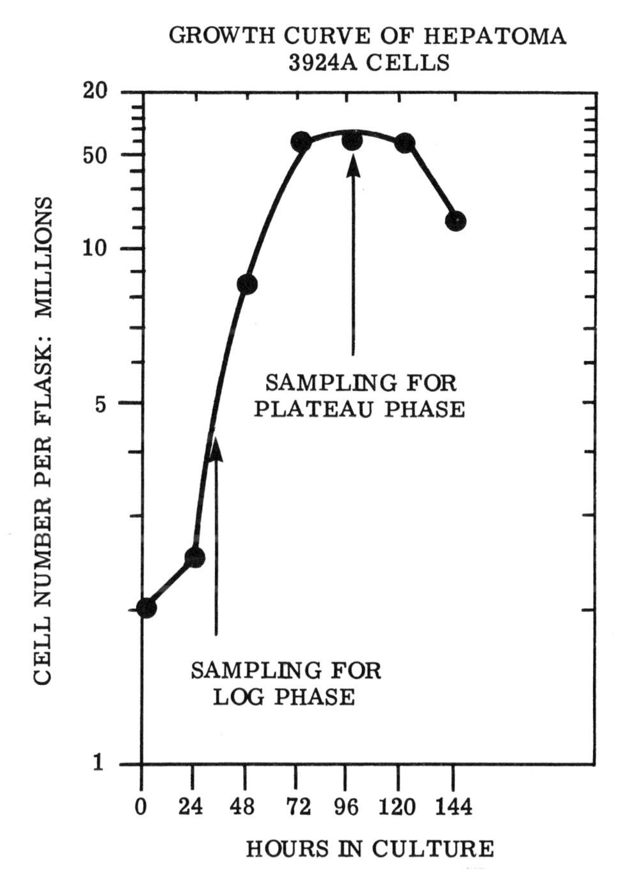

Fig. 3. Growth curve of hepatoma 3924A cells in culture. The time periods for sampling logarithmic and plateau phase cells are indicated by arrows.

phase-specific toxicity for PF which was reported in detail elsewhere (Olah and others, 1979). The effects of PF at 10 μM concentration are given in Fig. 4. The drug was equally toxic to early G_1 and early S phase cells with little or no effect on cells in the other phases of the cycle. These experiments throw new light on the cytological action of pyrazofurin by identifying the cell cycle phase selectivity of this drug.

Protection from action of pyrazofurin by nucleosides of tumor cells in tissue culture. Studies in this Laboratory in a wide spectrum of transplanted hepatomas and investigation of hepatoma 3924A growing in animals and in tissue culture showed that in addition to the increased activities of OMP decarboxylase (Table 2) and orotate phosphoribosyltransferase there was also an increase in the activities of the salvage enzymes, uridine phosphorylase, uridine-cytidine kinase and uracil phosphoribosyltransferase (Weber and others, 1978c). Therefore, it was important to elucidate whether some of the pyrimidine precursors such as uridine, cytidine or uracil provide protection against pyrazofurin under the tissue culture conditions employed in this study. A recent report suggested that PF might also inhibit an enzyme in the *de novo* biosynthesis of purines (Worzalla and Sweeney, 1978).

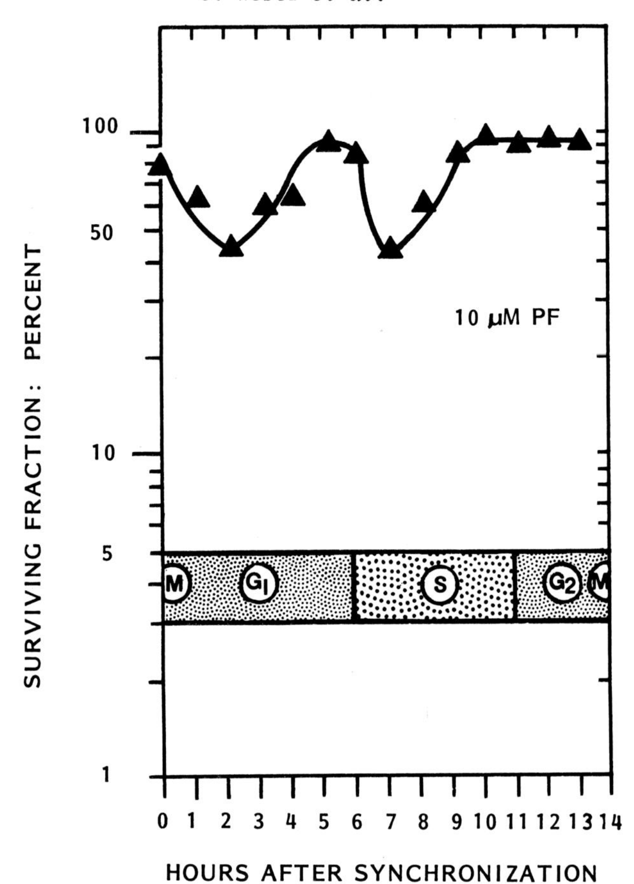

Fig. 4. Cell cycle phase specificity of pyrazofurin action in hepatoma 3924A cells in tissue culture.

Synchronized cells, 500 per flask, were exposed to the drug at 1-hr intervals after the mitotic cell action. PF was removed after 1 hr and the cells were washed with phosphate buffer saline solution. Six replicate flasks were used for each time period. The colonies were counted and the counts were expressed as percentages of colonies of untreated hepatoma cells.

It was of interest, therefore, to test whether hypoxanthine might prove to be effective in rescuing tumor cells from the action of PF. The results of these experiments showed that under our conditions only uridine or cytidine were able to protect the cell from the cytotoxic action of PF; by contrast, uracil or hypoxanthine had very little action as protective factors. These data are important because circulating or tissue levels of uridine or cytidine, depending on the presence of increased activities of the pyrimidine salvage enzymes in the different tumors or in the various critical normal tissues, might modify the cytotoxic action of PF which is inhibitory only to the de novo pathway of UMP biosynthesis. These results further support the assumption that the major anti-tumor action of PF is exerted through inhibition of OMP decarboxylase.

TABLE 2 Protection by Nucleosides and Bases against Pyrazofurin in Hepatoma 3924A Cells

Nucleosides and bases	Concentration (μM)	Percent survival
Uridine	1	0
	10	37
	30	54
	100	90
	1000	71
Cytidine	1	0
	10	43
	30	51
	100	66
	300	77
	1000	78
Uracil	1	0
	10	0.02
	100	0.04
	1000	0.10
Hypoxanthine	1	0
	10	2
	100	3
	1000	5

Cells were treated with 1 μM pyrazofurin and nucleoside or base for 7 days. Percentages were calculated from mean of 3 flasks for each treatment group.

Effect of PF on purine and pyrimidine ribonucleotide and deoxyribonucleotide concentrations in tumor cells. A further insight was gained into the mechanism of action of PF by determining the levels of purine and pyrimidine ribonucleotides and deoxyribonucleotides in tumor cells. Rapidly growing Novikoff hepatoma cells in tissue culture were treated for 16 hr with 1 μM PF and the nucleotide pools were determined in both treated and untreated cells. Table 3 shows that PF administration decreased only the concentrations of UTP (22%), CTP (9%), dTTP (50%) and dCTP (51%). PF treatment did not affect the pools of ATP, dATP, GTP or dGTP. The selective depression of the pools of uridylate, cytidylate, dTTP and dCTP supports the assumption of a primary action of PF on pyrimidine metabolism in the concentration used in this system (Weber and others, 1979).

Synergistic Action of Pyrazofurin and Galactosamine on Growth and Nucleotide Concentrations in Hepatoma Cells

Novikoff hepatoma cells were grown in the logarithmic phase and at the time of subcultivation the drugs were added to the medium. Incubation was continued for 48 hr at which time the cells were counted. As shown in Table 4, addition of pyrazofurin resulted in survival of 31% of the cells. In presence of galactosamine

TABLE 3 Action of Pyrazofurin on Ribonucleoside and Deoxyribonucleoside Triphosphate Concentrations in Novikoff Cells

Nucleotides	Control: No treatment	Pyrazofurin-treated (1 μM)
	nmol/10^9/cells	% of control
ATP	7235 ± 615	133
GTP	1510 ± 60	111
UTP	2585 ± 305	22*
CTP	2006 ± 352	9*
dATP	34 ± 13	139
dGTP	68 ± 1	107
dTTP	92 ± 2	50*
dCTP	15 ± 2	51

Means ± St. E. of 3 or more experiments were given. Treatment was for 16 hr.
*Significantly different from untreated ($p < 0.05$).

TABLE 4 Effect of Pyrazofurin and Galactosamine on Growth and Nucleotides in Novikoff Cells

Treatment	Cell count	UTP	CTP
	% of control	% of control	% of control
None, controls	100	100	100
Pyrazofurin	31*	51*	36*
Galactosamine	61*	92	96
Galactosamine + PF	14†	38*	25†

Mean ± St. E. of 3 experiments were given and expressed as % of control. PF (3×10^{-8} M), Gal_2NH (2×10^{-3} M) were added to medium at the time of subcultivation and were incubated for 16 hr for biochemical studies; cell counts were done at 48 hr.

*Significantly different from untreated cells ($p < 0.05$).
†Significantly different from pyrazofurin- or galactosamine-treated cells ($p = < 0.05$).

61% of the cells survived. Pyrazofurin and galactosamine together yielded a strong inhibition with only 14% of the hepatoma cells surviving.

Using a similar experimental set-up with a lower dose of pyrazofurin, samples were taken from Novikoff hepatoma cells at 16 hr after addition of the drugs and the concentrations of UTP and CTP were determined. As shown in Table 4, galactosamine alone had no effect; however, pyrazofurin decreased the UTP and CTP concentrations to 51 and 36%, respectively. Pyrazofurin and galactosamine together yielded a significantly more profound depression in UTP and CTP pools (38 and 25%, respectively) than pyrazofurin alone.

Enzyme-Pattern-Targeted Chemotherapy: In Vivo Selective Action on CTP Concentration in the Transplanted Hepatoma

Animals carrying the rapidly-growing, transplantable hepatoma 3924A were injected intraperitoneally with the drugs 17 and 2 hr before taking the tissue samples from the tumor and the host liver by the freeze-clamp method, as cited in Materials and Methods. The results summarized in Table 5 indicate that administration of galactosamine decreased UTP concentration in the liver to 25%, but did not significantly alter it in the hepatoma. Pyrazofurin did not cause a significant change in UTP concentration in either tissue. Combination chemotherapy with pyrazofurin and galactosamine resulted in a marked decrease of UTP pools in the liver (7%) and hepatoma (22%). Administration of galactosamine as well as of pyrazofurin and combination chemotherapy with galactosamine and pyrazofurin decreased the CTP concentration only in the tumor, not in the liver (Lui, Jackson, and Weber, 1979). The decrease in CTP levels in the hepatoma is attributed to the decline in UTP concentration which would curtail the substrate level for CTP synthetase activity. As reported earlier, CTP synthetase activity was 7-fold increased in hepatoma 3924A (Williams and others, 1978). Thus, limiting the concentration of the substrate, UTP, was more critical for the metabolic processes of this rapidly growing tumor than for those of the resting liver.

TABLE 5 In Vivo Selective Depression of CTP Concentration in Hepatoma 3924A in Rat

Treatment	UTP		CTP	
	Liver	Hepatoma	Liver	Hepatoma
None, controls	100	100	100	100
Galactosamine	25*	75	124	66*
Pyrazofurin	75	50	78	57*
Galactosamine + PF	7*	22*	145	48*

From Lui, Jackson and Weber (1979). Data were means of 3 or more rats. Nucleotide concentrations were calculated as nmol/g tissue and expressed as % of control untreated animals. Drugs were injected i.p. 17 and 2 hr before killing. Galactosamine (500 mg/kg); pyrazofurin (10 mg/kg).

*Significantly different from untreated ($p < 0.05$).

Theoretical and Experimental Approaches of Enzyme-Pattern-Targeted Chemotherapy

The theoretical approach of the enzyme-pattern-targeted chemotherapy is based on a need to obtain a more precise knowledge of the operation of organ and cell specific differentiated functions and key enzymes and isozymes in tumor cells. Then through application of multiple inhibition of the biosynthetic networks of tumor cells favorable sites for action of enzyme inhibitors may be utilized. It is particularly useful to achieve a depletion of the concentrations of key metabolites because this should enhance the effectiveness of blockers acting before and subsequent to metabolic utilization of that substrate.

Our approaches in enzyme-pattern-targeted chemotherapy also postulate that the identification of the selective advantages that enzymic imbalances confer to cancer cells should provide a biochemical rationale for drug treatment (Weber, 1977a, 1977b). It was pointed out in a discussion of the molecular correlation concept that marked elevations of activities of key enzymes should confer selective advantages to tumor cells and thus such enzymes should be sensitive targets for the design of anti-cancer chemotherapy. This argument is based on the consideration that the activities of most enzymes, even those that are classified as key enzymes, including most or all the rate-limiting enzymes, are present in tissues in an excess in comparison to the overall metabolic flux that occurs in those cells. A display of an increase in the activities of such enzymes indicates that the stepped-up metabolic capacity that the elevated enzyme activity confers to cancer cells should be important for neoplasia. Thus, marked elevation of enzyme activities reveals promising sites for design and utilization of anti-metabolites. Similarly, increased concentrations of substrate pools may indicate, among other possible interpretations, the need for an enlarged pool of these precursors in neoplastic cells.

In testing these reasonings in experiments we have been impressed by the marked rise observed in all types of cancer cells in the activity of CTP synthetase, the rate-limiting enzyme of CTP biosynthesis (Williams and others, 1978). It is further relevant to the present argument that the concentration of CTP was increased 5-fold in rapidly growing hepatoma 3924A, as determined by our freeze-clamp studies (Jackson and others, 1976). Earlier reports from this Laboratory demonstrated in the tissue culture system a synergistic interaction of 3-deazauridine with D-galactosamine. In such experiments, galactosamine was used to deplete UTP concentration in liver cancer cells and the further utilization of UTP by CTP synthetase was blocked by deazauridine which in its metabolite form as 3-deazauridine triphosphate is a competitive inhibitor for CTP synthetase. Similar studies in this Laboratory also successfully utilized galactosamine and cytosine arabinoside (ara-C) and subsequently combinations of methotrexate, 5-fluorodeoxyuridine and thymidine on rat hepatoma cells in vitro (Jackson and Weber, 1976).

Most recent investigations, that are also the subject of this report, concentrated on utilization of the liver-specific pattern of galactosamine metabolism that is retained in the hepatomas and on the action of pyrazofurin, an anti-metabolite of de novo pyrimidine biosynthesis (Lui, Jackson, and Weber, 1979). The rationale for the studies presented is visualized in Figure 5. Treatment with galactosamine was utilized as in our earlier studies because this compound is activated through a liver-specific pathway that is retained in hepatocellular carcinomas. D-Galactosamine is metabolized through activation by galactokinase (EC 2.7.1.6) and by utilization of UDPG which is synthesized from UTP by action of UDPglucose pyrophosphorylase (glucose-1-phosphate uridyltransferase, EC 2.7.7.9). Administration of galactosamine leads to depletion of UTP concentration in liver (Decker, Keppler, and Pausch, 1973). We tested the organ selectivity of galactosamine by treating rats with this compound and determining the concentrations of UTP and CTP in different tissues by the freeze-clamp method. We observed that a decrease in UTP

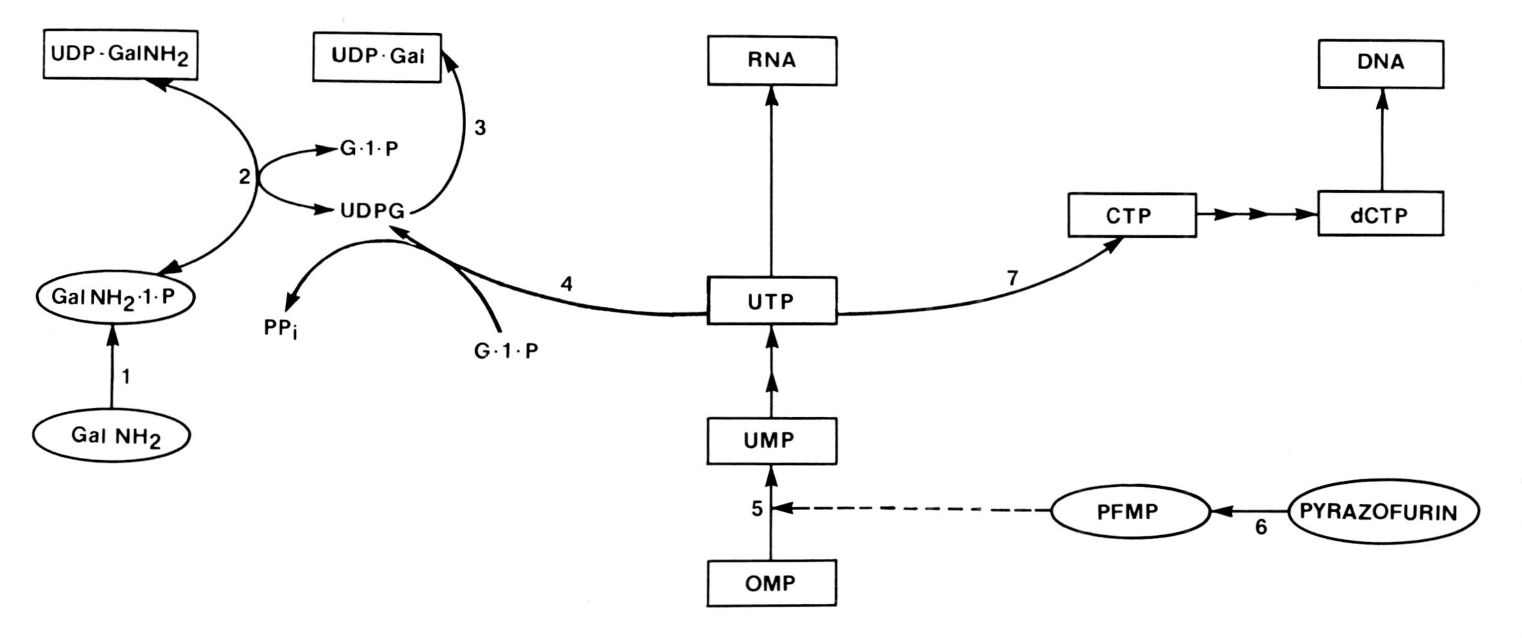

Fig. 5. Example of design of enzyme-pattern-targeted chemotherapy.

The liver-specific enzymic activities that are retained in the hepatoma and the increased enzymic activities in this liver tumor are used to selectively target chemotherapy. This simplified diagram shows relevant areas of pyrimidine biosynthesis in hepatoma and the sites of action of D-galactosamine and pyrazofurin. 1, Galactokinase; 2, UDPG: galactose 1-P uridyltransferase; 3, UDP: galactose 4-epimerase; 4, glucose 1-P uridyltransferase; 5, OMP decarboxylase; 6, adenosine kinase; 7, CTP synthetase. PF, pyrazofurin; PFMP, pyrazofurin 5'-phosphate; 3-DAU, 3-deazauridine; 3-DAUTP, 3-deazauridine triphosphate; Gal NH_2, D-galactosamine. Dashed line shows site of inhibition by pyrazofurin 5'-phosphate. Heavy arrow indicates enzymic activity increased in all hepatomas (OMP decarboxylase); heavy tapered arrow indicates enzymic activity elevated in parallel with tumor growth rate (CTP synthetase).

concentration took place exclusively in the liver (Weber and others, 1979). In the conceptual approach of the enzyme-pattern-targeted chemotherapy we postulated a selective and synergistic action for pyrazofurin and galactosamine (Fig. 5). Pyrazofurin should be activated by adenosine kinase which is retained by hepatoma cells (Jackson, Morris, and Weber, 1978). The metabolite, pyrazofurin 5'-phosphate, acts as an analog of orotidine 5'-phosphate and in appropriate dosages it should lead to a competitive inhibition of OMP decarboxylase. This approach is targeted primarily against OMP decarboxylase, which was increased in all the hepatomas we assayed and this should confer selective advantages to the cancer cells. Concurrent administration of galactosamine should further curtail UTP concentration, achieving a limitation in the concentration of the substrate for CTP synthetase activity which was increased in all the examined hepatomas (Williams and others, 1978). The prediction was tested that under these conditions pyrazofurin and galactosamine should synergistically depress the UTP concentration both in the liver and in the hepatoma. However, in the host the decline in UTP concentration and the resulting decrease in CTP synthetase activity may not be life-threatening to hepatic cells because sufficient CTP synthetase activity should be present to maintain necessary CTP levels in the resting liver. By contrast, in the rapidly growing tumors with an increased metabolic demand for CTP, a depletion in UTP level should curtail CTP synthetase activity, resulting in a critical drop in CTP pool which should lead to a fatal imbalance in the concentrations of nucleoside triphosphates required for the increased growth rate of these tumor cells. Thus, the marker for the synergistic action was the postulated depletion of the CTP pools and the prediction was that this should occur selectively in tumor cells.

This chain of reasoning was tested in a sequence of experiments and the outcome indicates that a selective depression of tumor CTP concentration was indeed achieved by the synergistic action of D-galactosamine and pyrazofurin.

Extensive studies are needed to elucidate the precise anti-cancer action of these and other combination drug treatments to test the practical results on the growth of tumors of the conceptual and experimental approaches of the molecular correlation concept and the enzyme-pattern-targeted chemotherapy.

ACKNOWLEDGEMENTS

The research work outlined in this paper was supported by grants from the United States Public Health Service, National Cancer Institute, Grants CA-05034 and CA-13526.

REFERENCES

Cadman, E. C., D. E. Dix, and R. E. Handschumacher (1978a). Clinical, biological, and biochemical effects of pyrazofurin. Cancer Res., 38, 682-688.

Cadman, E. C., F. Eiferman, R. Heimer, and L. Davis (1978b). Pyrazofurin enhancement of 5-azacytidine antitumor activity in L5178Y and human leukemia cells. Cancer Res., 38, 4610-4617.

Decker, K., D. Keppler, and J. Pausch (1973). The regulation of pyrimidine nucleotide level and its role in experimental hepatitis. Advances in Enzyme Regulation, 11, 205-230.

Frei, E., III, N. Jaffe, M. Gero, H. Skipper, and H. Watts (1978). Guest editorial. Adjuvant chemotherapy of osteogenic sarcoma: Progress and perspectives. J. Natl. Cancer Inst., 60, 3-10.

Gutowski, G. E., M. J. Sweeney, D. C. DeLong, R. L. Hamill, K. Gerzon, and R. W. Dyke (1975). Biochemistry and biological effects of the pyrazofurins (pyrazo-

mycins): Initial clinical trial. Ann. N. Y. Acad. Sci., 255, 544-551.
Holland, J. F., and E. Frei, III (1973). Cancer Medicine. Lea & Febiger, Philadelphia.
Jackson, R. C., T. J. Boritzki, H. P. Morris, and G. Weber (1976). Purine and pyrimidine ribonucleotide contents of rat liver and hepatoma 3924A and the effect of ischemia. Life Sci., 19, 1531-1536.
Jackson, R. C., and G. Weber (1976). Enzyme pattern-directed chemotherapy: The effects of combinations of methotrexate, 5-fluorodeoxyuridine and thymidine on rat hepatoma cells in vitro. Biochem. Pharmacol., 25, 2613-2618.
Jackson, R. C., J. C. Williams, and G. Weber (1976). Enzyme pattern-directed chemotherapy: Synergistic interaction of 3-deazauridine with D-galactosamine. Cancer Treatment Reports, 60, 835-843.
Jackson, R. C., H. P. Morris, and G. Weber (1978). Adenosine deaminase and adenosine kinase in rat hepatomas and kidney tumors. Brit. J. Cancer, 37, 701-713.
Lui, M. S., R. C. Jackson, and G. Weber (1979). Enzyme pattern-directed chemotherapy. Effects of antipyrimidine combinations on the ribonucleotide content of hepatomas. Biochem. Pharmacol., 27, in press.
Olah, E., D. Tzeng, M. S. Lui, and G. Weber (1979). Phase and cell cycle specificity of pyrazofurin action. Submitted for publication.
Sartorelli, A. C., and D. G. Johns (1974). Antineoplastic and Immunosuppressive Agents I & II. Springer-Verlag, New York.
Steel, C. G. (1977). Growth Kinetics of Tumours. Clarendon Press, Oxford.
Streightoff, F. J., J. D. Nelson, J. C. Cline, K. Gerzon, M. Hoehn, R. H. Williams, M. Gorman, and D. C. DeLong (1969). Antiviral activity of pyrazomycin. In Ninth Conference on Antimicrobial Agents and Chemotherapy, Washington, D. C., p. 8.
Sweeney, M. J., F. A. Davis, G. E. Gutowski, R. L. Hamill, D. H. Hoffman, and G. A. Poore (1973). Experimental antitumor activity of pyrazomycin. Cancer Res., 33, 2619-2623.
Sweeney, M. J., J. W. Parton, and D. H. Hoffman (1974). Biosynthesis of uridine-5'-monophosphate in rat liver and Morris hepatomas. Advances in Enzyme Regulation, 12, 385-396.
Weber, G. (1977a). Enzymology of cancer cells, Part I. New England J. Med., 296, 486-493.
Weber, G. (1977b). Enzymology of cancer cells, Part II. New England J. Med., 296, 541-551.
Weber, G., M. Stubbs, and H. P. Morris (1971). Metabolism of hepatomas of different growth rates in situ and during ischemia. Cancer Res., 31, 2177-2183.
Weber, G., R. A. Malt, J. L. Glover, J. C. Williams, N. Prajda, and C. D. Waggoner (1976). Biochemical basis of malignancy in man. In W. Davis and C. Maltoni (Eds.), Biological Characterization of Human Tumours. Excerpta Medica, Amsterdam, pp. 60-72.
Weber, G., R. C. Jackson, J. C. Williams, F. J. Goulding, and T. J. Eberts (1977). Enzymatic markers of neoplastic transformation and regulation of purine and pyrimidine metabolism. Advances in Enzyme Regulation, 15, 53-77.
Weber, G., F. J. Goulding, R. C. Jackson, and J. N. Eble (1978a). Biochemistry of human renal cell carcinoma. In W. Davis and K. R. Harrap (Eds.), Characterization and Treatment of Human Tumours. Excerpta Medica, Amsterdam. pp. 227-234.
Weber, G., H. Kizaki, D. Tzeng, T. Shiotani, and E. Olah (1978b). Colon tumor: Enzymology of the neoplastic program. Life Sci., 23, 729-736.
Weber, G., T. Shiotani, H. Kizaki, D. Tzeng, J. C. Williams, and N. Gladstone (1978c). Biochemical strategy of the genome as expressed in regulation of pyrimidine metabolism. Advances in Enzyme Regulation, 16, 3-19.
Weber, G., E. Olah, M. S. Lui, and D. Tzeng (1979). Enzymic programs and enzyme pattern-targeted chemotherapy. Advances in Enzyme Regulation, 17, in press.
Williams, J. C., H. Kizaki, H. P. Morris, and G. Weber (1978). Increased CTP synthetase activity in hepatomas. Nature, 271, 71-73.

Worzalla, J. F., and M. J. Sweeney (1978). Pyrazofurin inhibits de novo purine biosynthesis in rats. Proc. Am. Assoc. Cancer Res., 19, 56.
Zubrod, C. G. (1972). Chemical control of cancer. Proc. Natl. Acad. Sci. USA, 69, 1042-1047.

Transfemoral Hepatic Artery Infusion for Metastatic Carcinoma

J. Helsper*, T. Hall, J. Lance*** and W. Luxford*****

*Associate Clinical Professor of Surgery, University of Southern California School of Medicine, Los Angeles, California, U.S.A.
**American Cancer Society Research Professor
***Huntington Memorial Hospital, Pasadena, California, U.S.A.

ABSTRACT

We have infused the hepatic artery in 79 patients over a ten year period, with 5-Fluorouracil, FUDR, or Adriamycin. The catheter is placed by a transfemoral arteriotomy, under radiographic control, and threaded into the hepatic artery. A constant high volume technic is used. The infusion is continued for periods varying from one week to three weeks. Complications have been few. Toxicity has been minimal. Response rates, that is objective improvement for three months or more, are approximately 59%. After objective remission has been obtained, the patients have all been carried on systemic chemotherapy, to prolong the remission. Long-term remissions, up to five years, have been observed, although these are exceptional. The relative ease of the procedure, without the necessity of laparotomy, has increased the quality of survival in these patients. Tumor types have varied from cancer of the colon, cancer of the breast, small cell cancer of the lung, and carcinoid. Indications, objective measures of response, and technic will be presented.

KEYWORDS

Hepatic artery; Adriamycin; radiographic placement; high volume; remission; 5-FU

INTRODUCTION

The purpose of this report is to present our experience with high volume hepatic arterial infusion of chemotherapeutic agents for primary and metastatic cancer of the liver using a femoral artery approach to the hepatic artery. This was done

TABLE 1 INDICATIONS

1. Rapidly progressive hepatic malignancy
2. Potentially responsive tumor
3. Radiologic catheter placement possible
4. No life-threatening metastases elsewhere
5. Other methods have failed

over a three to twenty-one day hospitalization and was followed in all cases with systemic chemotherapy in the usual fashion. These patients were treated by various members of the hospital staff so that there were variations in adherence to the protocol; however, one of three radiologists placed all catheters.

As is well known, the average survival of patients with inoperable liver metastases is variously reported as 75.9 days by Purves and others (10); a 50% survival in 390 cases of 75 days by Jaffe (5); and 2.5 months by Kinami and Miyazaki (6).

PROTOCOL

1. All patients had histologic diagnoses of the primary.
2. All patients had positive liver scans.
3. Progressive metastases which were symptomatic.
4. Pathologic confirmation of liver metastases (about 60% achieved) by needle biopsy or laparotomy.
5. Confirmation of metastases by arteriogram at the time of catheter placement.
6. Routine laboratory: CBC, platelet count, SMA-12, prothrombin time, CEA, and alpha fetoprotein (if indicated).
7. Chest x-ray.
8. Bone scan and brain scan (where indicated).
9. Bone marrow study in anemic patients (or where indicated).
10. Informed Consent Form signed.
11. Alternates to therapy offered were reviewed with each patient:
    a. Systemic chemotherapy
    b. Radiation therapy combined with systemic chemotherapy
    c. Hepatic artery ligation at surgery or by various trans-catheter clotting agents
    d. Palliative care.

TABLE 2 CONTRAINDICATIONS

1. Hepatic failure (coma) or impending coma
2. Unresponsive and significantly decreased prothrombin time (less than 65% after Vitamin-K)
3. Significant metastases other than in liver
4. Severe arterial disease in lower extremities

PATIENT SELECTION

Since 1969, we have infused the hepatic artery in seventy four patients for progressive metastatic or primary cancer of the liver, a total of ninety three times. The largest group (forty patients) had metastatic cancer of the colon. In addition seven patients had metastatic pancreatic cancer. The remaining were scattered among cancers of the gallbladder, ileum, ovary, stomach, common duct, lung, melanoma and/or unknown primary (as shown in Table 1).

In breast cancer, the chemotherapy infusion was considered only after hormone manipulation or ablation was exhausted, and patient had progressive liver metastases.

TABLE 3 PATIENTS INFUSED

AVERAGE AGE:	60.7 years (26-77)
MALES	28
FEMALES	46
CAUCASIAN	69
BLACK	4
ORIENTAL	1

METHODS

The catheter was placed percutaneously through the femoral artery, with the majority of them having the tip of the catheter in the hepatic artery just beyond the superior pancreatico-duodenal artery. Percutaneous selective catheterization of the hepatic artery is usually relatively easy to accomplish. Once installed, the catheter can usually be maintained for several weeks without complication or loss of the catheter position. Entering the celiac axis is easier approaching from

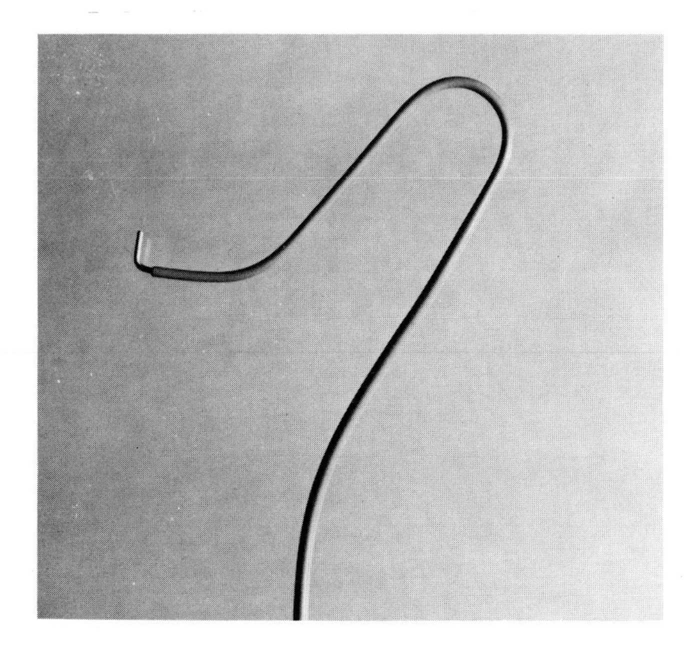

Fig. 1. Showing the curved tip #7 French Simmons #3 which we have found to be most successful (see text). (A stylet is in place.)

Figure 1. This catheter has a secondary U-shaped curve approximately 10 cm. from its tip which is reformed in the aortic arch. As the catheter is then retracted, the tip enters the celiac axis from above, giving the advantages of an antegrade approach. In the normal anatomy with the hepatic artery arising from the celiac axis, the catheter usually enters the hepatic artery without difficulty. The catheter has sufficient torque control to facilitate this maneuver. An effort was made to advance beyond the pancreatico-duodenal branch, but this has not seemed a significant factor. The U-shaped curve of the catheter seems to facilitate maintaining its position. The catheter was then secured to the skin at point of entry, and connected to a Magaw[R] volumetric pump. A moderate compression bandage was placed across the pubic area and the catheter taped to this lower belly wall bandage further to protect its position. Antibiotic ointment was placed on the arteriotomy site. The patients are allowed to be up in a chair and use bathroom facilities. The most difficult problem has been the obese patient. Heavy subcutaneous fat allowed shift of the skin to permit the catheter to be dislodged. These patients were maintained on complete bed rest.

There have only been three cases in which we were unable to place the catheter in at least one hepatic lobe artery.

We have confined the patients to a single nursing unit of our hospital, where the nursing staff is readily conversant with the problems of arterial infusion, and our complications have been kept to a minimum.

In the early years of this study, gravity was used as the propellant force with the I.V. container reservoir at two meters above the patient in order to overcome arterial pressure. Later, pumps such as Sigmamotor pumps* and the Fenwall pressure bag technic, and more recently, Magaw pumps* have been used.

We use a high volume technic: 1000 cc. of normal saline as the vehicle, to which the drugs were added (as in Table 2), with 1000 units of Heparin. The pumps were set at thirteen drops per minute, which averaged 1000 cc. in twenty-four hours. Repeat plain films, but not repeat arteriograms, were taken every second or third day to compare with the originals, to be certain that the catheter position had not changed. If there was any question, additional dye was given to be sure the catheter remained properly placed. Special attention was directed to the arteriotomy site where antibiotic ointment was applied with an occlusive dressing.

The majority of the patients were able to continue on a regular diet, and a flow sheet was used to record and to observe trends in the measured parameters. If signs of necrosis of the tumor appeared (temperature, tenderness, nausea, and/or vomiting), the chemotherapeutic agent was removed from the infusate, but the flow continued in order to avoid clotting. Once these signs cleared, the infusion would be re-started with the chemotherapeutic agent, usually at a lesser dosage. Following their infusions, all patients were carried on systemic chemotherapeutic agents, usually that which was successfully infused, such as 5-FU at 12 mg. per kilogram per week, providing weekly blood counts remained at satisfactory levels.

DRUGS

Seventy nine patients were treated with 5-FU, varying the dose from 500 mg. per day through 1500 mg. per day, for durations from five through twenty-one days (1,4). Adriamycin, up to 20 mg. per meter square, was given daily on each of three days to a total of 60 mg. per meter square to six patients. The remaining patients were treated with 5-FUDR or DTIC (12). All patients were monitored with daily white counts and platelet counts; those with elevated CEA tests had daily monitoring of CEA. Serum uric acid, SGOT, LDH, alkaline phosphatase and bilirubin were monitored less than daily, or daily if they were elevated at the beginning of the infusions.

TABLE 4

DRUGS USED	CASES	DAILY DOSE	TOTAL DOSE
5-FU	79	500 mg - 1500 mg/d	7.0 gm - 25 gm
Adriamycin	6	10 mg - 25 mg/d	50-75 mg/M^2
5-FUDR	6	0.3 - 0.5 mg/kg/d	500 mg
Mitomycin	1	5 mg/d	15 mg
DTIC	1	250 mg/M^2/d	1500 mg

* Multiple drugs were used in patients where no response was noted during the early infusion.

RESULTS

Anti-tumor and Toxicity

As shown in Table 5, the results in colon cancer were that twenty-five of forty infused patients responded onjectively and lived three months or longer, the longest at twenty-seven months, still doing well, but with a positive liver scan. This patient had laparotomy confirmation of metastases and continues on 12 mg. per kilogram of 5-FU weekly. Her liver enlarged slightly each time the systemic therapy was stopped, with enlargement of the metastases seen on liver scan. However, the average duration of survival has been 11.3 months after the first infusion. Seven of eight infused breast cancer patients responded, but the duration was less, averaging eight months. In pancreatic cancer, we saw only two

TABLE 5

AVERAGE SURVIVAL AFTER FIRST OR ONLY HEPATIC ARTERY INFUSION IN RESPONDERS

PRIMARY	CASES	RESPONDERS	AVERAGE SURVIVAL (MOS.)
Colon	40	25	11.3
Breast	8	7	8
Pancreas	7	2	5.5
Gallbladder	4	3	7.3
Biliary Tree	3	1	4.0
Stomach	3	2	7; 60*
Liver	3	1	11
Melanoma	2	0	-
Ovary	1	1	5
Lung	1	1	12
carcinoid	1	1	6
Unknown	1	0	-
Total	74	44	

* 1 case NED 60 months (positive liver scan only) no biopsy.

of seven patients respond, and only for 5.5 months, on the average. Considering that the pancreatic cancer patients had the primary still in place, we might expect such poor results. The remaining cases were in small numbers and can be reviewed from the table. One patient with gastric cancer is alive, without evidence of cancer at five years. However, she did not have biopsy confirmation of the liver metastases, only a positive liver scan two years after subtotal gastrectomy for gastric cancer. Her liver chemistries were also compatible with metastases. The patient at this time is asymptomatic, with normal studies.

Over half the patients had objective benefit, as evidenced by decreasing size of the liver, relief of pain, improvement in liver scan, decrease in liver chemistries particularly LDH and alkaline phosphatase, reduction in CEA levels, and improved appetite and weight gain.

TABLE 6

PATHOLOGY	NO. OF PATIENTS	OBJECTIVE IMPROVEMENT
1. Adenocarcinoma		
a) G.I. origin		
Colon	40	25
Stomach	3	2
Pancreas	7	2
Biliary	7	4
b) Papillary adenoca ovary	1	1
c) Duct adenoca breast	8	7
d) Unknown 1° adenoca	1	0
e) Hepatoma	3	1
2. Small cell CA lung	1	1
3. Carcinoid	1	1
4. Melanoma	2	0
Total	74	44

Complications included infection at arteriotomy site, one, thrombocytopenia, two, oozing at arteriotomy site, five. No patient had a significant vascular complication. At the time of placement, arteriograms of the hepatic artery were done and further studies of the liver metastases were made. If the metastases were highly vascular, we saw a more rapid, and at times overwhelming, response to the infusion, with signs of toxicity, such as elevated temperature, tender liver, nausea and vomiting (3,1). In these patients, we added the chemotherapeutic agent in small amounts, first 500 mg. of 5-FU per day, and monitored uric acid, and only increased the dosage if it was well tolerated.

In those patients with avascular primary tumors, which account for 90% of our cases we advanced the FU dosage more rapidly, up to 1500 mg. per day, and rarely saw signs of toxicity.

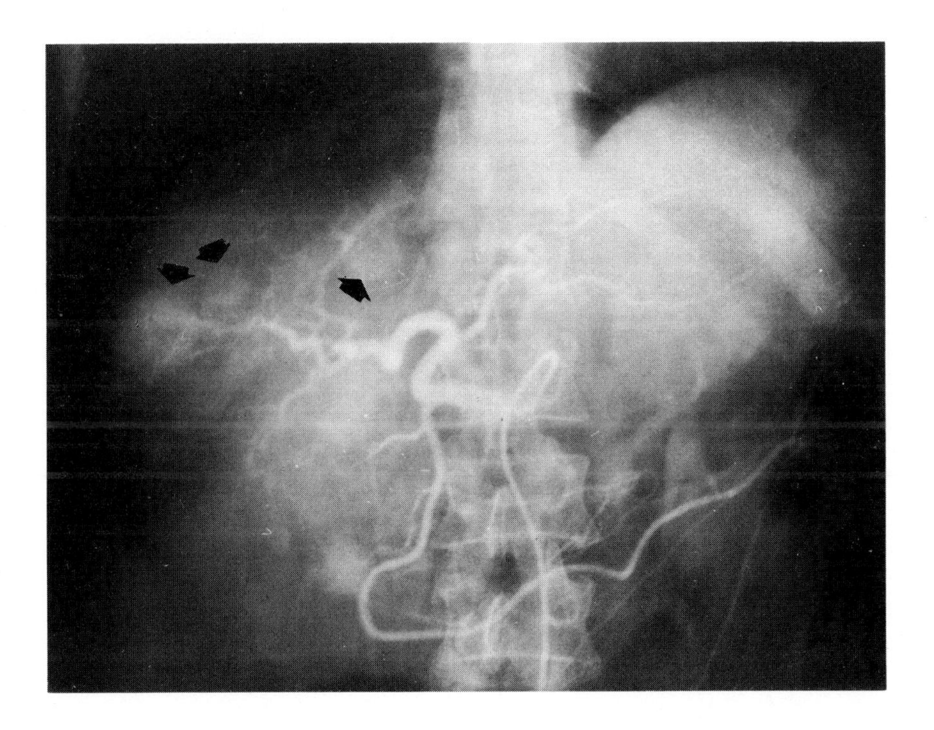

Fig. 2. A film showing the highly vascular tumors on arteriogram. Arrows mark a few of the metastases.

In several cases, we found good response in some of the metastases, but not in other masses. In these we assumed there was a lamellar flow in the vessel, but were not able to demonstrate it at arteriogram. We then added RISA to the infusate and scanned the liver. This showed unequal distribution, thought to be due to lamellar flow in these patients. Slight withdrawal of the catheter usually corrected this.

In a few cases, the hepatic artery was found to originate from the superior mesenteric artery, rather than from the celiac axis.

A few patients also had divided hepatic arteries, the left coming from the celiac axis, and the right coming from the superior mesenteric. In these cases, a single catheter was used, but the artery feeding the bulk of the malignancy was infused.

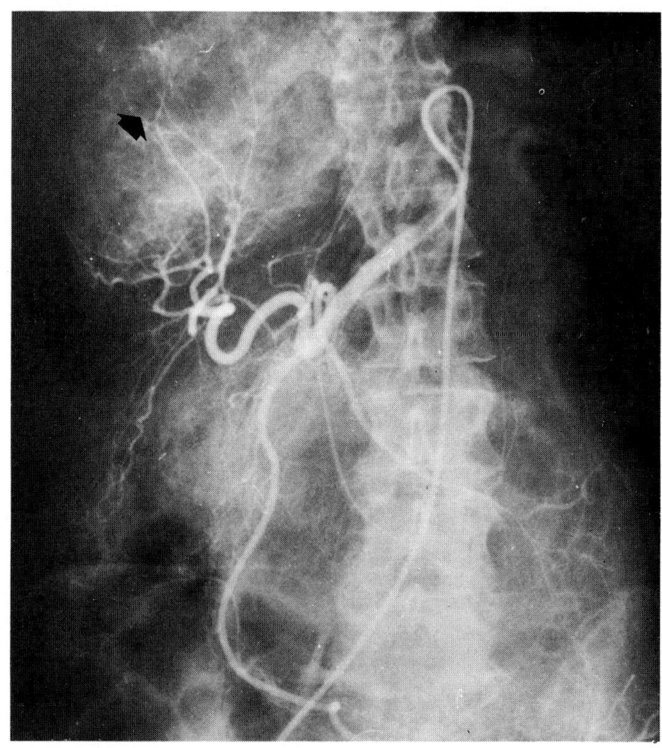

Fig. 3. A film showing the avascular tumors on arteriogram. In addition to the arrow, a huge avascular mass is seen in the left lower quadrant.

The majority of the patients tolerated the infusion well, with only 10% of the patients complaining of increasing pain in their liver, but without significant signs of toxicity, such as elevated temperature, nausea or vomiting. A few patients did develop other G.I. symptoms, such as diarrhea; however, this was seen principally in those in which the tip of the catheter could not be passed beyond the superior pancreatico-duodenal artery, so that a part of the G.I. tract was being infused, with probable local toxicity. Less than 5% of the patients showed evidence of hematologic toxicity in that their white count or platelet count did not go below 3500 or 100,000 respectively. If patients developed significant increases in uric acid, the infusion was diluted to 500 mg. per day of 5-FU, or the chemotherapeutic agent withdrawn altogether for one to three days. About 90% of the patients with elevated SGOT, alkaline phosphatase or bilirubin showed improvement during the infusion; however, in 5% of cases there was some increase in SGOT indicating some toxicity from the drug.

All six patients receiving Adriamycin showed significant drops in platelet counts, in one case requiring multiple platelet transfusions. However, four patients of six who were treated in this manner showed remarkable improvement both objectively and subjectively, and three of six of them lived more than one year following their arterial infusion, all of them dying of progression of disease at sites other than the liver. In three patients, evidence of hepatic artery thrombosis by subsequent arteriogram was demonstrated. Two of these had excellent anti-tumor results; one, however, had rapid deterioration due to massive necrosis, and expired during the procedure. The longest survivor is currently doing well five years following her hepatic artery infusion for gastric cancer; however, no biopsy was done, and only a positive liver scan, arteriogram, and abnormal liver chemistries were present before the infusion. Systemic therapy was stopped after two years.

SUMMARY

1. A total of 79 patients were infused in a high volume radiologically placed arterial catheter with chemotherapeutic drugs to the hepatic artery.

2. Drugs were well tolerated despite relatively high doses.

3. Infusions were given from three to twenty-one days without significant complications.

4. Significant improvements in subjective and objective parameters were seen in forty-four cases.

5. Complications were minimal.

6. This procedure seemed to offer patients a comfortable prolongation of life for a reasonable share of the health cost dollar.

NOTE: The authors wish to acknowledge the cooperation of numerous clinicians on the staff of the Huntington Memorial Hospital, Pasadena, California, for the use of their patients in preparing this study.

REFERENCES

1. Clarkson, B, Young, C., Dierick, W., Kuehn, P., Kim, M., Berrett, A., Clapp, P., and Lawrence, W. (1962) Effects of continuous hepatic artery infusion of antimetabolites on primary and metastatic cancer of the liver. Cancer 15: 472-488.

2. Collins, V.P., Loeffler, R.K., and Tivey, H. (1956) Observations on growth rates of human tumors. Amer. Jour. Roentgenology 76:988-1000.

3. Healey, J.E. (1965) Vascular patterns in human metastatic liver tumors. Surg., Gynecol. & Obstetrics 120:1187-1193.

4. Helsper, J.T. and deMoss, E.V. (1964) Regional intra-arterial infusion of 5-Fluorouracil for cancer. Surgery 56:340-348.

5. Jaffe, B.M., Donegan, W.L., Watson, F., and Spratt, J.S. (1968) Factors influencing survival in patients with untreated hepatic metastases. Surg., Gynecol. & Obstetrics 127:1-11.

6. Kinami, Yoshio, and Miyazaki, Izuo (1978) The superselective and the selective one shot methods for treating inoperable cancer of the liver. Cancer 41:1727.

7. Klopp, C.T., Alford, T.C., Bateman, J., Berry, G.N., and Winship, T. (1950) Fractionated intra-arterial cancer chemotherapy with methyl bis beta chloroethyl-amine hydrochloride; a preliminary report. Ann. Surg. 132:811-832.

8. Miller, T.R., and Griman, O.R. (1961) Hepatic artery catheterization for liver perfusion. Arch. Surg. 82:423-425.

9. Pestana, C., Reitemeier, R.J., Moertel, C.G., Judd, E.S., and Dockerty, M.B. (1964) The natural history of carcinoma of the colon and rectum. Amer. Jour. Surg. 108:826-829.

10. Purves, L.R., Bersohn, I., and Geddes, E.W. (1970) Serum alpha fetoprotein and primary cancer of the liver in man. Cancer 25:1261-1270.

11. Rochlin, D.B., and Smart, C.R. (1966) An evaluation of 51 patients with hepatic artery infusion. Surg., Gynecol. & Obstetrics 123:535-538.

12. Savlov, E.D., Hall, T.C., and Oberfield, R.A. (1971) Intra-arterial therapy of melanoma with dimethyl triazeno imidazole carboxamide (NSC 45388). Cancer 28:1161-1164.

13. Watkins, E., and Sullivan, R.D. (1964) Cancer chemotherapy by prolonged arterial infusion. Surg., Gynecol. & Obstetrics 118:3-19.

14. Wirtanen, G.W., Bernhardt, L.C., Mackman, S., Ramirez, G., Curreri, A.R., and Ansfield, F.J. (1968) Hepatic artery and celiac axis infusion for the treatment of upper abdominal malignant lesions. Ann. Surg. 168:137-141.

Investigation on Human Macrophage

Chang Yu-hui and Yao Chin-sheng

Department of Immunology, Institute of Cancer Research,
Chinese Academy of Medical Sciences, Peking, The People's Republic of China

ABSTRACT

In man, skin blister induced by cantharidin was a good source of macrophage. When cultured in the presence of 50% CFE[1], human macrophages were found to proliferate *in vitro*. They were also capable of exerting non-specific cytotoxic effect on human tumor cell lines as shown by an inhibition of ^{3}H-TdR incorporation into tumor cells. CFE did not abrogate the MMC.

INTRODUCTION

The fact that the immune system of the body is able to respond to cancer growth has been will established. It has been demonstrated that the immunity to cancer is mainly cell-mediated. T lymphocytes have been considered to be the chief effector cells whereas macrophages have been received less attention. Recent reports indicate that macrophages do play an important part in the rejection of tumor (Alexander, 1976; Keller, 1975). However, most studies have been limited to animal experiments, studies on human macrophages have been scanty. The role of macrophage in the immune defence mechanism against tumors in man remains to be elucidated. This article describes a simple method to collect human macrophages for *in vitro* cultivation and MMC assay.

MATERIALS AND METHODS

Induction of Skin Blister

Two pieces of filter papers, soaked with 10% alcoholic extract of cantharidin, were applied onto the skin of forearm. They were cover-

[1]Abbreviations used: CFE, cell-free exudate; CI, cytotoxic index; CP, C. parvum; CRBC, chicken red blood cell; EC-109, human esophageal cancer cell line; E/T, effector-to-target cell ratio; MA, human osteogenic sarcoma cell line; MMC, macrophage-mediated cytotoxicity; NBCS, newborn calf serum; OT, old tuberculin; PHA, phytohemagglutinin; RASC, rabbit antiserum against CRBC; ^{3}H-TdR, tritiated thymidine.

with a 2 cm^2 glass coverslip and held tight with adhesive tape. Four to 5 hrs later, the coverslip with the filter papers was removed. Local erythema and loosening of the superficial epidermis could be seen. Then, a circular plastic cap was placed over the newly forming blister to protect it from damage. At 48 hrs, the fluid was aspirated out under aseptic condition from the fully formed blister. A total white cell count and a differential cell count were performed from which the yield of macrophages could be estimated.

Intrablisteral Injections

PHA (200 µg), CP (0.35 mg) or OT (10^{-4} or 10^{-5}, 0.1 ml) was injected into blister cavity 24 hrs after its initiation. The blister fluid was collected at 48 hrs and processed as usual.

Cultivation of Macrophage

One ml of blister fluid was placed into a sterile culture vial with a 6 x 18 mm coverslip at its bottom and incubated at 37°C for 2 hrs. The coverslip was washed to remove non-adherent cells and the blister fluid centrifuged to obtain the CFE. Macrophages that had been attached on the coverslip were cultured in 50% CFE in TC 199 with or without NBCS.

Observation and Identification of Cultured Macrophage

Morphological studies. The coverslip was taken out from the vial, placed on a microscopic slide and observed under phase contrast microscope. It was also observed by ordinary light microscope after staining.

Phagocytosis. The coverslip was transferred into a clean vial to which were added 1 ml autologous CFE or 50% RASC in TC 199 and CRBC. After 30 min incubation at 37°C, the coverslip was washed and stained and examined microscopically for phagocytosis.

Demonstration of Fc receptors. The method described by Moskalewski and coworkers (1974) was adopted.

Quantitative estimation of cells adhered to glass surface. The method described by Nathan and others (1971) was used.

Incorporation of ^{3}H-TdR into macrophage. A sample of blister fluid was divided into 2 equal parts and each was incubated at 37°C in a culture vial for 6 hrs. The fluid was then aspirated out and centrifuged. The CFE so obtained was diluted with TC 199 to 50% concentration and introduced to one of the vials. TC 199 was added to the other as control. Both were incubated in 5% CO_2 in air for 6-10 days. Six hrs before termination 1 µCi of ^{3}H-TdR was added. At the end of incubation the fluid was decanted and the vials thoroughly washed 3x. The cells adhered to the bottom were treated with a mixture of 6% perchloric acid and H_2O_2 and incubated at 80°C for 1 hr. After cooling, scintillating cocktail was added and the radioactivity (counts/min) determined in a liquid scintillation counter.

Macrophage-mediated Cytotoxicity

In vitro MMC was performed according to Williams, Germain and Benacerraf (1975) with some modification. According to cell counts, appropriate amount of blister fluid was put into flat-bottomed culture tubes so that 5x, 10x and 20×10^4 macrophages were present in each tube, respectively. TC 199 was added to make up the fluid volume to 1 ml. After 24 hrs preincubation the supernatant was removed and each tube washed to get rid of nonadherent cells. Then 1 ml of 10^4 target cell (EC-109 or MA) suspension was added so that the E/T were 5, 10, and 20, respectively. Incubation was continued. At 56 hrs, 1 µCi of ^{3}H-TdR was added and the radioactivity determined at 72 hrs by the method mentioned above. The results were expressed as CI which was calculated from the counts/min of tumor cells cultured with macrophages as compared to that of tumor cells cultured alone. To study the possible influence of humoral factors on MMC, autologous CFE was added to some of the culture with a final concentration of 10 or 20%.

RESULTS

Cell Yield

The amount of blister fluid and its cell count varied. On the average, 1-1.5 x 10^6 macrophages could be harvested from each ml of exudate. Intrablisteral injection of OT, PHA or CP all led to an increase in cell counts but the percentage of macrophage did not change significantly as compared with the uninjected controls.

Morphological Observation of in vitro Cultured Macrophage

The in vitro cultured macrophages had the same morphological characteristics as those of uncultured ones but the cell size appeared much larger and the translucent ruffled membrane more conspicuous. Moreover, binucleated or multinucleated macrophages were more often observed. The latter might reach a size of 80-100 micron with innumerable nuclei. Multinucleated cells of Langhan's type or of foreign-body giant cell type were both seen. Under phase contrast microscope, one could see that macrophages were actively motile and the undulating membrane spread out in all directions with long and slender cytoplasmic processes. Numerous phase-lucent and -opaque granules could be seen oscillating in the cytoplasm. Irregular vermiform villi or evenly distributed projections were present on the cell surface. The in vitro cultured macrophages maintained their capacity to phagocytize opsonized CRBC. Multinucleated giant cells were actively phagocytic, too. Both showed on their cell surface Fc receptors.

Proliferation of Macrophage in vitro

When cultured in TC 199 with 50% CFE for 6-10 days, macrophages showed signs of proliferation, e.g., increase in cell count and protein content, mitosis and incorporation of ^{3}H-TdR. Proliferation of macrophage did not occur when cultured in TC 199 without CFE.

Macrophage-mediated Cytotoxicity

Altogether, 62 cases were examined for their MMC, of which 48 were patients with malignant tumors, including 33 cases of esophageal cancer and 14 were of benign diseases. The results indicated that, when different E/T were used, the response showed a linear relation to the dose of effector cell used. Using EC-109 as targets, 67% of the esophageal cancer patients gave positive MMC. The positive rate for patients with other malignant tumors was about the same (60%) while that for benign patients was somewhat lower (41.8%). Patients with different malignancies other than sarcoma also gave positive MMC against MA cells. Thus, MMC was nonspecific in nature. CFE, regardless of its concentration (10 or 20%) or its origin (from treated or untreated cases), did not abrogate MMC, and in one instance, the post-treatment CFE markedly enhanced the cytotoxicity.

DISCUSSION

Macrophages are now recognized as one of the cell types that participate in the immune response of the host against foreign antigens, including those on tumor cell surface. These have been proved in animal experimental systems. Our knowledge about human macrophage has largely been extrapolated from animal experiments, as hitherto it has not been easy to obtain sufficient number of human macrophages for *in vitro* studies. By the method herein reported, adequate number of human macrophages can be harvested from skin blister. Cantharis is a common herb used in traditional Chinese medical practice to induce skin blister and is highly blisterogenic. It has been used in our laboratory for years without untoward side effects.

According to several reports, macrophages of animal or human origin can be cultured *in vitro* for varying length of time but they seldom proliferate unless appropriate feeder cell layer or conditioned medium is used. In this study, it has been shown that, when the culture medium contains 50% CFE, human macrophages are able to proliferate as evidenced by an increase in cell number, mitosis and ^{3}H-TdR incorporation. However, it has been difficult to subpassage macrophages in such medium.

Employing the ^{3}H-TdR incorporation inhibition test we have been able to demonstrate that human macrophages are cytotoxic to human tumor cell lines *in vitro*. The results thus far obtained show that MMC is nonspecific in nature. This is in accord with many of the results from animal experiments. There is slight difference in the proportion of positive reactors between tumor-bearing (9/33) and tumor-free (13/33) patients. This difference is, however, not significant. The relatively low positive rate in tumor-bearing patients does not seem to be due to humoral factors since no humoral factors were present in our experimental system, nor did they show blocking effect when CFE was added to the system. In some of the negative cases, CI values were negative which might mean that macrophages, instead of being cytotoxic, actually promoted tumor growth. This phenomenon must be further studied.

REFERENCES

Alexander, P. (1976). The functions of the macrophage in malignant disease. Ann. Rev. Med., 27, 207.

Keller, R. (1975). Cytostatic killing of syngeneic tumor cells by activated non-immune macrophages. In R. van Furth (ED.), Mononuclear Phagocytes in Immunity, Infection and Pathology, Blackwell Scientific Publications, Oxford. pp. 857-867.

Moskalewski, S., W. Ptak, and J. Strzyzewska (1974). Macrophages in mouse placenta: morphologic and functional identification. J. Reticuloendoth. Soc., 16, 9.
Nathan, C. F., M. L. Karnovsky, and J. R. David (1971). Alteration of macrophage function by mediators from lymphocytes. J. Exp. Med., 133, 1356.
Williams, R. M., R. N. Germain, and B. Benacerraf (1975). Specific and non-specific anti-tumor immunity. I. Description of an in vitro assay based on inhibition of DNA synthesis in tumor cells. J. Nat. Cancer Inst., 54, 697.

Hormones and Cancer

Contraceptives and Cancer

A. González-Angulo* and H. Salazar**

**División de Patología, Subjefatura de Investigacíon Básica. Centro Médico Nal. Instituto Mexicano del Seguro Social. México City, Mexico*
***Department of Pathology, Magee-Womens Hospital, University of Pittsburgh, School of Medicine, Pittsburgh, Pa., U.S.A.*

ABSTRACT

It is well known that certain modifications of the normal morphological pattern of the female reproductive tract, breast and liver take place under contraceptive medication. A great interest has emerged concerning the possible carcinogenic properties of contraceptives, particularly in uterine cervix endometrium breast and liver. The presence of cytologic atypia in women under sequential therapy twice as high as expected in the general population and the findings of cervical atypias and squamous metaplasia in cervical biopsies have arisen the interest to find a cause-effect relationship. The now well known polypoid atypical glandular hiperplasia of endocervix has been proven to be initiated by oral contraceptives. The foundation of a registry to study those cases of carcinoma of endometrium associated with oral contraceptives therapy represent a true concern in determining how safe certains contraceptive drugs are. As to the breast, it is now widely accepted that the incidence of fibroadenoma and fibrocystic disease have not increased with contraceptive medication and that there is no clear cut evidence for an association to cancer. Many reports on benign hepatic tumors associated to oral contraceptives are on record. Perhaps this particular cell proliferation may truly be related to hormones administration. The tumors have been reported in young women in hospitals where no other benign hepatic tumors were found in many decades. As to the uterine devices recent studies are in keeping with earlier reports in demonstrating no increased risk of carcinoma of the cervix or endometrium with IUD's.

INTRODUCTION

Through the years it has been recognized that certain modifications of the normal morphological pattern of the female reproductive tract and other organs take place during contraceptive medications. While most pathologic changes include benign, non neoplastic modifications, a great interest has emerged concerning the possible carcinogenic properties of some steroids present in contraceptive formulas. Special attention has been paid to find a cause-effect relationship between contraceptive medication and the appearance of various types of tumors. This has been the result of several factors (Rinehart and Felt, 1978).

1.- Experimental evidence that ovarian hormones may alter the incidence of some tumors in laboratory animals.
2.- Epidemiology data suggesting a direct effect of endogenous hormones upon the development of mammary and endometrial neoplasia.
3.- Development of nodules in the liver of rats treated with synthetic estrogens.
4.- Reports in the literature dealing with premalignant and malignant lesions in women using estrogens and contraceptives.
5.- Immediate access to information concerning neoplasms in women using contraceptive medication (family planning registries, cancer detection programs, physicians in private practice etc.)
6.- Generalized public concern about cancer.

This review will deal with the most common concepts originated from recent publications on controversial aspects of contraceptives and cancer. A final answer to this problem has not been found not even with regard to the well documented cases of nodular hyperplasias and liver cell adenomas associated to contraceptive intake by women. Special attention will be given to changes in the uterus, breast and liver.

Uterus: In studying the pathologic effects iatrogenically induced on the uteru by the chronic administration of hormonal contraceptive preparations, we mus refer separately to the cervix uteri, to the endometrium and to the myometrium, mainly for presentation purposes, but also because each part of the organ responds differently to the medication.

Cervix: Normally the cervix uteri is a target organ to endogenous estrogen and progesterone. As such, it also responds with special predilection to the action of synthetic steroids used in contraceptive drugs (Fechner, 1971). The main effects of these drugs are morphologic changes of the endo and exocervix and modifications in the amount and the physicochemical characteristics of the cervical mucus.

Pincus (1965) was the first to indicate that the contraceptive drugs had a co siderable effect on cervical function. Very soon after the oral contraceptives were placed in the market, Dito and Batsakis (1961) made the first report o glandular hyperplasia and hypersecretion of the cervix in patients treated wit a combined preparation (Enovid) for more than three months. Similar studie were reported by Zañartu (1964), by Ryan and coworkers (1964) and by Maqueo and collaborators in larger groups of patients using various combinations of steroids for more than six cycles (Maqueo, 1966). They also found a close association between the use of the contraceptive drugs and the chang of the cervix, including glandular hyperplasia, increased secretory activity, inflammation, reserve cell hyperplasia, squamous metaplasia and progestational-like changes.

Taylor, Irey and Norris (1967) described the now classical "atypical polypoi glandular hyperplasia" of the endocervix in 13 patients treated with both com bined and sequential contraceptive preparations, and called attention to the frequent bizarre cytological and histological appearance of some of these lesions, susceptible of being misinterpreted as carcinoma. These lesions consist chiefly of an exophytic polypoid hyperplasia of the endocervical surface epithelium, with extensive squamous metaplasia and microglandular formation, cellular pleomorphism with focal atypical nuclear and cytoplasmic forms, severe chronic and acute inflammation, and scant number of mitosis. The glands usually appear enlarged with evidence of marked secretory activity. The stroma is usually edematous, with well developed vascularity and, occasionally, pseudodecidual reaction (Fig. 1). Other authors have reported aditional cases (Candy and Abell, 1968) (Govan and coworkers, 1969) (Nichols and Fidler, 1971) and others have used the term of microglandular hyperplasia (Kiriakos and coworkers, 1968).

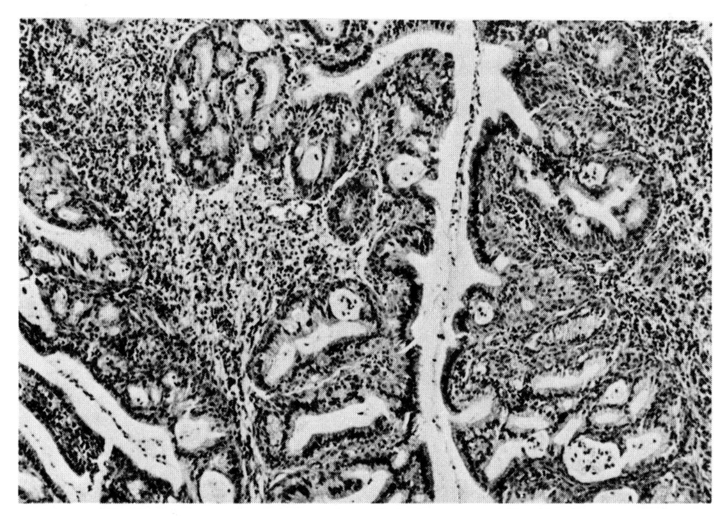

Fig. 1- Endocervix of a patient (microglandular hyperplasia) treated with estrogen-progestin combination. Note closely packed glands containing mucus. (H and E x 300)

On gross examination, these lesions may not be readily apparent sometimes but most commonly they present themselves as typical polyps, as friable polypoid excrescences, or as areas of superficial erosion in a diffusely enlarged cervix. Clinically, on the other hand, the majority of patients with these lesions are free of symptoms. In those with symptoms, the most prevalent are vaginal discharge, pruritis, postcoital or intermenstrual spotting or bleeding and dyspareunia. The lesions described above as atypical polypoid microglandular hyperplasia of the endocervix is a benign reversible process with no evidence of a pre-malignant or malignant nature. Many of these lesions may be and have been misdiagnosed as adenocarcinomas of the cervix, mainly when the microglandular pattern is associated to areas of squamous metaplasia and when the epithelial elements have a reticular arrangement or appear in solid cords displaying cellular atypia. As a matter of good practice, in most cases the probability of a true malignancy should be considered and ruled out based upon histologic and cytologic characteristics. Misdiagnosis of a polypoid lesion of this origin as malignant can lead to unnecessary and harmful treatments such as irradiation, radical surgery or chemotherapy. There is only a single case of malignancy in the endocervix. This was the case reported by Czernobilsky and coworkers (1974) of a 25 years old woman under combined contraceptive medication who developed a papillary adenocarcinoma of the endocervix. Being this the only case one has to regard it as a coincidental association. Regarding cytopatho logical studies, conflicting results on the incidence of abnormal smears in women using oral contraceptives have been reported. Most studies, however point out the lack of significant differences in the incidence (1-2%) of abnorm findings between treated and non-treated groups of patients, when retrospectively evaluated. The only important series studied suggestive of a higher incidence of cellular abnormalities, including carcinoma in situ, in the treated group, has been that of Melamed and coworkers (1969, 1973) who reported a two-fold increase in the users of oral contraceptives when compared with a control group using diaphragms. The same difference was found as well after one year as after five years of contraceptive therapy. These findings have been interpreted by the authors and by others, not as an increase in the incidence of abnormalities in the oral contraceptives group, but as a decrease in the diaphragm group, perhaps related to the blocking of penile/ cervical contact. Every other cytological study of the problem of possible association of hormonal contraceptives and cervical malignancy has proved a negative association, concluding that these drugs do not initiate or accelerate precancerous growths or overt cancer. Furthermore, some suggest that combinations with high doses of progestogen may even have cancer-inhibiting or regressive effects. In general, it has been concluded that there is no evidence at all to suggest that hormonal contraceptive preparations cause any increased incidence of premalignant or malignant diseases of the cervix. It has also been emphasized, nevertheless, that the follow-up of patients in long-term treatments should continue and that final valid conclusions should await accurate data on large samples over long periods of observation, certainly more than ten years, given the prolonged natural history of malignant disease of the cervix.

Endometrium: As the classical target organ to ovarian steroids, the endometrium is most responsive to both estrogens and progestogens, either endo genous or exogenous, showing a direct relationship between the magnitude and type of changes and the doses and proportion of the two steroids. It is also well known that the different portions of the endometrium do not react in the same manner, nor simultaneously under the same stimulus during the normal cycle. Furthermore, the endometrial response varies from cycle to cycle. These biological variations are also observed when exogenous steroids are administered, making more difficult the interpretation of specific patterns of response. In spite of the great variability in the dosage and nature of the synthetic steroids used in the different contraceptive preparations, the morphologic changes are, in the average, consistently observed in each of the cycles of a given contraceptive regime, regardless of the length of therapy. This is to say, virtually the same changes are seen in the first cycles of a patient, than after five or more years of therapy (Fechner, 1971). In general, the combination preparations produce lack of estrogenic stimulation of endometrial glands and stroma with premature but incipient secretory changes, arrested glandular secretory maturation followed by glandular atrophy, and periglandular pseudodecidual reaction of the stroma. The spiral arterioles are poorly developed and the venules and capillaries appear engorged. This general picture may very in the timing and intensity of specific changes depending upon the concentration of each steroid and the type of progestin given (either 19-nortestosterone or 17 -hydroxyprogesterone derivative), reflecting mainly the estrogenic or progestogenic activity of each compound (Fechner, 1971). After prolonged combined treatments the endometrium may become progressively atrophic, lower in height, the glands will appear scarcer and smaller and the stroma denser with minimal decidual response. The pattern of changes described above including glandular arrest and atrophy concommitantly with stromal maturation and pseudodecidualization is known as "glandular-stromal dissociation", a term used to designate succinctly the endometrial response to combined oral contraceptives. Occasionally, atypical cells with enlarged pleomorphic and hyperchromatic nuclei have been described, both in glands and stroma, similar to those seen in early pregnancy (Arias-Stella reaction). These atypical changes, however, have not a malignant character and disappear after withdrawal of the drugs. A dense cellular, or nodular sarcomatoid reaction of the stroma around atrophic glands has also been reported (Dockerty and coworkers, 1959) (Song and coworkers, 1970), including two cases diagnosed as questionable endometrial stroma sarcomas (Song and coworkers, 1959). The stromal changes seen in some patients are totally reversible soon after discontinuation of the hormones. Exfoliative cytology has also demonstrated atypical endometrial cells in the smears of aspiration of patients under combined therapy, always raising the possibility of a malignant interpretation. When the endometrium of women under a sequential contraceptive regime is studied at various times of the cycle, the pattern resembles that of a normal menstrual cycle, with slight variations. Depending upon the concentration of "unopposed" estrogen received during the first part of the cycle, the proliferative activity of the endometrium will vary and progress until the combined steroids are given in the second part. If the estrogen phase lasts for 15 days (i.e. up to day 20 of the cycle), there is no evidence of secretory activity during that time, confirming the lack of ovulation and progesterone production. When the second phase starts with estrogen-progestin combination, secretory activity of the glands is seen very soon in the form of cytoplasmic vacuolization (retarded secretion); there is no tortuosity of the glands and pseudodecidualization of the stroma is discrete. The spiral arterioles fail to coil up and venules and capillaries are not engorged. The endometrial arterioles and capillaries seem to develop endothelial hyperplasia and proliferation of the medial smooth muscle layer of arterioles, under both combined and sequential regimes (Blaustein and collaborators, 1968). This is perhaps due to estrogenic stimulation, not inhibited by progestins. When a progestogen alone, such as chlormadinone acetate or norethindrone, is given in continuous low doses, the endometrium shows a variable histology depending mainly upon the type of steroid and the dose used. Derivatives of 17 aceto-progesterone produce a more extensive deciduoid reaction after continuous therapy than derivatives of 19-nortestosterone. The magnitude of the decidual response is also directly proportional to the doses of any progestin used. The fact that these low-dosage progestins do not seem to inhibit ovulation in most cases (Martínez-Manautou and coworkers, 1967) would explain why the responses of the endometrium to simultaneous endogenous and exogenous steroids, are so variable, from a usually normal proliferative pattern during the first half of the cycle, up to irregular secretory or completely inactive patterns during the second half. Other types of contraceptives such as Depo-Provera when administered intramuscularly 200 mg. every six weeks, produces an almost atrophic endometrial pattern with a compact stroma (Ober, 1977).

A great interest has recently emerged concerning the possible carcinogenic properties of certain contraceptive methods. This is based upon several publications that call the attention to the development of tumors in cervix and vagina in daughters of women exposed to exogenous estrogens (Sandberg, 1976) (Greenwald and coworkers, 1977) and the appearance of adenocarcinoma of the endometrium in patients with Turner's syndrome (McCarroll and others, 1975) (Roberts and Wells, 1975). It seems therefore that there is a certain risk in developing adenocarcinoma of the endometrium when this type of medication is administered. (Smith and others, 1975) (Ziel and Finkle, 1975). A study of endometrial biopsies of nine patients with gonadal dysgenesis (Turner's syndrome) treated with estrogens disclosed foci of endometrial hyperplasia in two cases. All biopsies had active proliferative type of endometrium (González-Angulo and others, 1976) indicating a definite constant estrogenic effect upon this tissue. Silverberg and Makowski (1975) reported 21 cases of adenocarcinoma of the endometrium recorded in the Registry for Endometrial Carcinoma in Young Women taking Oral Contraceptive Agents. In eight of the 21 patients there were factors militating against a close relation between oral contraceptives and carcinoma and five of these eight patients had taken only or predominately combined agents. Eleven of the remaining 13 patients had taken sequential agents. Other reports are also in record pointing out an association of adenocarcinoma of the endometrium and contraceptives. (Lyon, 1975; Cohen and Deppe, 1977; Kaufman, 1976; Kelley and others, 1976; Silverberg and coworkers, 1977). A careful study of the material had revealed that quite frequently the women in these series were given this medication because abnormal uterine bleeding raising the question of a possible preexisting tumor. However in another series of 30 cases of endometrial cancer in women taking oral contraceptives, there were few individuals taking sequentials who had the common characteristics of premenopausal women prone to develop carcinoma (Rinehart and Felt, 1978 It appears then that there may be a meaningful relation between carcinoma and the sequential form of therapy. Analitical epidemiological studies however are needed to give a stronger support to this association (Rinehart and Felt, 1978).

Myometrium: The number of studies reported in the literature regarding the effects of hormonal contraceptives on the myometrium is relatively small, particularly those comprising adequate numbers of patients and significant clinico-pathologic correlations (González-Angulo and Salazar, 1973). Most publications refer to isolated or small groups of cases. Perhaps the most recent and complete review on the subject is that of Fechner (1971). The most significant effect of hormonal contraceptive preparations on the myometrium is that of changes observed in leiomyomas. It is well documented that preexisting leiomyomas can undergo changes of various kinds. Perhaps the most commonly seen is enlargement of the tumoral masses, frequently within a short period of time after initiation of the therapy and sometimes displaying severe clinical manifestations. It is considered that the increase in size is secondary to vascular congestion and interstitial edema, as well as to certain degree of hypertrophy of smooth muscle cells. The most dramatic clinical episodes are due to acute hemorrhage within the tumors associated to necrosis. These changes have been found in patients under both combination therapy and continuous low-dose progestins, particularly during the first two months of treatment. After the drug is withdrawn the size of the tumors regress even to become smaller than before treatment. All these data and previous knowledge on the effects of progesterone on uterine leiomyomas, suggest that the myometrial changes are probably induced by the progestin component of hormonal contraceptives. The lack of information regarding myometrial changes in patients under sequential therapy may be also of some significance in that respect. In addition to the enlargement and hemorrhagic necrosis of some leiomyomas under progestin-dominant therapy, histologically some of these tumors present increased cellularity and atypia, for which they have been called "atypical leiomyomas". The atypia consists of scattered foci of cells with bizarre nuclear changes including enlargement and lobulation with either vesiculation or hyperchromatism. Mitotic activity and degenerative changes are, however, minimal. The relationship of the cellular atypia of leiomyomas to the hormonal therapy is not clear. The importance of this lesion resides in the possibility of misdiagnosing it as malignant, given the morbidity associated to the treatment of leiomyosarcomas. The benign biological behavior of cellular atypical leiomyomas in general has led to disregard multinucleation and nuclear atypism as criteria for malignancy, leaving the rate of mitotic activity as the most reliable, but not infalible, parameter to predict a malignant course. The theory of malignant transformation of leiomyomas is not generally accepted and it is more commonly believed that leiomyosarcomas originate as such. Against a malignant interpretation of atypical changes in leiomyomas influenced by contraceptive hormones is the fact that the atypias are focal, discrete and not associated to increased mitotic activity (González-Angulo and Salazar, 1973).

Breast: It is well documented that breast changes occur in women currently taking contraceptives. These patients frequently note fullness of the breasts and often some nodularity and increase in size. The fullness and firmness of the breast tissue frequently appear as lobular and nodular, making the gross clinical distinction from a true tumor rather difficult. These modifications regress once the medication is withdrawn. When sequential type of therapy is used the symptoms are of a lesser magnitude but seem to persist for longer periods of time. All these effects are more marked with products containing larger amounts of estrogens especially when the treatment started shortly after delivery. Galactorrhea in variable degrees has been described in women under contraceptive therapy and also in nulliparous women who have prolonged amenorrheas following discontinuance of contraceptive medications. The histological changes observed in women taking oral contraceptives are chiefly lobular development with acinar formation and cytoplasmic secretory vacuoles resembling lactating breast. (Fig. 2) Taylor (1971) has encountered that the histologic picture of normal breast tissue in women under oral contraception is variable from woman to woman and from one part of the breast to another, and also that the changes, when present, bear no relationship to the type of hormone administered. It appears that estrogens are, at least during normal pregnancy, and perhaps during hormonal contraceptive therapy, the causative agent of breast hypertrophy and enlargement.

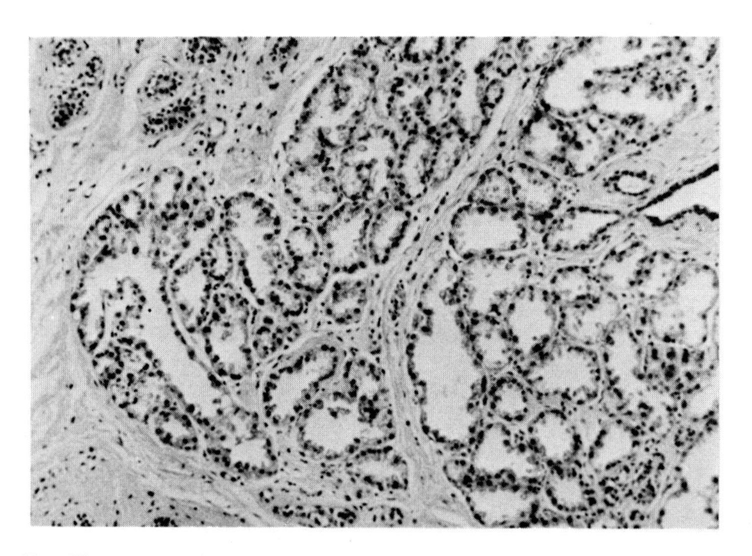

Fig. 2- Changes of lactating breast in a mammary nodule of a woman under combined contraception. (H. and E. x 300).

Histologic changes have been seen in abnormal mammary tissue under the influence of estrogen-progestin contraceptive therapy. Estrogen administration has resulted in a more florid form of a quiascent sclerosing adenosis but this has not been demonstrated with contraceptive preparations. Unusually complex duct formation was described by Brown (1970) and by Goldenberg and others (1968) in fibroadenomas of the breast without evidence of cytologic atypia. These findings, however, were not compared with corresponding control groups. Fechner (1970) made a comparative analysis of the appearance of fibroadenomas in 54 women who had been taking oral contraceptives and 54 who have never been under hormonal therapy, which revealed no morphologic differences related to the hormones, except for benign acinar hyperplasia in four nulligravid patients of the treated group. Prechtel (1970) also found no histologic differences between fibroadenomas in women taking contraceptives and those in patients who were not treated. Wiegenstein and coworkers (1971) reported the appearance of multiple fibroadenomas of the breast in 12 women on hormonal contraceptives, ranging in age from 17 to 30 years. In all but one case, the fibroadenomas were limited to one breast In three there were features of a florid type fibroadenoma with epithelial hyperplasia and secretory activity. Fechner (1970) found no specific structural differences between the breasts of 25 patients with fibrocystic disease (mammary dysplasia) under hormonal medication, and those of 25 women with the same benign changes and similar ages, who were not treated. Epithelial hyperplasia was observed in 24% of women under contraceptives and in 32% of the control group. These figures are very similar to those found by Davis and others, (1964) and Karpas and co-workers (1965) while studying hyperplasia and papillomatosis in surgical specimens of women not receiving oral contraceptives. Recent observations indicate that there is a reduced risk of benign breast diseases in women under oral contraceptives (Ory and colleagues, 1976). As to the potential risk of oral contraceptives in producing cancer of the breast, it can be stated at this time that there is no evidence indicating that these products cause malignant changes (Fechner, 1971). The known effects of estrogens on breast cancer produce concern, preoccupation and uncertainty, but the data available demonstrates that the use of oral contraceptives has no influence on the risk of breast cancer. The association of breast cancer and oral contraceptives have been rarely reported, except for sporadic cases (González-Angulo and Salazar, 1973). Ten cases of breast carcinoma studied by Fechner (1970) in women less than 35 years old who have been on oral contraceptives, represented 4.4 per cent of all breast cancers seen during the same period of three years. This proportion is practically the same (4.3%) found by Horseley and coworkers (1969) in their study of 1,564 breast cancers in the same age group, before the oral contraception period. In the same group studied by Fechner, three patients were less than 30 years of age, representing 2.1% of the breast cancers in that age group. This figure is comparable to the 1.8 to 2.0% incidence reported in patients of the same age group not under hormonal therapy. We do not have at present adequate data or direct knowledge about the pre-clinical stages of breast cancer in young women, or about the effects of prolonged estrogen therapy on the breast in that age group. However, the longest-term treatments (up to 12 years) with various steroid contraceptive drugs have failed to demonstrate any cause-effect association with the development of breast cancer.

Liver: Following the introduction of oral contraceptive drugs in 1960, and since the original report of Pérez-Meza and Shields (1960), increasing number of publications have appeared pointing out that contraceptive steroid hormones cause impairment of liver function, resulting in intrahepatic cholestasis, abnormal BSP excretion and jaundice. Information and discussion of these changes can be found elsewhere (González-Angulo and Salazar, 1973) (González-Angulo and colleagues, 1970). Tumors and non-neoplastic lesions composed of liver cells have been reported with increased frequency in women who are taking oral contraceptives (McAvoy and colleagues, 1976). Due to the fact that ethinil estradiol and estranol impair biliary secretion with a decrease in the bile flow, it has been thought that the retention of this steroids might produce toxic changes with secondary hyperplasia of hepatocites that lead to neoplastic formation (Check and colleagues, 1978) Baum and colleagues (1973) reported seven women with benign hepatic tumors who were taking oral contraceptive hormones. These cases were considered of significance because three of the patients were from a single hospital where no other benign hepatic tumors could be found since 1913. There are more than 100 cases reported in the literature and although few have been diagnosed as either hepatoma or hepatoblastoma, the majority of these lesions have been considered as benign and labeled focal nodular hyperplasia, liver cell adenoma, hamartoma and benign hepatoma (Fechner, 1977). Dramatic changes have been reported to occur in hepatic tumors in young women with contraceptive medication. The article by Mays and coworkers (1976), illustrates a large benign tumor in the right lobe of the liver that had ruptured producing massive intraperitoneal hemorrhage. The report of Mays deals with 13 young women who had been ingesting contraceptive steroids. Nine of these tumors were benign and four malignant. This increased frequency of reports has prompted a retrospective examination of liver lesions in various hospitals and institutions (Alcocer-Gregory, 1978) (Gold, Guzmán and Rosai, 1978). In a series of cases of focal nodular hyperplasia and liver cell adenoma found in the General Hospital of the National Medical Center in México City, 15 cases of focal nodular hyperplasia were found in women with no history of contraceptive medication, there was only a single woman to whom anabolic steroids were administered. There were six cases of liver all adenoma with no history of contraceptive administration. (Alcocer-Gregory, 1978) in this respect would tend to agree with Gold Guzmán and Rosai, (1978) in that it is not the time yet to draw any conclusion as to a true association to these benign lesions and contraceptives group. Experimental production of liver tumors in animals treated with oral contraceptives hormones produce variable results. These inconsistencies have not contributed to understanding the possible relationship of these hormones to hepatic tumors in humans. Focal nodular hyperplasia is characterized by a circumscribed lesion with a poor delineated capsule with central scarring and nodularity with bile ducts in the fibrous bands and at the periphery. (Fig. 3). There is no evidence of cholestasis and the glycogen is increased. Liver cell adenomas have a partial or complete capsule do not have scarring bands, nodules or bile ducts. The glycogen content is usually normal or increased. We believe as stated by Fechner (1977) that "the finding of hepatic tumors detected clinically in women under contraceptive medication is a rare event among the more than 20 million of women who have used or are using oral contraceptives".

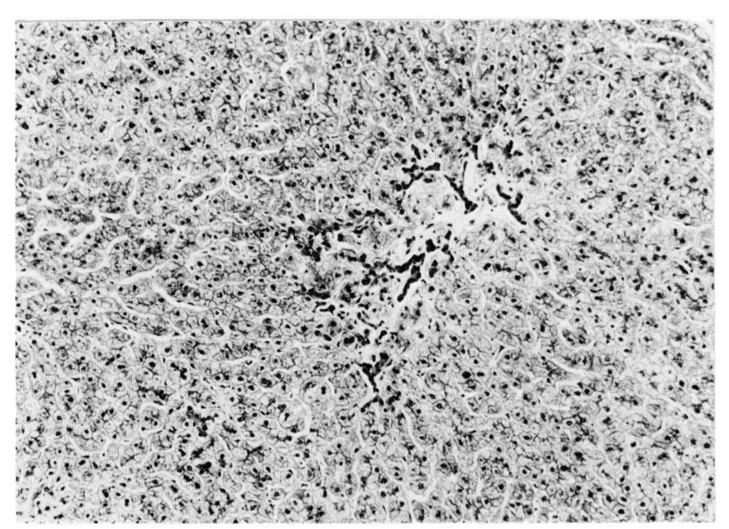

Fig. 3- Focal nodular hyperplasia in patient taking anabolic steroids. (H. and E. x 300) (Courtesy of Dr. J. Aguirre-García).

REFERENCES

Rinehart, W. and J.C. Felt (1978) Anticonceptivos orales. Population Report. Serie A, No. 4, Abril, pp A-81 - A-112.
Fechner, R.E. (1971) in Sommers, S.L. (ed), Pathology Annual 1971. Appleton-Century-Crofts, New York, p. 299.
Pincus, G. (1965) Acta Endocrinol. (Copenhagen) 18 (Suppl. 28)
Dito, W.R., and J.G. Batsakis, (1961) Obst. Gynec. 18: 1,
Zañartu, J. (1964) Int. J. Fertil. 9: 225.
Ryan, G.M., J. Craig, and D.E. Reid (1964) Amer. J. Obst. Gynec. 90: 715.
Maqueo, M., J.C. Azuela, J.J. Calderon, and J.W. Goldziehr (1966) Contraception 5: 177.
Taylor, H.B., N.S. Irey and H.J. Norris (1967) JAMA 202: 637.
Candy, J. and M.R. Abell (1968) JAMA 203: 323.
Govan, A.D.T., W.P. Black, and J.L. Sharp (1969) J. Clin. Path. 22: 84.
Nichols, T.M., and H.K. Fidler (1971) Amer. J. Clin. Path. 56: 424.
Kiriakos, M.R., L. Kempson, and N.F. Konikov (1968) Cancer 22: 99.
Czernobilsky, B., I. Kessler and M. Lancet (1974) Obst. Gynec. 43: 517.
Melamed, M.R., L.G. Koss, B.J. Flehinger, R.P. Kelisky, and H. Dubrow (1969) Brit. Med. J. 3: 195.
Melamed, M.R. and B.J. Flehinger (1973) Gynec. Oncol. 1: 290.
Dougherty, C.M. et al. (1970) Obst. Gynec. 36: 741.
Song, J., M.S. Mark, and M.P. Lawler (1970) Amer. J. Obst. Gynec. 107: 717.
Blaustein, A., L. Shenker, and R.C. Post (1968) In. J. Fertil. 13: 466.
Martínez-Manautou, J., J. Giner-Velázquez, and H. Rudel (1967) Fertil. Steril. 18: 57.
Ober, W.B. (1977) Human Path. 8: 513.
Sandberg, E.C. (1976) Amer. J. Obst. Gynec. 125: 777.
Greenwald, P., P.C. Nasca, T.A. Caputo and D.T. Janerich (1977) N.Y. State J. Med. 77: 1069.
McCarroll, A.M., D.A.D. Montgomery, J. McD. G. Harley, E.F. McKeown, and J.C. MacHenry, (1975) Brit. J. Obstet. Gynec. 82: 421-423.
Roberts, G. and A.L. Wells (1975) Brit. J. Obst. and Gynec. 82: 417-420.
Smith, D.C., R. Prentice, D.J. Thompson, W.L. Herrmann (1975) New Engl. J. Med. 293: 1164-1167.
González-Angulo, A., S. Armendarez-Sagrega, I. Ruiz de Chávez, H. Márquez-Monter, and R. Aznar. (1976) Proc. of the Thirty fourth EMSA Meet. pp. 148-149.
Ziel, H.K., W.D. Finkle (1975) New Eng. J. Med. 293: 1167-1170
Silverberg, S.G. and E.L. Makowski (1975) Obstet. Gynec. 46: 503-506.
Lyon, F.A. (1975) Am. J. Obstet. Gynec. 123: 299-301.
Cohen, C.J. and G. Deppe (1977) Obstet. & Gynec. 49: 390-392.
Kaufman, R.H., K.O. Reeves, and C.M. Dougherty (1976) J. Amer. Med. Assoc. 236: 923-926.
Kelley, H.W., P.A. Miles, J.E. Buster and W.H. Scragg (1976) Obstet. & Gynec. 47: 200-202.
Silverberg, S.G., E.L. Makowski and W.D. Roche (1977) Cancer 39: 592.
González-Angulo, A., and H. Salazar (1973) Uterine contraction. Josimovich, J.B., (ed) pp 343-380.

Taylor, H.B. (1971) Cancer 28: 1388.
Brown, J.M. (1970) Med. J. Aust. 1: 276.
Goldenberg, V.E., Wiegenstein and N.K. Mottet (1968) Amer. J. Clin. Pathol. 49: 52.
Fechner, R.E. (1970) Amer. J. Clin. Pathol. 53: 857.
Prechtel, K. (1970) Proc. of the 8th Intern. Cong. of the Inter. Acad. of Pathol. México City.
Wiegenstein, L., R. Tank, and V.E. Gould (1971) New Engl. J. Med. 284: 676.
Fechner, R.E. (1970) Cancer 25: 1332.
Davis, H.H., M. Simons, and J.B. Davis (1964) Cancer 17: 957.
Karpas, C.M., H.P. Leis, A. Oppenheim, and W.L. Mersheimer (1965) Ann. Surg. 162: 1.
Ory, H., P. Cole, B. MacMahon, and R. Hoover (1976) New Engl. J. Med. 294: 419-422.
Horseley, J.S. (1969) III Ann. Surg. 169: 839.
Pérez-Meza, R.A., and C.E. Shields, (1962) New Engl. J. Med. 267: 1137.
González-Angulo, A., R. Aznar-Ramos, H. Márquez-Monter, G. Bierzwinsky and J. Martínez-Manautou (1970) Acta Endocrinol. 65: 193.
McAvoy, J.M., R.K. Tompkins, W.P. Longmire Jr. (1976) Arch. Surg. 111: 761-767.
Check, J.H., L.C. King, A.E. Rakoff (1978) Obstet. Gynec. 52: 28s-29s.
Baum, J.K., J.J. Bookstein and F. Holtz (1973) Lancet 11: 926.
Fechner, R.E. (1977) Path. Ann. Part I Sommers S.C. and Rosen, P.P. (ed) pp 293.
Mays, E.T., W.M. Christopherson, M.M. Mahr, and H.C. Williams (1976) JAMA 253: 730.
Alcocer-Gregory, P. (1978) Receptional Thesis (Anatomic Pathology) National University of México, México City.
Gold, J.H., I.J. Guzmán, and J. Rosai (1978) Amer. J. Clin. Path. 70: 6-16.

Prediction of Drug Efficacy Potentialities and Limitations

St. Tanneberger, E. Nissen and W. Schälicke

Central Institute of Cancer Research, Academy of Sciences, German Democratic Republic

ABSTRACT

The result of every cancer chemotherapy depends on the individual tumor-host-drug relationship which should be analysed to improve the antineoplastic drug treatment by a more biology-adapted application of the drugs tailored to the patient.

Two fundamental elements of individualized cancer chemotherapy are prediction of drug efficacy before starting antineoplastic drug treatment and monitoring the drug action in the course of treatment. There are some biologically and pharmacologically acceptable approaches with improved predictive value for antineoplastic drug efficacy in experimental systems and also in man. But these approaches are complicated, time consuming and need special laboratory facilities.

In contrast to some overoptimistic evaluations in the past there is no assay for prediction of antineoplastic drug efficacy available now which could be recommended for general use in clinical practice. Most evidence for clinical usefulness has been accumulated for the hormone receptor analysis. The main obstacles we have to overcome in elaborating a biopsy-based drug prediction assay is the heterogenety of human tumors in space and time and a considerable lack of knowledge in pharmacokinetics of most antineoplastic compounds.

The data available now about monitoring of drug response using biological tumor markers are promising and mayby the situation in general is more favourable with respect to assays that can be recommended for use in clinical practice than with respect to pretherapeutic biopsy-based predictive tests. But further efforts are necessary, particularly to collect more data about the biological basis of tumor markers and, furthermore, to overcome methodical limitations. With respect to the last thermomonitoring could be a promising approach.

INTRODUCTION

A considerable number of antineoplastic drugs is now available, providing an arsenal of weapons for cancer chemotherapy.
The drugs are different in their chemical structure and the mechanism of action, the latter both in terms of biochemistry and cell kinetics. More or less the drugs are characterized according to the remission rate and survival time achieved by applying the drugs in different tumor localisations.

The target of antineoplastic drug action, as increasingly recognized during the last few years, is a dynamic inhomogenous population of 10^9 or more cells, inhomogenous in space and time and different from case to case with respect to the metabolism and, of course, to the drug sensitivity; in other words, human tumors behave individually both on cell biological and clinical level (Tanneberger, 1977). Tumor growth is the net result of a complicated balance between cell renewal, cell differentation and cell loss (Iversen, 1975).

The result of cancer chemotherapy depends on the tumor-host-drug relationship. Doubtless there is a high degree of individuality of the tumor-host-drug relationship which should be analyzed to improve the antineoplastic drug treatment by a more biology-adapted application of the drugs. Mihich (1976) formulated this demand by writing to the participants of the UICC/CICA meeting on - Human Tumor Sampling for Biochemical Pharmacological Studies of Target Determinants of Drug Action -: "The developement of chemotherapy regimens specific for an individual patient would be of major importance in view of the known limitations of treatments based on tumor-type sensitivities, limitations that are related to the variability of response among individual patients with the same clinical type of tumor."

Of course this does not mean that general chemotherapy protocols compiled in phase III studies would be of no value, particularly when considering institutions having no facilities for special biological-biochemical investigations but it does mean that efforts have to be stepped up in the future to introduce more than localisation, stage and histological type into the planning of antineoplastic drug treatment.

There are some different aspects of individualization of antineoplastic drug treatment. Two fundamental elements of indivualized cancer chemotherapy are prediction of drug efficacy before starting antineoplastic drug treatment and monitoring of the drug action during the course of treatment. A definition of the two terms, prediction and monitoring, as used in this paper is given in Fig. 1 showing the classical cell kill model as used by Skipper (1964), De Vita (1971) and others. However, it has to be mentioned that the term "monitoring" is used very often in a very wide sense to describe the follow-up of patients, regularly checking these for dissemination and relapse of the tumor.

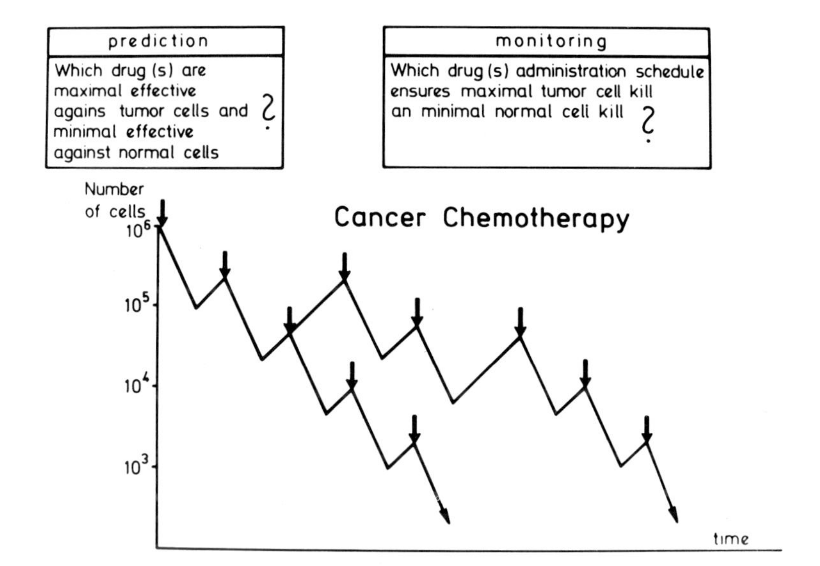

Fig. 1. Definition of prediction and monitoring of treatment response based on the cell kill model of cancer chemotherapy.

Prediction and monitoring are closely connected. Prediction is the best way of selecting the most effective drug, but if this will be not achieved for some reason or other, then monitoring of a time-limited application of the drug will give the information about their efficacy.

PREDICTION OF DRUG EFFICACY

The different approaches for antineoplastic drug prediction adopted in the last decade can be grouped in long-term tissue culture assays and short-term biopsy incubation assays. In these, the drug effects are detected either directly or indirectly by morphological or biochemical methods. Moreover cell kinetic data and also in vivo assays with diffusion chambers have been used to predict the drug sensitivity of human tumors. Also, biochemical assays are increasingly discussed which use the content of drug specific targets or intermediates in the tumor to predict the drug efficacy, particular in the case of antimetabolites (Klubes and others, 1978). Furthermore the assessment of drug distribution measured by labelled drugs could have a predictive value as demonstrated, for example, by Shani and others (1977) in an animal system.

The value of the drug prediction assays available at this time is assessed quite differently by the groups working in this field. Table 1 gives a survey about some of the main activities and about the evaluation of the assays by the authors.

TABLE 1 Prediction of Human Tumor Drug Sensitivity
Conclusion: + = valuable; - non-valuable

Approach	Method	Clinical Value patients	correl. (%)	concl.	Authors, year
long-term tissue cultures + drug	cell cultures/ morphology	102		(+)	Lickiss a.o., 1974
		188		(+)	Cobb a.o.,1964
		85	65	+	Limburg, 1973
		88	63	+	Tanneberger a.o., 1967
		41	46	+	Krafft a.o., 1973
		33	81	+	Terentieva a.o., 1976
		39		+	Izsak a.o., 1971
		201		+	Marzotko a.o., 1976
	cell cultures/ cell counting	36	92	+	Holmes a.o., 1974
	cell cultures/ autoradiography	48	100	+	Murphy a.o., 1975
		53	77	+	Zittoun a.o., 1975
	cell cultures/ DNA or RNA synthesis	450		(+)	Mitchel a.o., 1972
		55		+	Wheeler a.o., 1974
	organ cultures/ DNA synthesis	108		(-)	Tanneberger, 1977
	organ cultures/ histochemistry	10	45	+	Hecker a.o., 1976
	tumor-colony-assay/cloning efficieny	18		(+)	Salmon a.o., 1978
short-term tissue incubation + drug	cell suspension/ morphology	48	73	+	Dendy a.o., 1973
		148	55	+	Wright a.o., 1973
	biopsy/ autoradiography	25	27	-	Wolberg,1971
	cell suspension/ autoradiography	15	93	+	Thirlwell a.o., 1976
	cell suspension/ DNA or RNA synthesis	50	92	+	Andrysek, 1973
		23	50	(+)	Hirschmann,1973
		24	100	+	Possinger a.o., 1976
		24	40	+	Mattern a.o., 1976

(continued) TABLE 1 Prediction of Human Tumor Drug Sensitivity
Conclusion: + = valuable; - non-valuable

Approach	Method	Clinical Value patients	correl. (%)	concl.	Authors, year
short-term tissue incubation + drug	cell suspension/ DNA or RNA synthesis	125		+	Bastert a.o., 1975
		23	90	+	Volm, 1975
	biopsy/ DNA synthesis	21		(+)	Kaufmann a.o., 1971
	cell suspension/ enzyme assay	22	83	+	Kondo, 1971
		89	27	(+)	Di Paolo,1971
		60	70	+	Knock a.o., 1974
	tissue homogenate/ SH groups			+	Kulik, 1977
	biopsy/ DNA dependent RNA polymerase	23	45	+	Müller, 1977
	bone marrow + cell suspension/ DNA synthesis	11		(+)	Tisman, 1973
cell kinetic data	clumps/biopsy/ labeling index	34			Kirmiss a.o., 1976
	cell suspensions/ cytofluometry	23	82	+	Smets a.o., 1976
in vivo test	diffusion chamber/ cell suspension	47		+	Heckmann,1967

Our own activities in the field of antineoplastic drug prediction began approximately in 1960, and via a period of cell culture studies (Tanneberger and others, 1970) we developed an organ culture assay in which the drug action is determined by DNA synthesis measurement before and after 48 hours culture in the presence of the drugs (Tanneberger and other, 1973). To achieve a definite conclusion about the predictive value of this organ culture assay we started some randomized clinical trials, comparing no or non-predicted chemotherapy with predicted treatment with antineoplastic drugs (UICC registered trial No. 72-020; 72-053; 77-051). The most representative results were obtained with the ovarian cancer trial comparing a standardized surgical adjuvant therapy with Trenimon (one of the most recommended antineoplastic drugs when we started this trial) with the in vitro predicted long-term surgical adjuvant therapy (Nissen and others, 1978). Figure 2 shows the actual survival data of the trial evaluated according to the method of Peto (1977). There was no statisticaly significant difference between the two arms in this trial.

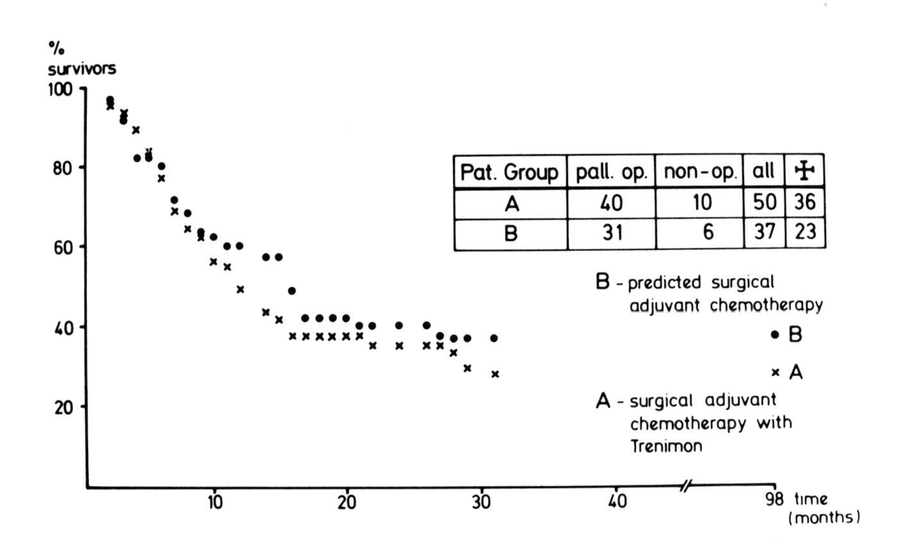

Fig. 2. Preliminary results:
Individualized postoperative chemotherapy in ovarian cancer (UICC registered trial No: 72 - 053)

A special problem of drug prediction is the pretherapeutic determination of the hormone sensitivity in breast cancer. Table 2 gives a survey about some of the most important activities in this field of research. Our own results are in good accordance with the international position concerning hormone sensitivity prediction. In a controlled clinical study involving 19 patients we found 8 tumors having a high concentration of estrogen receptors and a significant inhibition of the estrogen uptake by antiestrogens in vitro. When treated with tamoxifen, 4/6 of the receptor positive patients were responders.

TABLE 2 Prediction of breast cancer hormon sensitivity

Approach	Method	Clinical Value pats.	correl.	concl.	Authors year
suppression of prolactin level	L-dopa-test	43	80%	+	Sasaki a.o., 1976
glycolytic enzyme activity	LDH, NADP-isocitrate dehydrogenase, glucose-phosphate isomerase	30	50%	+	Hilf a.o., 1976
Barr body frequency	microscopical counting	93		-	Rajeswari, a.o., 1976
steroid hormone binding ability	dextran-charcoal-separation	580	54,4% ER^+ 6,7% ER^-	+	McGuire, 1975

(continued) TABLE 2 Prediction of breast cancer hormon sensitivity

Approach	Method	Clinical Value pats.	correl.	concl.	Authors year
steroid hormone binding ability	dextran-charcoal separation	423	55,8% ER^+ 9,6% ER^-	+	Leclercq a.o.,1977
		34	32 % ER^+	(+)	Heuson a.o. 1977
		19	66,7% ER^+	+	Heise a.o. 1977
	sucrose gradient sed.	35	66,7% ER^+ 5,9% ER^-	+	Westerberg a.o., 1978
		98	52,0% ER^+ 4,2% ER^-	+	Jensen, 1975
	gelfiltration	41	83,3% ER^+ 17,6% ER^-		

From all the data available at this time it has to be concluded that there is a positive correlation between the amount of specific estradiol binding in human mammary cancer and their response to subsequent antiestrogen therapy. However the statement of Heuson and other (1977) that all patients possibly are hormone dependent, though to a varying degree certainly is satisfied. In view of this fact, the result of the estrogen receptor assessment has to be considered as a guidline for treatment planning more than as an absolute order. Maybe, still the assessment of both estrogen and progesteron receptors as proposed by Horwitz and others (1975) will further qualify the prediction concept for hormone sensitivity.

To assess the clinical value of both antineoplastic drug prediction and measurement of hormone receptor content, we started a randomized clinical trial in 1974 comparing no surgical adjuvant chemotherapy and predicted hormone - and chemotherapy in breast cancer stage III with more than 3 lymph nodes involved.[1] Table 3 shows the preliminary results of this trial.

TABLE 3 Surgical Adjuvant Therapy of Operable Breast Cancer (Stage III, 3 + nodes; ICT-trial, ZIK 1974 - V/1978)

		N	treatment failure	local recurrence	distant metastasis	✝
O		56	21 (38 %)	18	17	4
I	resist.	11	7 (64 %)	1	7	4
	treated.	45	12 (27 %)	8	8	6(+1)
I	total	56	19 (34 %)	9	15	10(+1)

[1] Scientists involved: Dr.U.Peek, Dr.G.Marx, Dr.R.Gürtler, Dr.M.Gőrlich, Dr.E.Heise, Dr.B.Bodeck, Dr.D.Kunde

Up to now there is a clear tendency that patients being sensitive and receiving predicted treatment have the lowest rate of failure (27%), lower than patients receiving treatment and being resistant (64%) and also of the rate of failure in the group of patients with no surgical adjuvant chemotherapy (38%). Of course, if these preliminary results will be confirmed in a subsequent trial, predicted chemotherapy has to be compared with a non-predicted surgical adjuvant treatment regimen, because in the meantime Fishers (1977) and Bonadonna's (1978) results demonstrated a high probability for prolonged survival also by non-predicted surgical adjuvant chemotherapy.

Considering the whole field of pretherapeutic prediction of human tumor drug sensitivity the following conclusion seems to be justified:
There are some biologically and pharmacologically acceptable approaches with improved predictive value for antineoplastic drug efficacy in experimental systems and also in man. But these approaches are complicated, time consuming and need special laboratory facilities. In contrast to some overoptimistic evaluations in the past, there is no assay for prediction of antineoplastic drug efficacy available at present which could be recommended for general use in clinical practice. Most evidence for clinical usefulness we have for the hormone receptor analysis. The quantitative assessment of the estrogen receptor content in breast cancer should be used more often in planning breast cancer treatment. The main obstacle we have to overcome in elaborating a biopsy-based drug prediction assay is the heterogenety of human tumors in space and time and a considerable lack of knowledge in pharmacokinetics od most antineoplastic compounds.

MONITORING OF ANTINEOPLASTIC CHEMOTHERAPY RESPONSE BY MEASUREMENT OF BIOLOGICAL TUMOR MARKERS AND THEIR PREDICTIVE VALUE

As already mentioned earlier, monitoring of response is of importance in antineoplastic drug treatment in a double sense: Individualized cancer treatment demands an individual schedule for drug administration, based on monitoring of the tumor-host-drug interaction and furthermore, the monitoring of tumor response will indicate the drug efficacy within a short period of time.

During the last few years there has been a rapid development in the field of biological tumor markers and many efforts have been made to use these biological markers for monitoring treatment response (Rees 1977). Table 4 gives a survey about some of the most important approaches in this field, where clinical application particularly for cancer chemotherapy monitoring was the principle for selection.

TABLE 4 Monitoring of Treatment Response in Human Cancer

Approach	Biological/ marker	Clinical Value patients	correl. (%)	concl.	Authors, year
assay of cancer-related antigens	CEA	60		+	Klvaňa a.o.,1977
		147	80	+	Munjahl a.o.,1976
		38		(+)	Zamchek, 1976
		72		+	Pompecki a.o.,197[illegible]
	Circulating immune complexes	22		(+)	Hoffken a.o.,1977

(continued)
TABLE 4 Monitoring of Treatment Response in Human Cancer

Approach	Biological/ marker	Clinical Value patients	correl. (%)	concl.	Authors, year
assay of cancer-related antigens	CEA, AFP, HCG			+	Franchimont a.o., 1977
	AFP	26	75	+	Bourgeaux a.o., 1976
		153	94	+	Grigor a.o.,1977
assay of proteins and polypeptid-hormones	ACTH, Calcitonin			+	Franchimont a.o., 1977
	FSH	51		+	Coombes a.o.,1977
	α_1-acid glucoprotein	38	94	(+)	Rapp a.o.,1975
	fucose - protein ratio	150		(+)	Waalkes a.o.,1978
assay of key enzymes	LDH, Aldolase, Hexokinase	600		+	Klvaňa a.o.,1977
	AP (alkaline phosphatase)	51		+	Coombes a.o.,1977
	LDH Isoenzyme			+	Ziegenbein a.o., 1971
	Aldolase	200	80	+	Dallüge a.o.,1975
	RN-ase	22	95	+	Sheid a.o.,1977
	Sialyltransferase	51	80	+	Ganzinger a.o., 1977
		65	54	+	Kessel a.o.,1975
	Phosphohexoisomerase, LDH, x-glutamyl transpeptidase	147	80	+	Munjahl a.o.,1976
assay of metabolites	Hydroxyproline	27		+	Klvaňa a.o.,1977
	Creatinine	181		+	Gielen a.o.,1976
		31		+	Powles a.o.,1975
assay of polyamines	Polyamines:	56	60	+	Durie a.o.,1977
	Putrescin,Spermin,Spermidine	42	63	+	Russel a.o.,1975
				+	Jänne a.o.,1978
assay of free DNA	^{125}J Desoxyuridine	173		+	Leon a.o., 1977

Doubtless several of the biological markers described are of clinical importance. In spite of this, the remark of Lokich (1978) is quite correct. He underlined that tumor markers as sequential monitors or quantitative measures of disease and response to therapy have similarly demonstrated limited utility, related to the heterogeneity of the tumors and to the lack of a precise quantitative relationship between circulating levels of tumor markers and tumor volume or mass. It has also to be mentioned that for example in case of CEA the marker behaviour seems to reflect a reduction of tumor mass only if the pretherapeutic level is highly abnormal. This is shown in Table 5, demonstrating one of our studies on the CEA serum level in colorectal cancer(Drs.K.Jacobasch,G.Pilgrim,W.Seifart,R.Ziegenbein)

TABLE 5 Dependency of post-operative CEA-level behaviour from the pre-operative CEA value in colorectal cancer

CEA level	Number of patients	post-operative decrease of CEA-level	post-operative increase or no change of CEA level
> 2.5 ng/ml	21	14 (66.7 %)	7
> 5.0 ng/ml	13	13 (100 %)	-
< 5.0 ng/ml	8	1 (12.5 %)	7

The data about monitoring of treatment response available to date are promising, and maybe the situation in general is more favourable with respect to assays that can be recommended for use in clinical practice than with respect to pretherapeutic biopsy-based predictive tests. But further efforts are necessary, particularly to collect more data about the biological basis of tumor markers and furthermore to overcome methodical limitations. Another field of efforts should be data processing. Here the quantitative approach to determine disease response during therapy by Woo and others (1978) seems to be a very promising one.

PERSPECTIVES FOR PREDICTION OF DRUG EFFICACY

As underlined several times in this paper the main obstacles to drug prediction are

- the heterogeneity of human tumors with respect to the cell composition and the tumor cell behavior limiting the chance of developing biopsy-based in vitro assays
- the lack of knowledge about biological markers which reflect tumor growth and tumor destruction suitable for monitoring of treatment response
- the methodical limitations particularly in human tumor sampling, routine determination of biological markers etc. In particular, the complete lack of methods for continuous measurment of drug response has to be mentioned.

Further efforts should be taken to overcome these obstacles. There is a considerable number of intelligent approaches to do so. One approach now under investigation in our institute should be mentioned at the end of this paper.

Basing on the idea that maybe not only biochemical markers but also biophysical markers should be used to try monitoring of tumor response we started a programme on thermomonitoring in 1977. Thermomonitoring means continous measurement of skin temperature and body core temperature before, during and after treatment with antineoplastic drugs. The tentative hypothesis for this approach consists in the assumption that cell multiplication and cell kill is accompanied by thermic phenomenons particularly by release of pyrogenic substances. The pyrogenic substances with high probability influence the position of the homoiothermy regulation mechanism. It seems possible that a high sensitive thermomonitoring considering frequency and amplitude of the thermoregulation and taking into account the circadian rythm, performed under strictly standardized conditions could reflect cell proliferation and cell kill. Of course it is necessary in such an experiment to differentiate which cells are proliferating and which cells are killed. Figure 3 shows schematically our system for measurement.

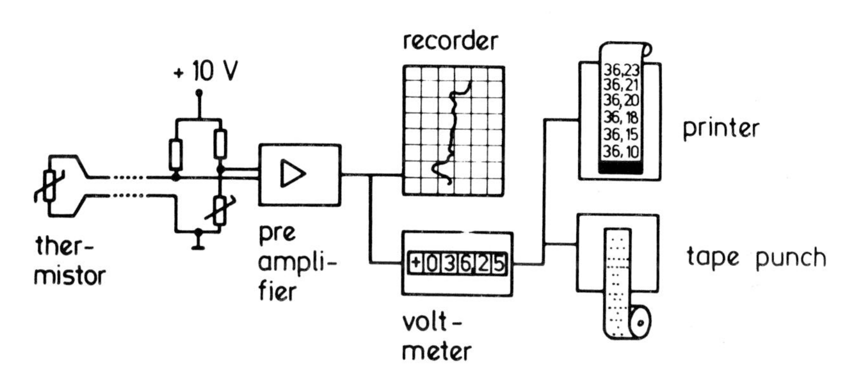

Sensitivity : 1/100 °C ≙ 1/100 V
measuring time : 30 - 60 min
interval of measurements : ~ 20 sec.

Fig. 3. Measuring arrangement for thermo-monitoring

Of course, thermomonitoring is in a very early stage of development but maybe this approach will contribute to the solution of the prediction and monitoring problem. One of the main advantages we see now in the countinous way of working provided by thermomonitoring.

REFERENCES

Andrysek, O. (1973). Sensibilitätstestungen von Zytostatika an menschlichen Tumoren mit ^{125}J-Jododeoxy-uridin. In G.Wüst (Ed.), Aktuelle Probleme der Therapie maligner Tumoren, Georg Thieme, Stuttgart, pp. 80-85.

Bastert, G., H.Schmidt-Matthiesen, R. Gerner, R.T. Michel, D. Nord, and G. Leppien (1975). In vitro Testung der Sensibilität von Mamma Karzinomen gegen Zytostatika. Dtsch. med. Wschr., 100, 2035-2043.

Bonadonna, G., P. Valagussa, A. Rossi, R. Zucali and U. Veronesi (1978). Actual data on efficacy of surgical chemotherapy with CMF in breast cancer. Arch. Geschwulstforsch., 48 (in press).

Bourgeaux, C., N. Martel, P. Sizaret, J. Guerrin (1976). Prognostic value of Fetoprotein Radioimmunoassay in surgical treated patients with embryonal cell carcinoma of the testis. Cancer, 38, 1658-1660.

Cobb, J.P., D.G. Walker (1964). Studies on human melanoma cells in tissue culture. II. Effects of several cancer chemotherapeutic agents on cytology and growth. Unio internationalis contra

cancerum Acta, 20, 206-211.
Coombes, R.C., J. Powles, J.C. Gazet, H.T. Ford, A.G. Nash, J.P. Sloane, C.J. Hillgard, P. Thomas, J.W. Keyser, D. Marcus, N. Zinberg, W.H.Stimson, and A.M. Neville (1977). A biochemical approach to the staging of human breast cancer. Cancer, 40, 937-944.
Dallüge, K.H., H.-J. Eichhorn, R. Ziegenbein, J. Hüttner und R. Jacob (1975). Die Wertigkeit von Serumfermentuntersuchungen für die Erkennung von Rezidiven und Metastasen bösartiger Geschwülste. Dt. Gesundh.-Wesen, 30, 1090-1096.
Dendy, P.P., M.P.A. Dawson, and D.J. Honess (1973). Studies on the drug sensitivity of human tumor cells in short term culture. In G. Wüst (Ed.), Aktuelle Probleme der Therapie maligner Tumoren, Georg Thieme, Stuttgart, pp. 34-34.
De Vita, V.T.(1971). Cell kinetics and the Chemotherapy of cancer. Cancer Chemother. Rep., 2, Part. 3, 23-33.
Di Paolo, J.A. (1971). Analysis of an individual chemotherapy assay system. Natl. Cancer Inst., Monogr. 34, 240-247.
Durie, B.G.M., S.E. Salmon, and D.H. Russell (1977). Polyamines as markers of response and disease activity in cancer chemotherapy. Cancer Res., 37, 214-221.
Fisher, B., A.Glass, C. Redmond, E.R. Fisher and B. Borton (1977). L-Phenylalanin Mustard (L-PAM) in the management of primary breast cancer. Cancer, 39, Supplement, 2883-2903
Franchimont, P., P.F. Zangerle (1977). Present and future: Clinical relevance of tumour marker. Europ. J. Cancer, 13, 637-646
Ganzinger, U., F. Dorner, F.M. Unger, K. Moser, und K. Jentzsch (1977). Erhöhung der Serum-Sialyltransferase bei menschlichen Malignomen. Grundlage eines neuen Diagnostikums ? Klin.Wschr., 55, 553-555.
Gielen, F., J. Dequeker, A. Drochmans, J. Wildiers, and M. Merlevede (1976). Relevance of hydroxpyroline excretion to bone metastasis in breast cancer. Br. J.Cancer, 34, 279-285.
Grigor, K.M., S.I. Detre, J. Kohn, and A.M. Neville (1977). Serum α-Fetoprotein levels in 153 male patients with germ cell tumors Br.J.Cancer, 35, 52-58.
Hecker, D., G. Saul und G. Wolf (1976). Untersuchungen zum Einsatz von Zell- und Organkulturen bei Zytostatikasensibilitätstestungen an menschlichen Ovarial-Karzinomen. Arch. Geschwulstforsch., 46, 34-43.
Heckmann, U. (1967). Neue Möglichkeiten einer Resistenztestung menschlicher Karzinomgewebe gegen Zytostatika im in vitro Test. Dtsch. med. Wschr., 92, 932-943.
Heuson, J.C., E. Longeval, W.H. Mattheiem, M.C. Deboel, R.J. Sylvester, and G. Leclercq (1977). Significance of quantitative assessment of estrogen receptors for endocrine therapy in advanced breast cancer. Cancer, 39, 1971-1977.
Hirschmann, W.D. (1973). Experimentelle Ergebnisse der Testung von zytostatischen Substanzen an menschlichen Leukämiezellen in Relation zur in vivo Sensibilität. In G. Wüst (Ed.), Aktuelle Probleme der Therapie maligner Tumoren. Georg Thieme, Stuttgart, pp. 108-115.
Hoffken, K., I.D. Meredith, R.A. Robins, R.W. Baldwin, Ch.J. Davies, and R.W. Blamey (1977). Circulating immune complexes in patients with breast cancer. British Medical Journal, 2, 218-220.
Holmes, H.L., and J.M. Little (1974). Tissue culture microtest for predicting response of human cancer to chemotherapy. Lancet II, 985-986.

Horowitz, K.B, W.L. McGuire, O.H. Pearson, and A. Segaloff (1975), Predicting response to endocrine therapy in human breast cancer: A hypothesis. Sci., 189, 726-727.

Izsak, F.Ch., E. Eylan, A. Gazith, J. Shapiro, S. Nattarin, and Ch. Raanani (1971). Growth inhibiting effects of cytotoxic agents on human tumor bearing normal tissue in vitro. Europ.J.Cancer, 7, 33-39.

Iversen, O.H. (1975). Kinetic characterization of malignant tumors. In Excerpta Medica International Congress Series No. 375, Proceedings of the sixth Int.Symposium on the biological characterization of human tumors, Copenhagen, May.

Jänne, J., H. Pösö and A. Raina (1978). Polyamines in rapid growth and cancer. Biochim. Biophys. Acta, 473, 241-293.

Kaufmann, M., M. Volm, und K. Goertler (1971). Zur Sensibilitätstestung maligner Tumoren des Menschen gegenüber Zytostatika. Klin. Wschr., 49, 219-224.

Kessel, D., and J. Allen (1975). Elevated sialyltransferase in the cancer patients. Cancer Res., 35, 670-672.

Kirmiss, K., R. Gürtler, P. Langen, R. De Heureuse und K. Arndt (1976). Die Bestimmung des ^{3}H-Labeling Indexes von Knochenmarkzellen zur Kontrolle von Therapieerfolg und Krankheitsverlauf bei akuten Leukosen. Dt. Gesundh.-Wesen, 31, 2056-2060.

Klubes, Ph., I. Cerna, K. Conelly, G.W. Geelhoed, and H.G. Mandel (1978). Effects of 5-Fluorouracil on human colon carcinoma and solid rat walker 256 carcinosarcoma: Evaluation as in vitro predictors of clinical response. Cancer Treatment Reports, 62, 1065-1073.

Klvaňa M., I.Pokorná, L. Jurga, M. Wágnerová(1977). Monitoring problems in chemotherapy of cancer. III. Congressus Oncologicus Cechoslovacus cum Participatione Internationali, Bratislavae, 26.-29.X.

Knock, F.E., R.M. Galt, Y.T. Oester, and R. Sylvester (1974). In vitro estimate of sensitivity of individual human tumors to antitumour agents. Oncology, 30, 1-22.

Kondo, K. (1971). Prediction of response of tumor and host to cancer chemotherapy. Natl. Cancer Inst. Monogr. 34, 251-256.

Krafft, W., W. Preibsch und F. Marzotko (1973). Erfahrungen mit dem Onkobiogramm bei gynäkologischen Karzinomen. Arch.Geschwulstforsch., 41, 241-247.

Kulik, G.I., W.S. Korol und R.I.Kadetzki (1976). Bestimmung der individuellen Chemotherapie gegenüber antineoplastischen Pharmaca (russ.) Mitteilungen der Akademie der Wissenschaften der UdSSR - Ministerium für Gesundheitswesen.

Leon, S.A.. B. Shapiro, D.M. Sklaroff, and M.J. Yaros (1977). Free DNA in the serum of cancer patients and the effect of therapy. Cancer Res., 37, 646-650.

Lickiss, J.N., K.A. Cane, and A.G. Baikie (1974). In vitro drug selection in antineoplastic chemotherapy. Europ. J. Cancer, 10, 809 -815.

Limburg, H. (1973). Selektion von Zytostatika bei gynäkologischen Tumoren in der Gewebekultur. In Aktuelle Probleme der Therapie maligner Tumoren, G. Wüst (Ed.), Georg Thieme, Stuttgart, pp. 7-17.

Lokich, J.J. (1978). Tumor markers: Hormones, antigens, and enzymes in malignant disease. Oncology, 35, 54-57.

Mattern, J., M. Kaufmann, K. Wayss, M. Volm, M. Kleckow, M. Mosthagi, and I.Moykopf (1976). Clinical correlates of in vitro effect of adriamycin on advanced lung carcinoma. Klin.Wschr., 54, 665-673

Marzotko, F., W. Krafft, W.Preibsch und M. Schröder (1976). Die in

vitro Testung und klinische Erfahrung mit 5-Fluorouracil bei gynäkologischen Tumoren. Arch. Geschwulstforsch., 46, 140-149.
Mihich, E., D.J.R. Laurence, D.M. Laurence and S. Eckardt (1978). UICC Workshop on Human Tumor Sampling for Biochemical Pharmacological Studies of Target Determinants of Drug Action. UICC Technical Report (in press).
Mitchel, J.S., P.P. Dendy, M.P.A. Dawson, and T.K. Wheeler (1972). In vitro testing of anticancer drug. Lancet, I, 955-958.
Müller, W.E.G. (1977). Bleomycin-sensitivity test-application for human cell carcinoma. Cancer, 40, 2787-2792.
Munjahl, D., P.L. Chawla, J.L. Lokich, and N. Zamcheck (1976). CEA and Phosphohexoisomerase, gamma glutamyl transpeptidase and Lactatdehydrogenase levels in patients with and without liver metastase. Cancer, 37, 1800-1807.
Murphy, W.K., R.B. Livingston, V.G. Ruiz, F.G. Gercovich, S.L. George, J.S. Hart, and E.J. Freireich (1975). Serial labeling index determination as a predictor of response in human solid tumors. Cancer Res., 35, 1438-1446.
Nissen, E., St. Tanneberger und A. Projan (1978). Prediction in human tumor chemotherapy. Arch. Geschwulstforsch., 48, (in press)
Peto, R., M.C. Pike, P. Armitage, N.E. Breslow, D.R. Cox, S.V. Howard, N. Mantel, K. McPherson, J. Peto and P.S. Smith (1977). Design and analysis of randomized clinical trial requiring prolonged observation of each patient. II. Analysis and examples. Br.J.Cancer, 35, 1-39.
Pompecki, R., G. Schröder, M. Garbrecht und H. Frahm (1978). Carcinoembryonales Antigen (CEA) im Plasma bei Patienten mit metastasierendem Mammakarzinom unter endokriner und zytostatischer Therapie. Dtsch. med. Wschr. 103, 620-622.
Possinger, K., R. Hartenstein, H. Ehrhart (1976). Resistenztestung von menschlichen Tumoren gegenüber Zytostatika. Klin.Wschr. 54, 349-361.
Powles, T.J., C.L. Leese, and P.K. Bondy (1975). Hydroxyproline excretion in patients with breast cancer and response to treatment. Br. med. J., II, 164-168.
Rapp, W., M. Heim, J.G. v.Mukulicz-Radecki, and R. Ludwig (1975). α_1-acid glucoproteins in gastric cancer juice. Klin. Wschr., 53, 139-141.
Rees, L.H. (1977). Laboratory testing for cancer.In G. Mathé (Ed.), Tactics and strategy in cancer treatment. Springer, Berlin, Heidelberg, New York, pp. 39-40.
Russel, D.H., B.G.M. Durie, and S.E. Salmon (1975). Polyamine as predictors of success and failure in cancer chemotherapy. Lancet, II, 797-799.
Salmon, S.E., A.W. Hamburger, B. Soehnlein, B.G.M. Durie, D.S. Alberts, and Th.E. Moon (1978). Quantitation of differential sensitivity of human-tumor stem cells to anticancer drugs. New England Journal of Medicine, 298, 1321-1327.
Shani, J. and W. Wolf (1977). A model for prediction of chemotherapy response to 5-Fluorouracil based on the differential distribution of 5- [^{18}F] Fluorouracil in sensitive versus resistant Lymphocytic leukemia in mice. Cancer Research, 37, 2306-2308.
Sheid, B.S., T. Lu, L. Pedrinan, and J.H. Nelson (1977). Plasma ribonuclease - a marker for the detection of ovarian cancer. Cancer, 39, 2204-2208.
Skipper, H.E., F.M. Schabel, and W.S. Wilcox (1964). Experimental evaluation of potential anticancer agents. XIII. On the criteria and kinetics associated with "curability" of experimental leukemia. Cancer Chemotherapy Reports, 35, 1-111.

Smets, L.A., E. Mulder, F.C. de Waal, F.J. Cleton, and J. Blok (1976). Early response to chemotherapy detected by pulse cytophotometry. Br. J. Cancer, 35, 153-161.

Tanneberger, St. und G. Bacigalupo (1967). Die Benutzung von Zellkulturen zur Ermittlung der Sensibilität menschlicher Tumoren. Dt.Gesundh.Wesen, 22, 11-19.

Tanneberger, St. und G. Bacigalupo (1970). Einige Erfahrungen mit der individuellen zytostatischen Behandlung maligner Tumoren nach prätherapeutischer Zytostatika-Sensibilitätsprüfung in vitro (Onkobiogramm). Arch. Geschwulstforsch., 35, 44-53.

Tanneberger, St. und A. Mohr (1973). Biological characterization of human tumours by means of organ culture and individualized cytostatic cancer treatment. Arch. Geschwulstforsch., 42, 307-315.

Tanneberger, St. (1977). Individual cell biology: Basses and prospects for a rational cancer chemotherapy. Arch. Geschwulstforsch., 47, 755-765.

Terentieva, T.G., T.G. Bukhny, A.F. Durnov, and M.A. Ivanitzkaja (1976). On the role of individual sensitivity of children with neuroblastoma to various cytostatics (russ.). Antibiotiki, 21, 1011-1015.

Thirlwell, M.P., R.B. Livingston, W.K. Murphy, and J.S. Hart (1976). A rapid method in vitro labelling index - for prediction response of human solid tumors to chemotherapy. Cancer Res., 36, 3279 -3283.

Tisman, G., V. Herbert, and H. Edlis (1973). Determination of therapeutic index of drugs by in vitro sensitivity tests using human host and tumor cell suspensions. Cancer Chemotherapy Reports, 57, 11-19

Volm, M. (1975). Onkobiogramm und Krebschemotherapie: Realität oder Hoffnung? Langenbecks Arch. Chirurgie, 339, 4-12.

Waalkes, T.Ph., Ch.W. Gehrke, D.C. Tormey, K.B. Woo, K.C. Kuo, J.S. Snyder, and H.Hansen (1978). Biologic markers in breast carcinoma. IV. Serum fucose-protein ratio. comparisons with carcinoembryonic antigen and human chorionic gonadotrophin. Cancer, 41, 1871-1882.

Wheeler, T.K., P.P. Dendy, and A. Dawson (1974). Assessment of an in vitro screening test of cytotoxic agents in the treatment of advanced malignant disease. Oncology, 30, 362-376.

Wolberg, W.H. (1971). The realtion of thymidine labeling index in human tumors in vitro to the effectivness of 5-fluorouracil chemotherapy. Cancer Res. , 31, 448-453.

Wright, J.C. and D. Walker (1973). Tissue culture as a target test model for sensitivity of chemotherapeutic agents on tumors. In G. Wüst (Ed.), Aktuelle Probleme der Therapie maligner Tumoren. Georg Thieme, Stuttgart, pp. 17-28.

Woo, K.B., T.Ph. Waalkes, D.L. Ahmann, D.C.Tormey, Ch.W. Gehrke, and V.T. Oliverio (1978). A quantitative approach to determining disease response during therapy using multiple biologic markers. Cancer, 41, 1685-1703.

Zamcheck, N. (1976). A summary of the present status of CEA in diagnosis, prognosis and evaluation of therapy of colonic cancer. Bulletin du Cancer, 63, 463-472.

Ziegenbein, R. und K. Rieche (1971). LDH-Isoenzyme im Serum in der Verlaufskontrolle bei Tumorkranken. Dt.Gesundh.Wesen, 26, 1337.

Zittoun, R., M. Bonchard, J.Pacquet-Danis, M. Percia-Du-Sert, and J.Bousser (1975). Prediction of response to chemotherapy in acute leucemia. Cancer, 35, 507-514.

Hormone Receptors as Indicators of the Biological Properties of Neoplastic Tissues

C. Levy, P. Robel, J. P. Wolff, J. C. Nicolas and E. E. Baulieu

Laboratoire Hormones, 94270 Bicêtre, France

ABSTRACT

Biochemical tests of hormone responsiveness are definitely preferable to single measurements of hormone receptors. Endometrial cancer is taken as example. Estradiol and progesterone receptors, ornithine-decarboxylase and 17β-hydroxysteroid dehydrogenase activities have been measured in small biopsy samples of the tumor before and after administration of an antiestrogen. The responses observed, in particular the increase of progesterone receptor, allow a more rationale approach to hormone therapy.

Keywords : Endometrial adenocarcinoma. Estradiol receptor. Progesterone receptor. Ornithine-decarboxylase. 17β-hydroxysteroid dehydrogenase. Tamoxifen.

INTRODUCTION

In the recent years, a renewal of interest in endocrine therapy of potentially hormone dependent cancers has resulted from basic investigations which have led to the development of assays that hope to determine with confidence those patients which will or will not respond to endocrine therapy. This advance is largely the result of the major efforts to understand the subcellular biochemical pathways of hormone action on both normal and neoplastic tissues.

Normal target tissues contain specific intracellular receptors for steroid hormones. These receptors sites are responsible for the initial interaction between the hormone and the cell, and function to trigger the biochemical chain of events characteristic of the particular hormone (Jensen and DeSombre, 1972 ; Baulieu and co-workers, 1975).

Hormone responsive tumors always contain receptors whereas hormone negative tumors are generally autonomous (McGuire, Raynaud and Baulieu, 1977). At present the most well defined and clinically relevant studies are focused on breast cancer. However the presence of both estradiol and progesterone receptors in mammary tumor tissue is not sufficient to identify the 20-40 % of breast cancer patients who will actually benefit from endocrine therapy, since at most two thirds of receptor positive cases will be responsive (Horwitz and co-workers, 1975).

Knowing the present limitations of single receptor assays, we now propose to get a direct demonstration of responsiveness <u>in vivo</u>, by observing the biochemical changes provoked in the tumors by the administration of hormone or anti-hormone.

We have selected endometrial adenocarcinoma for such investigation. What are the reasons for that choice : First, although this disease occurs most frequently in postmenopausal women, epidemiological studies have shown an increased risk in women who have been exposed to unopposed estrogens for a long period of time. Hormone related risk factors (like obesity and diabetes) which are associated with increased levels of circulating estrone, have been identified (Gusberg, 1976 ; McDonald and Siiteri, 1974). Also, several reports indicate that advanced disease can undergo remissions after treatment with high doses of progestagens (Kohorn, 1976).

In addition, it is known that endometrial carcinomas contain estradiol and/or progesterone receptors (Crocker, Milton and King, 1974 ; Evans, Martin and Hähnel, 1974 ; Gustafsson and co-workers, 1977) and quantitative estimates have been published (Gurpide, Gusberg and Tseng, 1976 ; Pollow and co-workers, 1975c ; Young, Ehrlich and Cleary, 1976).

In most patients, two biopsy samples were obtained within acceptable ethical and practical conditions, the first one to confirm the diagnosis and the second during surgery or immediately before radium therapy. Hormone and antihormone (tamoxifen) was administered for the 5-7 days between biopsies and its effects on the second biopsy evaluated. In addition to receptor concentrations, the activities of 17β-hydroxysteroid oxidoreductase and ornithine-decarboxylase, were measured. The latter enzyme provided a convenient marker (Tabor and Tabor, 1976) for the possible inhibitory effect of tamoxifen on tumor growth.

MATERIALS AND METHODS

All patients were postmenopausal women and were studied at the department of gynecology of the Institut Gustave-Roussy. Twenty-five patients with histologically confirmed adenocarcinoma and 1 case of mulleroblastoma were studied.

Estradiol and progesterone receptors were measured in fresh biopsy samples, collected in ice cold medium and processed less than 60 min after removal, by an exchange technique allowing measurement of total (empty and filled) estradiol and progesterone receptor sites in the cytosol and nuclei (Bayard and co-workers, 1978). The measurement of nuclear receptor sites has been improved recently by the use of a glass fiber filter exchange technique (C. Levy, B. Eychenne and P. Robel, submitted) and a complete study can be performed on $\leqslant$ 50 mg wet weight specimens.

Seventeen-β-hydroxysteroid oxidoreductase activity was measured in the 105,000 g pellet according to Pollow and co-workers (1975a) and ornithine-decarboxylase was determined in the cytosol according to Jänne and Williams-Ashman (1971).

The endometrial cancer cases were classified according to their state of differenciation as : Grade I - well differenciated ; Grade II - moderately well differenciated ; Grade III - poorly differenciated ; Grade IV - undifferenciated tumors.

After the first biopsy 14 patients took orally 40 mg of tamoxifen and 3 patients received 250 mg of medroxyprogesterone acetate i.m. daily for 5 to 7 days. The second biopsy was performed 12-18 h after the last dose of hormone.

RESULTS

Estradiol and Progesterone Receptors Concentrations in Endometrial Carcinoma

Before any hormone administration all biopsy samples contained measurable though variable concentrations of estrogen receptor (Table 1), in the same range as in the

TABLE 1 Sex Steroid Receptors and 17β-Hydroxysteroid Oxidoreductase

Grade	Case N°	Age	Estradiol receptor T	Estradiol receptor C/N	Progesterone receptor T	Progesterone receptor C/N	17β
I	65	57	.5	.4	.2	1.3	3
	23	55	.6	> 10	.8	ND [X]	ND
	59	51	.9	2.8	1.8	16.3	13
	69	64	1.7	0	2.7	0	1
	67	76	7.4	3.8	1.7	> 10	99
	32	62	8.1	7.1	1.4	1.4	273
mean			3.2	3.0	1.5	5.8	78
± s.d.			± 3.5		± .9		± 116
II	28	69	.5	1.5	.4	.3	4
	18	71	.7	4.6	.1	> 10	0
	21	54	.7	1.5	1.2	1.9	ND
	57	66	.8	.7	.5	6.4	1
	55	67	1.0	.5	.1	ND	11
	71	43	1.2	.6	1.4	.3	7
	4	49	1.5	1.9	1.3	2.0	0
	26	73	1.6	> 10	.8	8.0	7
	16	76	1.6	> 10	1.0	1.4	7
	53	60	1.8	1.1	.7	> 10	67
	51	65	2.5	> 10	2.1	> 10	74
	61	69	4.0	3.5	3.1	> 10	62
	35	53	4.4	> 10	.9	3.6	21
	46	67	5.7	3.2	14.4	> 10	20
	45	47	5.7	4.2	12.0	> 10	49
mean			2.3	3.4	2.7	14	[illegible]
± s.d.			± 1.8		± 4.4		± 28
III + IV	17	64	.8	4.5	.3	> 10	8
	40	58	1.1	6.1	1.2	8.6	11
	30	56	3.1	> 10	.7	.6	1
	34	75	3.2	> 10	.2	3.8	3
mean			2.1	9.5	.6	2.7	6
± s.d.			± 1.3		± .4		± 5
Late proliferative (mean ± s.d.)			2.2 ± 1.3	1.1	3.1 ± 1.4	5.5	
Late secretory (mean ± s.d.)			.9 ± 1.0	1.5	1.1 ± .6	2.6	

Estradiol and progesterone totals (T, pmol/mg DNA) were the sum of cytosol (C) and nuclear (N) receptors. The 17β-hydroxysteroid oxidoreductase activity (17β) was expressed in pmol of estrone produced/min/mg protein.

X Not determined.

endometrium of normal women during the late proliferative phase, a remarkable result considering the low amount of circulating estrogen in postmenopausal women. Progesterone receptor was also found in all the samples investigated, although at very low concentrations, comparable to those found in the late secretory phase of normal menstrual cycle. The majority of both estradiol and progesterone receptors were found in the cytoplasmic fraction. It was not determined whether or not nuclear receptor sites were occupied by endogenous hormones.

No clearcut differences in estradiol and progesterone receptor concentrations were observed according to the degree of tumor differenciation, although the more differenciated tumors tended to have higher receptor concentrations especially of the progesterone receptor. The tumors which contained the higher concentrations of one receptor often had rather elevated concentrations of the other receptor. The 17β-hydroxysteroid oxidoreductase (Table 1) and ornithine-decarboxylase activities (data not shown) seemed to be directly correlated with the total concentrations of steroid receptors and not to depend upon the degree of tumor differenciation.

In Vivo Effects of Hormones and Antihormones

Estradiol and progesterone receptor levels and 17β-hydroxysteroid oxidoreductase activity were measured in 3 patients before and after 5-7 days of medroxyprogesterone acetate. A large decrease in estradiol and progesterone receptors concentrations and an important increase in 17β-hydroxysteroid oxidoreductase activity followed progestagen administration. Similar results have been obtained in laboratory animals (Gorski and co-workers, 1968 ; Brenner, Resko and West, 1974) and normal human endometrium (Tseng and Gurpide, 1975a).

In 14 patients who received tamoxifen an increase in estradiol receptor was induced in about half of the cases, and an increase in progesterone receptor in all but one case (Fig. 1). Ornithine-decarboxylase activity was measured in 7 cases. The activities measured corresponded to published values for actively growing cells and were unchanged by tamoxifen (Tabor and Tabor, 1976 ; Kaye, Isakson and Lindner, 1971).

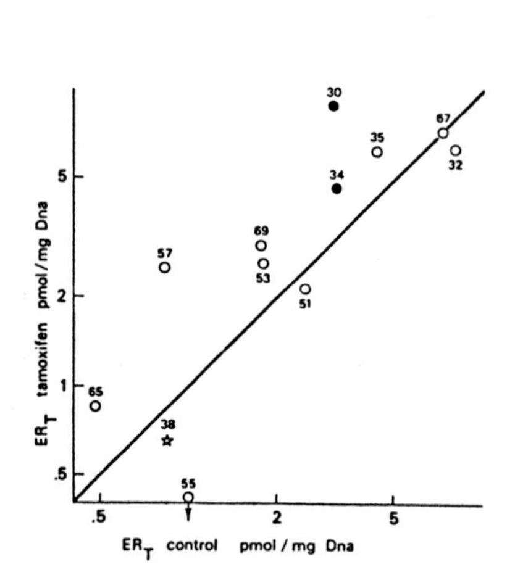

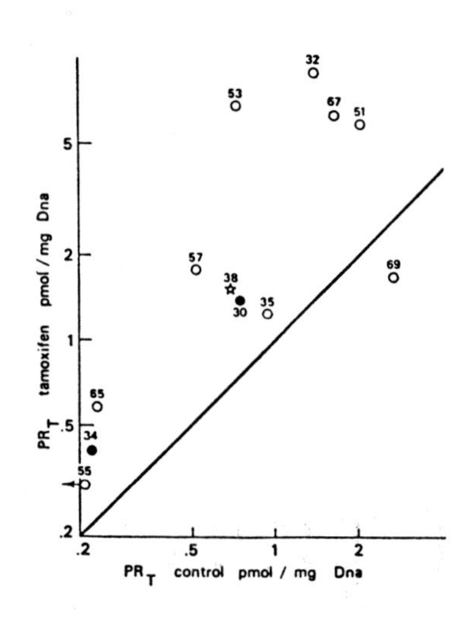

Fig. 1 Effects of tamoxifen on sex steroid receptors. Total concentrations (pmol/mg DNA) of estradiol receptor (ER_T) and of progesterone receptor (PR_T), were measured before and after tamoxifen (40 mg/day for 5 to 7 days). o well differenciated adenocarcinoma ; (Grades I and II), ● poorly differenciated adenocarcinoma ; (Grade III and IV), ✫ mulleroblastoma.

No obvious differences were observed in the histology of biopsy samples taken before and after hormone or anti-hormone.

DISCUSSION

Widespread concentrations of estradiol and progesterone receptor have been found in endometrial adenocarcinomas. It is not known to what extent they reflect the cellular heterogeneity of tumor samples. In this study all cases investigated contained measurable concentrations of estradiol receptor (> 0.5 pmol/mg DNA), which were definitely higher than expected in post menopausal women whereas very low and probably insignificant concentrations ($\sim$ 0.1 pmol/mg DNA) of progesterone receptor were found in several cases. The levels of both receptors correlate poorly with the degree of tumour differenciation (Gustafsson and co-workers, 1977 ; Gurpide, Gusberg and Tseng, 1976 ; Young, Ehrlich and Clearly, 1976). The similar lack of correlation found for the 17β-hydroxysteroid oxidoreductase activity agrees with the observations of other investigators (Gurpide, Gusberg and Tseng, 1976).

The presence of hormone receptors is a known prerequisite for hormone action. However the relationship between their concentrations and the extent to which a tumor will respond to hormone treatment has not yet been established. The question then arises as to whether or not such cases would respond to hormone challenge. With this in mind, the changes of biochemical parameters after administration of hormone or anti-hormone were taken to be a proof of the _in vivo_ sensitivity of the tumor. It is well established that positive and negative control of uterine estradiol and progesterone receptors is exerted by estradiol and progesterone respectively (Bayard and co-workers, 1978 ; Gorski and co-workers, 1968 ; Brenner, Resko and West, 1974 ; Baulieu and co-workers, 1975). Progestagens have been shown to counteract the estradiol induced increase of estradiol receptor (Tseng and Gurpide, 1975a) and to induce 17β-hydroxysteroid oxidoreductase activity in human endometrium, both _in vivo_ and _in vitro_ (Pollow and co-workers, 1975b ; Tseng and Gurpide, 1975b). Results reported here using medroxyprogesterone acetate confirm and extend other authors' published observations (Gurpide, Gusberg and Tseng, 1976 ; Pollow and co-workers, 1975c).

Since estrogens have been implicated in the onset and evolution of endometrial cancer, antiestrogens have a potential therapeutic value. Tamoxifen shows estrogenic or anti-estrogenic properties in different animal systems (Sutherland, Mester and Baulieu, 1977 ; Jordan and co-workers, 1977 ; Nicholson and Golder, 1975), it increases progesterone receptor concentration in the rat uterus (Koseki and co-workers, 1977), and has been used with good results in the treatment of advanced human breast cancer (Lerner and co-workers, 1976 ; Morgan and co-workers, 1976). In endometrial cancer, _the most marked effect of tamoxifen is the increase of progesterone receptor_, observed in all cases but one. The increase of estrogen receptor is less constant and rather small. Ornithine-decarboxylase activity measured in the cytosol before and 4 hours to 7 days after tamoxifen is unchanged. Therefore one can tentatively interpret tamoxifen effects on human endometrium cancer as "partially estrogenic". This type of response is in principle very favorable since, without apparently increasing growth related parameter, tamoxifen induces an increase in progesterone receptor and presumably makes the tumor more responsive to progestagen.

The fact that repeated biopsies of the same tumor may give different result in terms of state of tumor differenciation, relative amounts of epithelial and stromal cells and necrosis, and that complex hormonal changes may be elicited at the pituitary, adrenal and ovarian level means that hormonal tests should be interpreted with caution. Nonetheless we _consider the biochemical tamoxifen test to be denititely preferable to a single measurement of estradiol and progesterone receptors for predicting endometrial cancer hormonal responsiveness_.

Finally, progestagens have proven therapeutic value in endometrial carcinoma (Kohorn, 1976). Unfortunately, one drawback to their use alone is the rapid desensitization of tumor progesterone receptor which may be responsible for subsequent tumor hormone resistance. Tamoxifen has a promoting effect on progesterone receptor Therefore, once no adverse effect of tamoxifen on tumor growth has been demonstrated, the use of tamoxifen, in combination with progestagen, may be considered.

REFERENCES

Baulieu, E.E., M. Atger, M. Best-Belpomme, P. Corvol, J.C. Courvalin, J. Mester, E. Milgrom, P. Robel, H. Rochefort, and D. DeCatalogne (1975). Steroid hormone receptors. Vitamins and Hormones, 33, 649-736.

Bayard, F., S. Damilano, P. Robel and, and E.E. Baulieu (1978). Cytoplasmic and nuclear estradiol and progesterone receptors in human endometrium. J. Clin. Endocrinol. Metab., 46, 635-648.

Brenner, R.M., J.A. Resko, and N.B. West (1974). Cyclic changes in oviductal morphology and residual cytoplasmic estradiol binding capacity induced by sequential estradiol progesterone treatment of spayed rhesus monkeys. Endocrinology, 95, 1094-1104.

Crocker, S.G., P.J.D. Milton, and R.J.B. King (1974). Uptake of 6,7-^{3}H oestradiol-17β by normal and abnormal human endometrium. J. Endocrinol., 62, 145-152.

Evans, L.H., J.D. Martin, and R. Hähnel (1974). Estrogen receptor concentration in normal and pathological human uterine tissues. J. Clin. Endocrinol. Metab., 38, 23-32.

Gorski, J., D. Toft, G. Shyamala, D. Smith, and A. Notides (1968). Hormone receptors : studies on the interaction of estrogen with the uterus. Rec. Progr. Hormone Res., 24, 45-72.

Gurpide, E., S.B. Gusberg, and L. Tseng (1976). Estradiol binding and metabolism in human endometrial hyperplasia and adenocarcinoma. J. Ster. Biochem., 7, 891-896.

Gusberg, S.B. (1976). The individual at high risk for endometrial carcinoma. Am. J. Obst. Gynecol., 126, 535-542.

Gustafsson, J.A., N. Einhorn, G. Elfström, B. Nordenskjöld, and O. Wrange (1977). Progestin receptor in endometrial carcinoma. In W.L. McGuire, J.P. Raynaud, and E.E. Baulieu (Eds.), Progesterone Receptors in Normal and Neoplasic Tissue Raven Press, New-York. p. 299.

Horwitz, K.B., W.L. McGuire, O.H. Pearson, and A. Segaloff (1975). Predicting response to endocrine therapy in human breast cancer : a hypothesis. Science, 189 726-727.

Jänne, J., and H.G. Williams-Ashman (1971). Ornithine decarboxylase : on the purification of L-ornithine decarboxylase from rat prostate and effects of thiol compounds on the enzyme. J. Biol. Chem., 246, 1725-1732.

Jensen, E.V., and E.R. DeSombre (1972). Mechanism of action of female sex hormones. Ann. Rev. Biochem., 41, 203-230.

Jordan, V.C., C.J. Dix, L. Rowsby, and G. Prestwich (1977). Studies on the mechanis of action of the antiestrogen tamoxifen (ICI 46474) in the rat. Mol. Cell. Endocrinol., 7, 177-192.

Kaye, A.M., I. Isakson, and H.R. Lindner (1971). Stimulation by estrogens of ornithine and S-adenosylmethionine decarboxylase in the immature rat uterus. Biochim. Biophys. Acta, 252, 150-151.

Kohorn, I. (1976). Gestagens and endometrial cancer. Gynecol. Oncol., 4, 389-411.

Koseki, Y., D.T. Zava, G.C. Chamness, and W.L. McGuire (1977). Progesterone interactions with estrogen and antiestrogen in the rat uterus. Receptors effects. Steroids, 30, 169-177.

Lerner, H. Jr., P.R. Band, L. Israel, and B.S. Leung (1976). Phase II study of tamoxifen : report of 74 patients with stage IV breast cancer. Cancer Treatment Reports, 60, 1431-1435.

McDonald, P.C., and P.K. Siiteri (1974). Relationship between extraglandular production of estrone and the occurence of endometrial neoplasia. Gynecol. Oncol., 2, 259-263.

McGuire, W.L., J.P. Raynaud, and E.E. Baulieu (1977). Progesterone receptors in normal and neoplastic tissues. In Progress in Cancer Research and Therapy, Vol. 4. Raven Press, New-York.

Morgan, L.R., P.S. Schein, P.V. Woolley, D. Hoth, J. McDonald, M. Lippman, L.E. Posey, and R.W. Beazley (1976). Therapeutic use of tamoxifen in advanced breast cancer : correlation with biochemical parameters. Cancer Treatment Reports, 60, 1437-1443.

Nicholson, R.I., and M.P. Golder (1975). The effect of synthetic anti-oestrogens on growth and biochemistry of rat mammary tumours. Eur. J. Cancer, 11, 571-579.

Pollow, K., H. Lübbert, R. Jeske, and B. Pollow (1975a). Studies on 17β-hydroxysteroid dehydrogenase in human endometrium and endometrial carcinoma. 2. Characterization of the soluble enzyme from secretory endometrium. Acta Endocrinol., 79, 146-156.

Pollow, K., E. Boquoi, H. Lübbert, and B. Pollow (1975b). Effects of gestagen therapy upon 17β-hydroxysteroid dehydrogenase in human endometrial adenocarcinoma. J. Endocrinol. 67, 131-132.

Pollow, K., H. Lübbert, E. Boquoi, G. Kreuzer, and B. Pollow (1975c). Characterization and comparison of receptors for 17β-estradiol and progesterone in human proliferative endometrium and endometrial carcinoma. Endocrinology, 96, 319-328.

Sutherland, R., J. Mester, and E.E. Baulieu (1977). Tamoxifen is a potent "pure" antiestrogen in chick oviduct. Nature, 267, 434-435.

Tabor, C.W., and H. Tabor (1976). 1,4-diaminobutane (putrescine) spermidine and spermine. Ann. Rev. Biochem., 45, 285-306.

Tseng, L., and E. Gurpide (1975a). Effects of progestins on estradiol receptor levels in human endometrium. J. Clin. Endocrinol. Metab., 41, 402-404.

Tseng, L., and E. Gurpide (1975b). Induction of human endometrial estradiol dehydrogenase by progestins. Endocrinology, 97, 825-833.

Young, P.C.M., C.E. Ehrlich, and R.E. Cleary (1976). Progesterone binding in human endometrial carcinomas. Am. J. Obst. Gynecol., 125, 353-360.

Role of Nutrition in Changing the Hormonal Milieu and Influencing Carcinogenesis

K. K. Carroll* and G. J. Hopkins**

Department of Biochemistry, University of Western Ontario, London, Ontario N6A 5C1, Canada

ABSTRACT

Epidemiological data and studies with experimental animals have provided evidence that nutrition can have an important influence on carcinogenesis. High fat diets promote the development of mammary tumors in animals, and breast cancer mortality in humans shows a strong positive correlation with the amount of fat available for consumption. Total dietary fat is more positively correlated with breast cancer mortality than either animal fat or vegetal fat alone, and results of experiments with animals are in accord with these observations. Dietary fat may enhance mammary carcinogenesis by stimulating prolactin production and increasing the prolactin-estrogen ratio. Non-lipid components of the diet also appear to have some influence on mammary carcinogenesis. In addition to breast cancer, the epidemiological data show strong positive correlations between dietary fat and cancer at certain other sites, such as colon and prostate. High fat diets have been shown to promote development of intestinal tumors in animals and this could be related to increased production of bile acids, some of which are capable of acting as tumor-promoting agents.

KEY WORDS

Breast cancer, colon cancer, dietary fat, experimental models of mammary cancer, geographical distribution of cancer, estrogens, promoting agents, prolactin, bile acids.

INTRODUCTION

Two lines of evidence indicate that nutrition has an important influence on carcinogenesis (Nutrition in the Causation of Cancer, 1975; Nutrition and Cancer, 1977). One stems from epidemiological data, which show correlations between cancer incidence and the type of diet available in different countries. The other is based on experimental studies, which have demonstrated that the susceptibility of animals to tumorigenesis can be altered by changes in the diet.

---

* Medical Research Associate of the Medical Research Council of Canada.
** Recipient of a Fellowship from the Medical Research Council of Canada.

Although epidemiological data provide clues to interactions between diet and carcinogenesis, the existence of correlations between particular dietary variables and certain types of cancer does not necessarily mean that one is influenced by the other. Experiments with animals therefore provide a valuable source of additional evidence with which to test the significance of observed correlations. Dietary variables can be altered more readily in such experiments and the effects of the alterations on carcinogenesis can be observed within a reasonable time period.

DIETARY FAT AND CARCINOGENESIS

Epidemiological data show a strong positive correlation between the amount of fat available for consumption in different countries and mortality from cancers of the breast and colon, which account for a large percentage of cancer deaths in many parts of the world. These correlations are supported by experimental data showing that animals on high fat diets develop mammary and intestinal tumors more readily than those on comparable diets containing low levels of fat (Carroll, 1977; Wynder and Reddy, 1977)

In our experiments with mammary tumors induced in rats by 7,12-dimethylbenz(α)anthracene (DMBA), unsaturated fats enhanced tumorigenesis more effectively than saturated fats (Carroll and Khor, 1971, 1975). Similarly, mice fed unsaturated fat developed tumors more readily than those fed saturated fat, following injection of cells from a transplantable mammary tumor (Hopkins and West, 1977). In contrast, epidemiological data for human populations showed that animal fat intake was more positively correlated with breast cancer mortality than vegetal fat intake, although vegetal fat is likely to be more unsaturated than animal fat (Carroll, 1975).

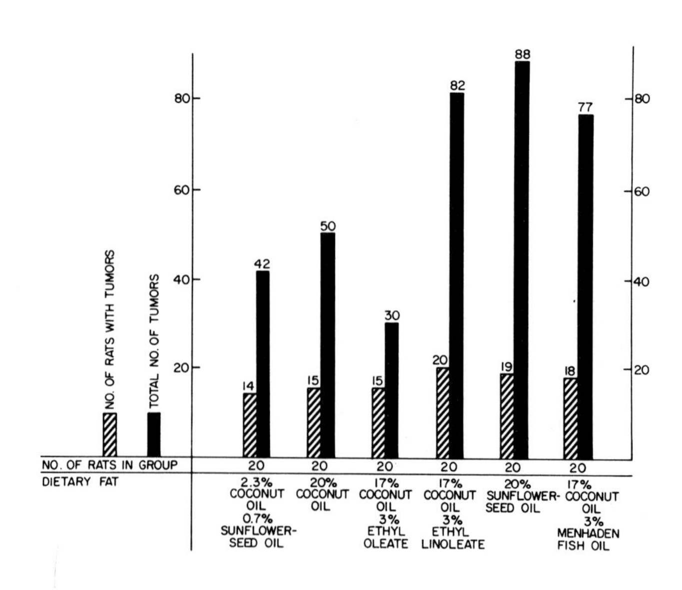

Fig. 1. Effects of saturated versus polyunsaturated fat on development of mammary tumors induced in rats by DMBA. The rats were maintained on Purina Chow and were given 5 mg of DMBA in 0.5 ml of sesame oil by stomach tube at 50 days of age. One week later, they were transferred to semipurified diets similar to those used in earlier studies (Carroll and Khor, 1971), containing fats as indicated. The rats were autopsied 4 months after receiving the DMBA.

This apparent discrepancy appears to have been resolved by recent experiments in our laboratory, the results of which are shown in Fig. 1. Rats on a high saturated fat diet, containing 20% coconut oil, developed only a few more tumors than those on a low fat diet containing 2.3% coconut oil and 0.7% sunflowerseed oil. A high fat diet containing 17% coconut oil and 3% ethyl oleate also gave a low tumor yield. Each of these diets contained about 0.5% linoleate. In contrast, a diet containing 17% coconut oil with 3% ethyl linoleate produced a marked increase in tumor yield comparable to that obtained by feeding a 20% sunflowerseed oil diet. This effect does not seem to be specific for linoleate, since a similar result was obtained with menhaden oil, which is highly unsaturated but contains only small amounts of fatty acids belonging to the linoleate family.

Although it is evident from these experiments that a certain amount of polyunsaturated fat is required to demonstrate the enhancement of mammary tumorigenesis by a high fat diet, the requirement is so small that it can probably be satisfied by most human diets. In countries where fat intake is low, the fat tends to be more unsaturated, since a larger percentage is normally derived from plant sources. Thus, if there is a threshold requirement for linoleate in the development of human breast cancer, this requirement is probably provided by most human diets. One would therefore not expect to see any relationship between breast cancer mortality and degree of unsaturation of the dietary fat. Total fat intake, however, could still be positively correlated with breast cancer mortality. The results of our experiments with animals are thus consistent with the data for human populations.

These findings suggest that if dietary fat is causally related to incidence of certain types of cancer in humans, attempts to lower the incidence by alterations in dietary fat should be concerned with reduction in total fat intake rather than changes in the type of fat consumed. In countries such as Canada, where fat intake is relatively high, data published by FAO (1971) indicate that about 90% of the available dietary fat is derived from meats, dairy products and visible fats and oils, such as margarines, shortenings and salad oils (Fig. 2). It is evident that the same types of food account for most of the available dietary fat in Argentina and Chile as well, but the amounts are smaller (Fig. 2). No data for cancer mortality in Argentina were included in the compilation of Segi and his associates

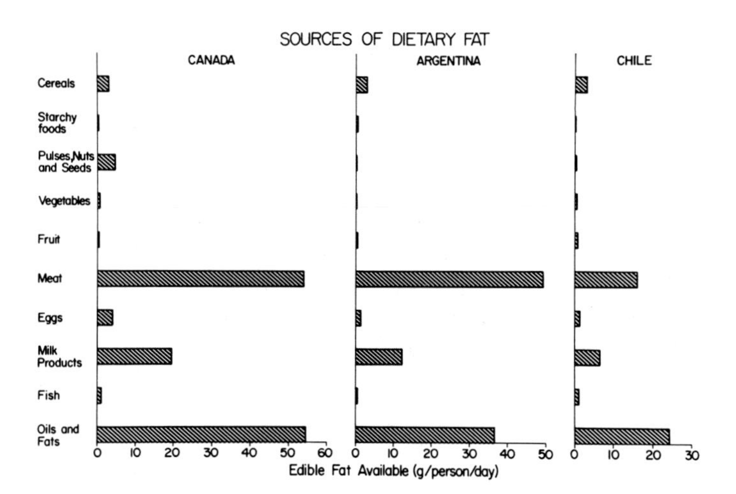

Fig. 2. Sources of available dietary fat in Canada, Argentina and Chile, based on data published by FAO (1971).

(1972), but mortality from breast cancer in Canada was reported to 2.5 times as high as in Chile.

In our studies with mammary tumors induced by DMBA, it was found that high fat diets enhanced the yield of mammary tumors only when fed after administration of the carcinogen (Carroll and Khor, 1970, 1975). Furthermore, experiments in a number of different laboratories have shown that the enhancing effect of the dietary fat can be demonstrated with tumors induced in various ways. These findings have led to the concept that dietary fat acts as a promoting agent, rather than affecting initiation of the tumors (Carroll and Khor, 1975). If this concept is applicable to human breast cancer, it seems possible that breast cancer incidence could be reduced substantially by simply decreasing the intake of dietary fat, without regard to the carcinogenic agents responsible for initiation. It may also be possible by decreasing the fat intake to reduce the possibility of recurrence of cancer after surgical removal of the primary tumor. In this connection, it is of interest that Japanese women, who eat a much lower fat diet than American women, show a lower recurrence of breast cancer following surgery than American women (Wynder and co-workers, 1963; Morrison and co-workers, 1976; Nemoto and co-workers, 1977).

Characteristics of Animal Models of Mammary Cancer relative to Human Breast Cancer

Because dietary fat appears to exert an important influence on mammary cancer, both in animals and humans, it is of interest to examine in more detail the characteristics of animal models for mammary cancer in relation to human breast cancer, and to consider possible mechanisms by which dietary fat may exert its influence on mammary cancer.

In our experiments, the mammary tumors were induced in young animals (Carroll and Khor, 1970, 1971) at a time when they are most susceptible to the carcinogen (Janss and Ben, 1978). Other investigators have also generally used young animals (Carroll and Khor, 1975). In contrast, the differences in breast cancer mortality which show a positive correlation with dietary fat are greater in post-menopausal women (De Waard, 1969; Hems, 1970). Benson and co-workers (1956) investigated the effect of dietary fat on spontaneous tumors in older rats, but most of the mammary tumors in their animals were fibroadenomas rather than adenocarcinomas.

A large proportion of the mammary tumors induced by DMBA are hormone-dependent and they also share many of the histological characteristics of human malignant breast tumors. Unlike human tumors, however, they seldom metastasize. Mammary tumors induced by N-nitrosomethylurea (NMU) may provide a better experimental model, since they do metastasize (Gullino and co-workers, 1975) and like other mammary tumors, they are influenced by the level of dietary fat (Chan and co-workers, 1977b).

In our experiments with DMBA-induced tumors, the tumor incidence was nearly always greater in rats on high fat compared to those on low fat diets, but the major difference was in the number of tumors which developed. Although experiments with other animal models may show a larger differential in tumor incidence, it is seldom as great as the 5 to 10-fold difference in breast cancer mortality between countries with high and low fat intake (Carroll, 1975).

Mechanism of Action of Dietary Fat in Mammary Carcinogenesis

In our studies with DMBA-induced mammary tumors, the effect of dietary fat could be demonstrated by transferring rats to a high fat diet 1 or 2 weeks after they were given the carcinogen. However, when the transfer was delayed until 4 weeks after giving the DMBA, the diet produced no enhancement in tumor yield (Carroll and Khor, 1975). Although, as suggested previously, these results provide evidence that dietary fat acts as a promoting agent, they seem to differ from the classical observations with skin tumors, in which application of the promoting agent could still produce tumors many weeks after exposure to the carcinogen (Berenblum and Shubik, 1947). Perhaps the effect of dietary fat on mammary tumorigenesis involves some physiological characteristic, such as hormonal balance, which changes as the animals grow older so that the effect can no longer be observed. It should be noted that Chan and co-workers (1977a) were able to demonstrate the high fat effect in rats ovariectomized and then started on a high fat diet 3 months after administration of DMBA.

Hopkins and West (1976) reviewed some of the possible mechanisms by which dietary fat may affect mammary tumorigenesis. One of the more appealing hypotheses is that of Chan and Cohen (1975), who suggested that dietary fat enhances mammary tumorigenesis by stimulating prolactin secretion and increasing the ratio of prolactin to estrogen in the circulation. They showed that treatment with an anti-prolactin drug lowered tumor incidence and abolished the difference between groups on high and low fat diets. In contrast, treatment with an anti-estrogen drug did not eliminate the difference between high and low fat diets (Chan and Cohen, 1974). Chan and co-workers (1975) also reported that rats on a high fat diet exhibited significantly higher prolactin levels during the proestrus and estrus phases of the cycle. In addition, they found that the ratio of prolactin to estrogen in rats treated with NMU was greater on a high fat compared to a low fat diet (Chan and co-workers, 1977a). Other evidence that the ratio of prolactin to estrogen is important has been provided by Meites and co-workers (1971) and Chan and co-workers (1976). This may be related to the observation of Kledzik and associates (1976) that high levels of estrogen decrease the binding of prolactin by mammary tumor membranes.

Studies with humans (Hill and Wynder, 1976) showed that the nocturnal peak in serum prolactin levels was significantly decreased in nurses after transfer from a Western-type diet to a vegetarian diet. Other studies (Hill and co-workers, 1976; Kwa and co-workers, 1976; Pike and co-workers 1977) have failed to provide clear evidence of an association between elevated plasma prolactin levels and increased risk of breast cancer, but such studies are complicated by the circadian nature of prolactin secretion. Since prolactin can act as a cancer promoter in animals (Welsch and Nagasawa, 1977), it seems reasonable to think that it could also do so in humans. Prior to the menopause, this effect may be counteracted by high levels of circulating estrogen, which could tend to minimize differences in breast cancer in different countries. The decline in estrogen levels after the menopause might explain the much greater disparity in breast cancer incidence among post-menopausal women in different countries.

It is not known by what mechanism dietary fat may influence prolactin secretion, but there is evidence that prolactin acts as a major liporegulatory hormone, affecting the utilization and storage of fat in animals (Meier, 1977). Prolactin has also been shown to affect lipoprotein lipase activity (Zinder and co-workers, 1974; Garrison and Scow, 1975). Since prolactin acts as a regulator of lipid metabolism, it seems reasonable to suspect that the reverse may also be true.

The emphasis in this review has been on promotional effects of the diet, rather than effects on tumor initiation. As suggested above, dietary fat may act as a promoting agent by providing a more favorable environment for development and proliferation of tumor cells. This could provide an explanation for the relatively high incidence of breast cancer in populations with high fat intake compared to those with low fat intake. However, other possibilities for explaining the geographic variations in cancer incidence should not be ignored. For example, the relatively low incidence of breast cancer in Eskimos may be difficult to explain on the basis of their dietary fat intake. An alternative possibility is that peoples living in the polar regions may be less exposed to carcinogenic stimuli than those living in the industralized countries of the temperate regions.

Role of Dietary Fat in Cancers other than Breast Cancer

As in the case of breast cancer, both epidemiological data and experiments with animals have provided evidence that dietary fat promotes the development of intestinal cancer (Wynder and Reddy, 1977). With intestinal tumors, however, there is little reason to think that endocrine factors are involved. Hill and co-workers (1971) suggested that high fat diets may facilitate growth of intestinal bacteria, which are capable of converting steroids to carcinogenic hydrocarbons. A more plausible theory, advanced by Reddy and co-workers (1975), is that high fat diets enhance the production of bile acids, some of which have been shown to act as cancer promoters.

The epidemiological data also show positive correlations between dietary fat and certain other endocrine-dependent tumors, notably cancer of the prostate (Carroll and Khor, 1975). To our knowledge, the significance of such correlations have not yet been tested in experimental animals. This may be related in part to a lack of suitable animal models. Speculation on possible mechanisms by which dietary fat may influence development of such tumors, therefore seems unwarranted at the present time.

EFFECTS OF NON-LIPID COMPONENTS OF THE DIET

The possibility that dietary factors other than fat can exert an influence on carcinogenesis also deserves consideration. Intake of animal protein, for example, shows a strong positive correlation with mortality from breast cancer (Carroll and Khor, 1975). This led us to compare the effects of diets containing either an animal protein, casein, or a plant protein, isolated soy protein, on the yield of mammary tumors induced by DMBA. However, no significant difference in tumor incidence was observed (Carroll, 1975).

Breast cancer mortality is also positively correlated with intake of sugar and is negatively correlated with intake of complex carbohydrate (Carroll, 1977). In this case, support for a causative relationship has been obtained from experiments with the DMBA-tumor model, which have shown that rats develop more tumors on diets containing simple sugars than on corresponding diets containing starches (S.K. Hoehn and K.K. Carroll, unpublished experiments). This difference may be related to endocrine factors. Dietary sucrose has been reported to give higher plasma concentrations of insulin and adrenal corticosteroid hormones than dietary starches (Yudkin, 1972).

SUMMARY

In this brief review, an attempt has been made to show that diet can have an important influence on carcinogenesis by promoting the development and proliferation of tumor cells. This may occur through various indirect mechanisms, some of which may involve endocrine factors. These concepts suggest that a substantial reduction in the incidence and recurrence of some of the more common types of cancer, such as breast cancer, could be achieved by alterations in the diet and specifically by a reduction in the intake of dietary fat.

ACKNOWLEDGMENTS

This work was supported by the National Cancer Institute of Canada and the Medical Research Council of Canada. Technical assistance by Cathy Frank and a gift of menhaden oil from A.P. Bimbo of Zapata Haynie Corp., Reedville, VA, U.S.A., are gratefully acknowledged.

REFERENCES

Benson, J., M. Lev, and C. G. Grand (1956). Enhancement of mammary fibroadenomas in the female rat by a high fat diet. Cancer Res., 16, 135-137.

Berenblum, I., and P. Shubik (1947). A new, quantitative, approach to the study of the stages of chemical carcinogenesis in the mouse's skin. Br. J. Cancer, 1, 383-391.

Carroll, K. K. (1975). Experimental evidence of dietary factors and hormone-dependent cancers. Cancer Res., 35, 3374-3383.

Carroll, K. K. (1977). Dietary factors in hormone-dependent cancers. In M. Winick (Ed.), Current Concepts of Nutrition, Vol. 6, Nutrition and Cancer, John Wiley & Sons, New York, pp. 25-40.

Carroll, K. K., and H. T. Khor (1970). Effects of dietary fat and dose level of 7,12-dimethylbenz(α)anthracene on mammary tumor incidence in rats. Cancer Res., 30, 2260-2264.

Carroll, K. K., and H. T. Khor (1971). Effects of level and type of dietary fat on incidence of mammary tumors induced in female Sprague-Dawley rats by 7,12-dimethylbenz(α)anthracene. Lipids, 6, 415-420.

Carroll, K. K., and H. T. Khor (1975). Dietary fat in relation to tumorigenesis. In K. K. Carroll (Ed.), Prog. Biochem. Pharmacol., Vol. 10, Lipids and Tumors, S. Karger, Basel. pp. 308-353.

Chan, P.-C., and L. A. Cohen (1974). Effect of dietary fat, antiestrogen, and antiprolactin on the development of mammary tumors in rats. J. Natl. Cancer Inst., 52, 25-30.

Chan, P.-C., and L. A. Cohen (1975). Dietary fat and growth promotion of rat mammary tumors. Cancer Res., 35, 3384-3386.

Chan, P.-C., F. Didato, and L. A. Cohen (1975). High dietary fat, elevation of rat serum prolactin and mammary cancer. Proc. Soc. exp. Biol. Med., 149, 133-135.

Chan, P.-C., J. F. Head, L. A. Cohen, and E. L. Wynder (1977a). Effect of high fat diet on serum prolactin levels and mammary cancer development in ovariectomized rats. Proc. Am. Ass. Cancer Res., 18, 189.

Chan, P.-C., J. F. Head, L. A. Cohen, and E. L. Wynder (1977b). Influence of dietary fat on the induction of mammary tumors by N-nitrosomethylurea: Associated hormone changes and differences between Sprague-Dawley and F344 rats. J. Natl. Cancer Inst., 59, 1279-1283.

Chan, P.-C., J. Tsuang, J. Head, and L. A. Cohen (1976). Effects of estradiol and prolactin on growth of rat mammary adenocarcinoma cells in monolayer cultures. Proc. Soc. exp. Biol. Med., 151, 362-365.

De Waard, F. (1969). The epidemiology of breast cancer; review and prospects. Int. J. Cancer, 4, 577-586.

Food and Agriculture Organization of the United Nations, Rome (1971). Food Balance Sheets, 1964-66 Average.

Garrison, M. M., and R. O. Scow (1975). Effect of prolactin on lipoprotein lipase in crop sac and adipose tissue of pigeons. Am. J. Physiol., 228, 1542-1544.

Gullino, P. M., H. M. Pettigrew, and F. H. Grantham (1975). N-Nitrosomethylurea as mammary gland carcinogen in rats. J. Natl. Cancer Inst., 54, 401-414.

Hems, G. (1970). Epidemiological characteristics of breast cancer in middle and late age. Br. J. Cancer, 24, 226-234.

Hill, M. J., B. S. Drasar, V. Aries, J. S. Crowther, G. Hawksworth, and R.E.O. Williams (1971). Bacteria and aetiology of cancer of large bowel. Lancet, 1, 95-100.

Hill, P., and E. Wynder (1976). Diet and prolactin release. Lancet, 2, 806-807.

Hill, P., E. L. Wynder, H. Kumar, P. Helman, G. Rona, and K. Kuno (1976). Prolactin levels in populations at risk for breast cancer. Cancer Res., 36, 4102-4106.

Hopkins, G. J., and C. E. West (1976). Possible roles of dietary fats in carcinogenesis. Life Sci., 19, 1103-1116.

Hopkins, G. J., and C. E. West (1977). Effect of dietary polyunsaturated fat on the growth of a transplantable adenocarcinoma in C3HA^{vy}fB mice. J. Natl. Cancer Inst., 58, 753-756.
Janss, D. H., and T. L. Ben (1978). Age-related modification of 7,12-dimethylbenz-(a)anthracene binding to rat mammary gland DNA. J. Natl. Cancer Inst., 60, 173-177.
Kledzik, G. S., C. J. Bradley, S. Marshall, G. A. Campbell, and J. Meites (1976). Effects of high doses of estrogen on prolactin-binding activity and growth of carcinogen-induced mammary cancers in rats. Cancer Res., 36, 3265-3268.
Kwa, H. G., F. Cleton, M. De Jong-Bakker, R. D. Bulbrook, J. L. Hayward, and D. Y. Wang (1976). Plasma prolactin and its relationship to risk factors in human breast cancer. Int. J. Cancer, 17, 441-447.
Meier, A. H. (1977). Prolactin, the liporegulatory hormone. In H.-D. Dellman, J. A. Johnson, and D. M. Klachko (Eds.), Adv. Exp. Med. Biol., Vol. 80, Comparative Endocrinology of Prolactin, Plenum Press, New York. pp. 153-171.
Meites, J., E. Cassell, and J. Clark (1971). Estrogen inhibition of mammary tumor growth in rats; counteraction by prolactin. Proc. Soc. exp. Biol. Med., 137, 1225-1227.
Morrison, A. S., C. R. Lowe, B. MacMahon, B. Ravnihar, and S. Yuasa (1976). Some international differences in treatment and survival in breast cancer. Int. J. Cancer, 18, 269-273.
Nemoto, T., T. Tominaga, A. Chamberlain, Z. Iwasa, H. Koyama, M. Hama, I. Bross, and T. Dao (1977). Differences in breast cancer between Japan and the United States. J. Natl. Cancer Inst., 58, 193-197.
Nutrition and Cancer (1977). M. Winick (Ed.), Current Concepts of Nutrition, Vol. 6, John Wiley & Sons, Inc., New York.
Nutrition in the Causation of Cancer, Proceedings of a Symposium held at Key Biscayne, Florida, (1975). Cancer Res., 35. No. 11, Pt. 2.
Pike, M. C., J. T. Casagrande, J. B. Brown, V. Gerkins, and B. E. Henderson (1977). Comparison of urinary and plasma hormone levels in daughters of breast cancer patients and controls. J. Natl. Cancer Inst., 59, 1351-1355.
Reddy, B. S., T. Narisawa, R. Maronpot, J. H. Weisburger, and E. L. Wynder (1975). Animal models for the study of dietary factors and cancer of the large bowel. Cancer Res., 35, 3421-3426.
Segi, M., and M. Kurihara, in collaboration with T. Matsuyama, M. Ito, Y. Nagano, and K. Yamamoto (1972). Cancer Mortality for Selected Sites in 24 Countries, No. 6 (1966-1967), Japan Cancer Soc., Nagoya.
Welsch, C. W., and H. Nagasawa (1977). Prolactin and murine mammary tumorigenesis: A review. Cancer Res., 37, 951-963.
Wynder, E. L., T. Kajitani, J. Kuno, J. C. Lucos, Jr., A. De Palo, and J. Farrow (1963). A comparison of survival rates between American and Japanese patients with breast cancer. Surgery, Gynecol. Obstet., 117, 196-200.
Wynder, E. L., and B. S. Reddy (1977). Diet and cancer of the colon. In M. Winick (Ed.), Current Concepts of Nutrition, Vol. 6, Nutrition and Cancer, John Wiley & Sons, New York. pp. 55-71.
Yudkin, J. (1972). Sugar and disease. Nature, Lond., 239, 197-199.
Zinder, O., M. Hamosh, T.R.C. Fleck, and R. O. Scow (1974). Effect of prolactin on lipoprotein lipase in mammary gland and adipose tissue of rats. Am. J. Physiol., 226, 744-748.

The Incidence of Cancer Following Long-Term Estrogen Therapy

Benjamin F. Byrd, Jr.*, and William K. Vaughn**

*Vanderbilt University School of Medicine, *Department of Surgery, **Department of Biostatistics, Nashville, Tennessee, U.S.A. Surgical Service, St. Thomas Hospital, Nashville, Tennessee, U.S.A.*

We previously reported a series of 1016 women who have been placed on estrogen support following hysterectomy. About one half of the women in the study group had residual ovarian tissue after hysterectomy.[2] The other women had all ovarian tissue removed at the time of hysterectomy. In general, the initiation of estrogen support was delayed until the patient developed some of the evidences of estrogen deficiency in those instances in which ovarian tissue was retained. Women in the study ranged from 22 to 78 years of age at the time of admission to the study. These women have been on estrogen for at least three months. If some major complication related to the study protocol developed after more than three months of estrogen therapy, these were included. In the absence of such complications women were not included in the study unless they had undergone at least three years of estrogen therapy.[2]

The entire group of 1016 women was followed for five or more years,but the special study reported at this time relates to that group of women who had been on estrogen support therapy for 15 or more years. This group of women received an average daily dose of 1.5 mgm. of conjugated equine estrogen. There were 402 patients who were followed from the beginning of their 15th year of estrogen support onward. The total study follow-up comprised 2030 patient/years after the 15th year of medication. The current average age of the women in the study is 68. The effect of environmental carcinogens is generally acknowledged to require from five to ten years following exposure. For this reason, we picked the minimum delay of 15 years following initiation of therapy with conjugated equine estrogen to evaluate this group. The group was evaluated for mortality from all causes, mortality from heart disease, mortality from cerebrovascular accidents, the instance of cancer and the mortality from cancer during the study period. The figures have all been corrected for the absence of the pelvic reproductive organs, including the ovaries, in studying cancer experience and mortality.

An equivalent group of women from the reports of the National Cancer Survey[3] and from the Vital Statistics of the State of Tennessee[1] would anticipate 16.5 instances of cancer. In reality there were only 11 experienced. These were spread through the various systems. Cancer of the lung and breast occurred at the anticipated level. The principal improvement was in cancers of the gastrointestinal tract, experiencing less than the anticipated number of cancers of the colon and rectum, but the figures are not statistically significant because of the small numbers involved. In reference to the mortality rate from all causes, it is seen that the leading cause of cancer death in women, breast cancer, was the basis for one death

TABLE 1 POPULATION CHARACTERISTICS OF WOMEN TAKING ESTROGENS

Total Patient Years on Estrogen	13,666
Total Patient Years Follow-Up	14,318
Average Number of Years on Estrogen	13.5
Average Age at Entrance into Study	44.0
Average Age at Current Follow-Up	57.7
Percent with Ovaries Removed	63.8
Percent with No Children	20.3

TABLE 2 CHARACTERISTICS OF WOMEN TAKING ESTROGEN* FOR MORE THAN 15 YEARS — 402 PATIENTS

Total Patient Years on Estrogen	7959
Total Patient Years Follow-Up After 15 Years	2030
Average Number Years on Estrogen	19.8
Average Age at Entrance into Study	42.5
Average Age at Current Follow-Up	63.0

*Conjugated Equine Estrogen

in the 402 patients and that there were six deaths from all cancer in the study period which was slightly less than the total anticipated. The remarkable improvement in the realized over the anticipated number of deaths from heart disease and from cerebrovascular accident completely outweighs the significance of alteration in other types of mortality. The figures show a statistical improvement in the mortality from both heart disease and cerebrovascular accident over that which was anticipated in a group of women of similar ages and social background.

In summary, there is a modest improvement in the occurrence of all types of cancer in a group of women who have been on supportive estrogen therapy for 15 or more yea There is a similar diminished incidence of death from all cancers in this group of

TABLE 3 MORTALITY EXPERIENCE OF SERIES ON ESTROGEN 402 PATIENTS

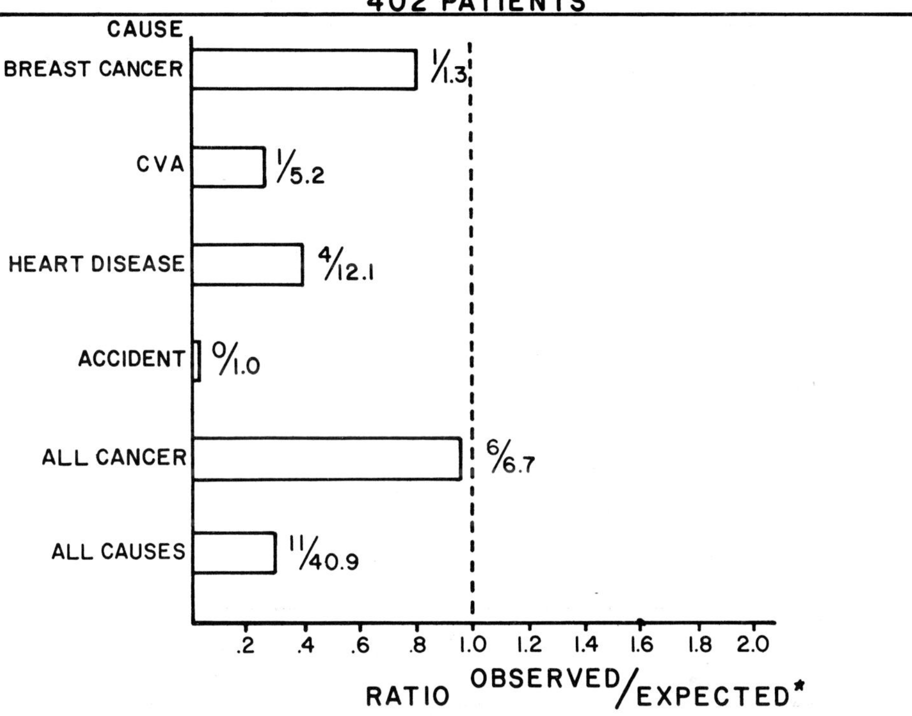

*VITAL STATISTICS OF TENN.

women. There is a dramatic improvement which is statistically valid in the mortality rate from cerebrovascular accident and from heart disease. This accounts for the significant diminution in mortality in this group of women, which outweighs any evidence of other side reactions from long-term maintenance from conjugated equine estrogen.

TABLE 4 CANCER INCIDENCE FOR SELECTED SITES OF SERIES
402 PATIENTS

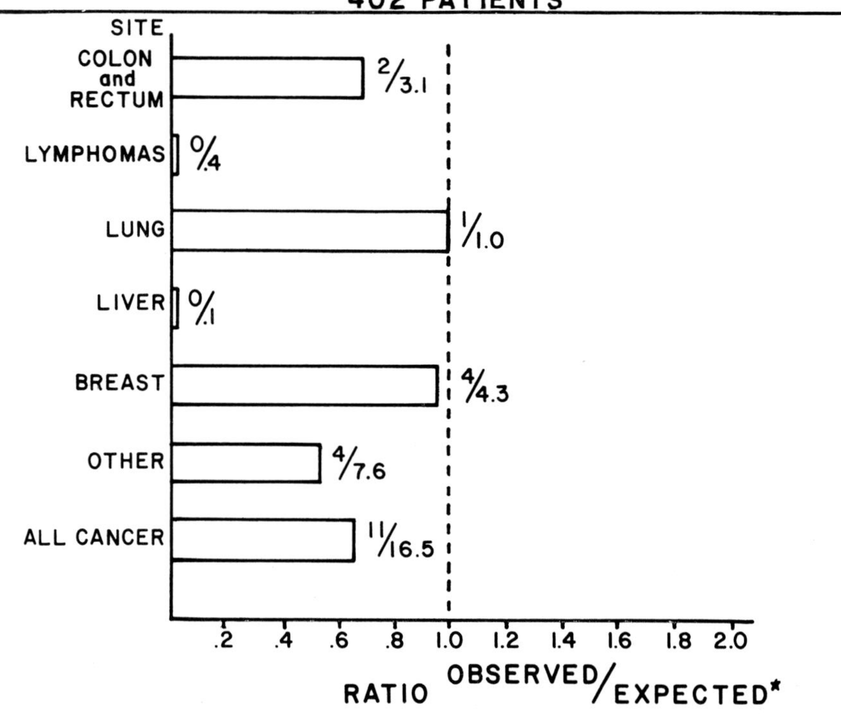

*THIRD NATIONAL CANCER SURVEY

*Aided by a grant from Ayerst Laboratories

REFERENCES

1. Annual bulletin of vital statistics for the year 1960. State of Tenn., Dept. of Public Health, Nashville, Tenn.

2. Byrd, B.F., J. C. Burch, and W. K. Vaughn (1977). The impact of long-term estrogen support after hysterectomy - a report of 1016 cases. Annals of Surgery, 185, 574-580.

3. Third national cancer survey advanced three year report. 1969-1971 incidence. DHEW Publication, No. NIH 74-637, Biometry Branch of NCI

Symposium No. 6 — Hormones and Cancer

Pietro M. Gullino

Dr. A. González-Angulo reviewed the evidence concerning oral contraceptives and increased risk of neoplastic transformation, giving special attention to the uterus and breast. Animal experiments demonstrating the appearance of carcinoma of the cervix in mice fed a liquid diet rich in estrogens and progestogens were described. The cause-effect relationship between oral contraceptive treatment and presence of cellular atypia or squamous neoplasia of the cervix was briefly discussed and the evidence linking polypoid glandular hyperplasia of the endocervix with oral contraceptives was emphasized. The risk of carcinoma of the cervix or endometrium due to intrauterine devices used as contraceptives was considered negligible. For the breast, data showing the lack of association between carcinoma and contraceptive medication was analyzed and accepted as a documented observation. The incidence of fibrocystic disease or fibroadenomas was also considered as not being substantially modified in subjects on oral contraceptive medication. The reports of almost 100 cases of hepatomas in young women on contraceptives were considered of epidemiologic significance particularly because they came from hospitals where no other benign hepatic tumors had been found in many decades. Brief considerations on the similarity of ultrastructural features of mitochondria in hepatoma cells and in hepatocytes of pregnant women concluded the presentation.

The epidemiologic and clinical evidence that estrogens used for replacement therapy at menopause alter the risk of neoplastic transformation of tissues of the female genital system and breast, was reviewed by Dr. D. B. Thomas. He concluded that strong evidence exists showing exogenous estrogens as the cause of endometrial carcinoma. On the contrary, the evidence that such estrogens increase the risk of carcinomas in the breast, ovary or cervix is rather weak and unconvincing. In the discussion that followed, emphasis was placed on the need of tailoring the use of hormones in replacement therapy at menopause to a minimal dose sufficient to control the symptoms.

Dr. Nandi discussed the role of hormones in enhancing virus production. He focused attention on the increment of virus elimination by cells injected with polyoma, EBV, and HSV-2 when treated with glucocorticoids. The glucocorticoid effect on polyoma infected cells was shown to be correlated with the binding of steroids on specific receptors. Production of murine mammary tumor virus in particular was shown to be enhanced 5- to 40-fold by glucocorticoids acting either on normal or neoplastic mammary cells *in vitro*. Estrogen, insulin, prolactin, and dibutyryl cyclic AMP act synergistically with glucocorticoids to enhance virus release. Murine mammary tumor virus expression also occurs via a specific interaction with a glucocorticoid receptor followed by increased transcription of the viral genome. The effects of hormones on leukemogenesis *in vivo* and C-type virus production from murine cells *in vitro* were discussed and the variety of effects of steroids on leukemia incidence in mice was briefly analyzed.

Dr. C. Levy spoke on the presence of hormone receptors as a prerequisite for hormone action. He discussed the data obtained in his laboratory on the concentration of estradiol and progesterone receptors determined in cytosol and nuclear fractions from postmenopausal endometrial carcinomas. The effects of tamofixen on the receptor levels as well as the activities of 17β-hydroxysteroid oxidoreductase and ornithine decarboxylase of the cytosol were analyzed as an approach to assess hormone responsiveness of the neoplasm.

Dr. K. K. Carrol discussed the role of nutrition in changing the hormonal milieu of the host and possibly influencing carcinogenesis. He first analyzed evidence showing that animals on high fat diets have consistently been found to develop mammary tumors more readily than animals on low fat diets. When the tumor was produced by a chemical carcinogen the increment in tumor frequency was obtained only when the high fat diet was started after carcinogen treatment. This kind of evidence was interpreted to suggest that dietary fat affects the promotional stage of carcinogenesis. Since mammary tumor development is susceptible to the action of various hormones, the evidence suggesting that dietary fat may alter the hormonal environment, thereby influencing mammary carcinogenesis, was discussed. The observations in this area were contrasted with the role that high fat diet has in enhancing the yield of colon tumors induced in rats by chemical carcinogens. The presentation was concluded by a critical analysis of the epidemiological data showing a positive correlation between intake of dietary fat and mortality from breast and colon cancer in different countries of the world.

The last presentation was by Dr. Gullino in substitution for Dr. Nora who was unable to attend the symposium. The topic discussed was mammary tumor regression after hormonal deprivation of the host. The speaker presented evidence to show that tumor regression is due not only to an arrest of cell division but also to an acceleration of cell destruction. This destruction results from an endocellular digestion requiring high energy consumption and new formation of lytic enzymes. The cyclic AMP system is involved in triggering the regression process in the sense that ovariectomy induced a decrease in estrogen receptors and increased the cyclic AMP receptor levels. The same increment of cyclic AMP receptors may be induced by injections of dibutyryl cyclic AMP and in both cases, a cyclic AMP+receptor complex is found in the nuclei of the regressing tumors. This complex coexists in the nuclei with a phosphorylation of non-histone proteins but the role of this event in triggering endocellular digestion is unknown.

Modern Trends and Prospectives in Cancer Surgery

Surgical Oncology as a Specialty. The Making of the Surgical Oncologist

Ronald W. Raven

Marie Curie Memorial Foundation, 124 Sloane Street, London SW1, U.K.
and
Royal Marsden Hospital and Institute of Cancer Research, Fulham Road, London SW3, U.K.

ABSTRACT

Oncology is a multidisciplinary subject composed of arts and sciences with subdivisions of research oncology, clinical oncology and social oncology. Clinical oncology is divided into medical, surgical, radiation and nursing oncology according to the main treatment modality. Team work is essential in joint consultation clinics for the investigation, diagnosis and treatment of patients with oncological diseases. Team members must speak the same scientific and clinical language and have a sound knowledge of basic sciences built into the theory of oncology. Trainees in surgical oncology (as in the other divisions of clinical oncology) must be educated in basic sciences forming the theory of oncology, and this is a continuous learning process throughout professional life. These sciences include pathology (including haematology); biochemistry of human cancer (including carcinogenesis); endocrinology and metabolic medicine; tumour biology and immunology; and clinical pharmacology; cell kinetics; radiobiology; epidemiology and genetics.

The training of the surgical oncologist in the art of surgery is arranged so that sound technical skill is acquired in general surgery, followed if desired by a period of training in a surgical specialty as gynaecology, urology, head and neck, or thoracic surgery. During training a short period should be spent in medical and radiation oncology to learn the indications, value and complications of medicines and radiations, alone or in combination. In the practice of oncology combination treatment is expanding continually so the surgical oncologist must keep abreast of these developments.

It is emphasized that the education and training of the surgical oncologist must be a continuous learning process for which the trainee must acquire solid foundations.

Keywords: surgical oncologist, education - theory - basic sciences; training - surgical art

INTRODUCTION

The synthesis of different arts and sciences to form oncology is of profound importance for the coordination of the clinical and research

efforts to control many diseases which have been grouped together under the general term of cancer and to elucidate their causation. This term is unscientific, and because it is associated with so much suffering and high mortality it generates considerable fear. Consequently the term oncological diseases is more appropriate and should now be substituted for cancer.
Oncology is constantly developing by the acquisition of new knowledge which is applicable to the management of patients and the care of the community. Many different disciplines and practical skills are required for all the work which is encompassed by oncology. (Table 1)

TABLE 1 Divisions of Oncology

Divisions of Oncology
Research Oncology
Laboratory
Clinical
Epidemiology
Clinical Oncology
Medical oncology
Surgical oncology
Radiation oncology
Nursing oncology
Social Oncology
Education and Training
Rehabilitation
Prevention

Clinical oncology is carried out by a team of different experts working in joint consultation clinics dealing with the diagnosis and treatment of patients with oncological diseases. It is essential that they have easy access to special departments including pathology, radiology and nuclear medicine. There are added benefits from an association with research laboratories and research oncologists to facilitate the feedback of new discoveries for patient care. The members of the team must speak the same scientific and clinical language in addition to possessing background knowledge of the various scientific disciplines, which is essential not only for their clinical work, but for the advancement of oncology. It is agreed that an oncologist cannot embrace all these sciences in depth, for a professional lifetime is necessary for this scientific education.

TRAINING THE SURGICAL ONCOLOGIST

The majority of oncological diseases require surgical treatment, either alone or in combination. which gives the surgical oncologist an important role in their management, so that his training is long and arduous embracing both the theory and practice of oncology.

The Theory of Oncology

A system of knowledge has been established with derivations from different scientific disciplines where laws are emerging for guidance and doctrine for teaching. In this system much scientific data gives certainty, while there is ample scope for research and speculative ideas. The theory of oncology must be related with its practice to bring immediate help to patients and provide a constant feedback from laboratory to clinic where more patient-orientated research is necessary.

Trainees in Surgical Oncology require guidance in their education and training, and the object of my paper is to outline such a scheme, which includes an educational programme in the theory based upon different medical sciences, a knowledge of which is essential for their work.

Pathology

A sound knowledge is necessary to understand the natural history of oncological diseases, including their development and progression. Tumour pathology influences methods of investigation, diagnosis and treatment, so the trainee will benefit from an attachment of several months in a pathology department handling plenty of tumour material, with access to a department of haematology. Knowledge will thereby be gained of morbid anatomy from autopsy material; dissection of surgical specimens with tissue block selection; a frozen section biopsy service; histopathology with tumour grading; and tumour spread, metastases and systems of staging. Clinico-pathological correlations are learnt by attending seminars, out-patient clinics and operation theatres.

Biochemistry of Human Cancer

This growing discipline is of great academic clinical importance. An increasing number of investigations are available in routine patient investigations and management where biochemical monitoring can be life-saving.

A key subject is carcinogenesis and carcinogens, which I compare in importance with inflammation and infections in general medicine. The mechanism of carcinogenesis, which includes the activation and binding of carcinogens at cellular level, is under constant investigation. Increasing attention is rightly given to transplacental and intra-uterine carcinogenesis which is responsible for paediatric oncological diseases, and perhaps for some which affect adult too. We are beginning to recognize the contribution which developmental biology can make to oncology.

The surgical oncologist must be familiar with general cancer metabolism, steroid chemistry, enzymes and the biochemistry of individual tumours.

Endocrinology

An intricate and significant relationship exists between a number of oncological diseases and the endocrine system. We are constantly impressed by the spectacular results in patients with carcinoma of the breast and prostate by manipulating in various ways hormonal control systems. The reversal of carcinomatous tissues to normal tissues by such manipulations as removing the ovaries, adrenals or hypophysis, or by the administration of synthetic hormones is one of the most significant discoveries in oncology. A big research effort is now essential to explain these mechanisms and to discover others, for this treatment modality remains empirical at present.

Many tumours have endocrine manifestations and others secrete ectopic hormones which cause clinical syndromes. Their recognition leads to tumour diagnosis and the appropriate treatment. Some of these secretory tumours are occult and are diagnosed by biochemical measurements. The metabolic abnormalities can be lethal unless they are rectified quickly by the removal or destruction of the primary tumour and the treatment of such disorders as hypercalcaemia, dilutional hyponatraemia, hypokalaemic alkalosis and the argentaffinoma syndrome. The surgical oncologist is involved in all this work which includes the recognition and rectification of electrolyte imbalances and endocrine replacement therapy.

Immunology

This subject continues to attract considerable attention from research and clinical oncologists who seek an understanding of the role of immunological reactions in modifying or controlling oncological diseases. There are problems to solve concerning immunological deficiencies, and immunopotentiation which may lead eventually to immunotherap Already immunological techniques are being used in diagnosis, includin alpha fetoprotein in hepatic tumours and carcinoembryonic antigen in colonic and other carcinomas. We cannot yet speak of immunotherapy, and remember that such techniques may even be harmful to the patient.

Clinical Pharmacology

In the rapidly expanding subject of chemotherapy which is used alone or combined with surgery and radiotherapy, an understanding of the pharmacology of many chemicals is essential. Thus dosage schedules must be studied in addition to drug metabolism and elimination. Many chemicals in clinical use are toxic causing severe side effects, and should be used only by those with a sound knowledge of their indications, limitations, cytotoxicity, contraindications and drug interactions. New chemicals are needed which are non-toxic to normal tissues and easy to administer over the long term. Clinical trials are expected to demonstrate their value; the surgical oncologist should be familiar with the organization of these trials, for which a knowledge of statistics is necessary for their evaluation.

Cell Kinetics

A knowledge of this subject is valuable in oncological work, which includes the characterization of normal and neoplastic cells; the cell cycle and cellular division; cell loss and multiplication and cellular repair. This is specially important for the radiotherapist and chemotherapist.

Radiobiology

Important discoveries in radiobiological research have found clinical application which is necessary, especially for radiotherapists. Other oncologists find much of interest in tumour responses to irradiation; cell sensitivity correlated with the cell cycle; radiation dose fractionation; modification of radiosensitivity and the oxygen effect which influences tumour response to radiotherapy. The search for radiosensitizers continues; there is some evidence that Bleomycin may possess this property, especially in squamous carcinomas as in the oesophagus.

Epidemiology

This subject is of interest to all oncologists, and a necessary part of their education. They are concerned with the world pattern of cancer incidence and its geographical distribution; migrant studies; aetiological clues and the identification of risk factors, which are necessary in prevention programmes. Additional studies include cancer incidence according to sex, age, habits, occupation and race. The solution of certain problems requires epidemiological methods with the use of cancer registries, questionnaires, case-records and organizing population survery. New epidemiological clues are needed which are comparable with the classical observation made by Percivall Pott (1775) on chimney sweeps' cancer.

Genetics

The importance of a knowledge of genetics and understanding genetic mechanisms in oncology is recognized. Information is constantly growing about the genetic code and protein synthesis; chromosomes and chromosomal disorders; chromosome analyses; sex chromatin; and marker chromosomes.

This work has considerable practical value, for more people are asking for guidance where there seems to be inherited disease in their families. Thus genetic counselling is developing and is made available on an increasing scale in special clinics.

THE PRACTICE OF ONCOLOGY

Oncology is practised by surgeons, physicians, radiotherapists and chemotherapists, usually working in joint clinics where patients are seen in consultation for their investigations, diagnosis and treatment planning. The final responsibility is usually taken by the clinician to whom the patient was initially referred.

The surgical oncologist must be skilled in technical surgery for which he has undergone a prolonged period of training lasting about 8 years. The basic training is in general surgery, followed if desired by a period of training in a special branch of surgery like gynaecology, urology, head and neck, or thoracic surgery. The majority of patients with oncological diseases require some form of surgical treatment so that the general or specialized surgeon plays an important part in the team and understands the indications, value and complications of radiotherapy and chemotherapy used singly or in combination. An increasing number of patients have combination treatment, which highlights the necessity of team work. Many of these operations for malignant tumours are major procedures, so that the surgical oncologist must be a well-trained and competent operator with special experience of malignant disease.

THE ART OF SURGERY

The development of the operations for malignant diseases has occurred during a period of less than a century and forms a brilliant part of the whole history of surgery. (Table 2) Many of these early operations were performed without the essential supportive care we enjoy today. The major operations are designed on a precise knowledge of anatomy and pathology, and patient management owes much to the principles of physiology and biochemistry. The good results obtained witness to the high standard of skilled anaesthesia and nursing, antibacterial chemicals, rehabilitation techniques and all the modern facilities provided in our postoperative and intensive care units.

Table 2 History of Major Oncological Operations

Surgeon	Year	Operation
Billroth	1881	Subtotal gastrectomy
Halsted	1890	Radical mastectomy
Schlatter	1897	Total gastrectomy
Von Mickulicz	1898	Oesophagogastrectomy
Wertheim	1900	Radical hysterectomy
Miles	1908	Abdominoperineal excision of rectum
Torek	1913	Oesophagectomy
Trotter	1913	Partial pharyngectomy
Graham and Singer	1923	Pneumonectomy

SURGICAL DEVELOPMENTS

Since these major operations were introduced decades ago many changes have occurred so that they are carried out with a minimum of operative risk and their morbidity has been decreased by reconstructive and plastic surgery. Some operations have become more radical with consequently improved prognosis. For example, the present author has a large series of cases with advanced carcinoma of the hypopharynx and cervical oesophagus uncontrolled by radiotherapy for whom he performed the operation of laryngo-oesophago-pharyngectomy with the construction of a skin hypopharynx and cervical oesophagus to restore perfect deglutition. The operative mortality is around 6 per cent when all patients are included who died in hospital within 2 months of the operation.

Surgeons must always be critical of their methods and end-results and be prepared to change them when this is indicated. The place of radic surgery for malignant diseases is now being challenged in favour of more local operations, and many ask the question whether lymph-node block dissections should be discarded. Thus the treatment of breast carcinoma is the centre of much controversy with considerable indecision about the value of an axillary dissection, and even the removal of the breast. I do not propose to discuss these subjects here but use them to stress the importance of a thorough knowledge of basic sciences, especially pathology, to guide the surgical oncologist in this responsible work. Education is a continuing process throughout our professional lives, and a constant critical examination of our end-results is essential. In addition, we can learn much from the experience of others, and today a great deal of help is expected from controlled clinical trials and closer collaboration with research oncologists.

CONCLUSION

In this paper an outline is given of the education and training of the Surgical Oncologist. The basic requirement in training is a sound surgical technique in general surgery with fine surgical judgement.

Later the surgical oncologist may be trained in a surgical speciality and become well acquainted with radiotherapy and chemotherapy. In the education knowledge of the basic medical sciences is essential, which is a continuing process throughout professional life, and constant contact with research oncology is very beneficial in this work.

Surgical Management of Regional Lymph Nodes

A. Kulakowski

Department of Surgery, Maria Sklodowska-Curie Memorial Institute of Oncology, Warsaw, Poland

The notion, that for obtaining cure of cancer, the primary tumor with regional lymph nodes should be removed, first appeared in the early XXth century. Halsted, Crile, Miles, and many other surgeons developed and put into practice some extensive procedures called radical operations for cancer.
The survival rate analysis have shown that even very sophisticated and supraradical operations are of limited curative value. Basic research and experimental data as well as growing clinical knowledge, have indicated that in most instances clinicians are treating patients with a systemic cancer. Even the supraradical, but local or loco-regional treatment, is under these circumstances obviously ineffective. In such cases the systemic adjuvant chemotherapy has proved promising. The integrated cytoreductive therapy becomes a new goal in cancer treatment. According to the stage of disease this therapy applies the optimal loco-regional treatment with most effective drugs in various time combinations of systemic chemotherapy.
Many experimental and clinical data have led to the conclusion that the presence of metastases in the lymph nodes reflects the agressivenes of primary tumor and/or the impairment of host defence mechanisms. As the contemporary indications for lymphangectomy for staging or curative purposes are the subject of aur present discusion, we should try to answer some very practical questions as follows:

1. When, in the course of neoplastic disease, the lymph nodes become metastatically involved?
2. How the tumor grows in the lymph nodes?
3. What is the role of the lymph nodes in the immune response against the tumor cells?
4. Is it possible to discover metastases in unenlarged lymph nodes?

5. Is it necessary to remove metastatically involved lymph nodes?
6. Is it harmful to remove histologically negative lymph nodes?

1. It is generally accepted that as the tumor mass increases, so does the time it takes to double its volume. A clinician might be able to recognise 0,5 - 1 cm minimal tumor mass. We would assume that the tumor was discovered quite early. Until then, the tumor had undergone 30 doublings, which is equivalent to 3/5 of its entire life span. This fact has several important clinical implications. Metastatic disease may well hed occured long before obvious evidence of the primary tumor. We should be aware that in most instances we are dealing with regional or systemic metastatic disease. In cases in which the diagnosis can be obtained in the "preclinical stage" of the disease, metastases are very rare. There are clinical observations proving this fact - Urban series of mastectomy patients with xeromammographically discovered lesion less than 1 cm in size, or Frazier and others 93,2 twenty year survival rate for minimal (less than 5 mm) breast cancer, results of treating early melanomas (Clark stage II) or minimal gastric cancer discovered by gastroscopy.

2. The lymph node represent a filtrative structure. The electronmicroscopic research provides details of the structure of the lymph node and the possibilities of lymphocytes and cancer cells migration. The mechanism of cancer cell inoculation and growth in a lymph node is complicated and still not clear. The number of cancer cells (the tumor mass) which enter to the node plays an important role. After inhibition of phagocytosis, the growing tumor tissue increases the intranodal lymph pressure, opens new lymph paths and allows futher spread of cancer cells. Some observations proved the direct lymphaticovenous communications (Pressman). In this way nodal metastases can give futher bloodborn metastases.

3. The experimental data show that tumor specific sensitization start in the regional lymph nodes. However, prior to the time when the primary tumor becomes palpable, the immune response become systemic. Tumor immunity extends to other lymphoreticular sites, thereby making the host less dependent on regional nodes to elicit a specific immune response. Thus the regional nodes operate as the only specific site of immune function for a brief period only during an early tumor development. When the tumor is clinically detectable the regional lymph nodes have already initiated tumor immunity and have transferred this immunity to other lymphoreticular sites. The regional lymphadenectomy after the tumor is clinically detectable is therefore not likely to decrease tumor immunity.

4. Unfortunately we do not have a precise diagnostic method for

evaluation of lymph node metastases. Lymphography and fine needle biopsy is helpful but not decisive in not enlarged, clinically negative nodes. The only precise diagnostic method still remains the same - histological examination of excised nodes.

5. The presence of regional lymph node metastases is associated with about 80% chance of distant metastases and in their absence there is only 20% recurrence rate. We also do know that metastases can give metastases. The prognostic information is very important. When the nodes are positive, surgery alone is often inadequate and additional therapy may be required. It allows the selection of those patients who have a poor prognosis for clinical trials.

6. Knowing the immune role played by the lymph nodes which is important only in a very early stage of disease - the excision of nodes when there is a reasonable suspition of involvement with metastatic tumor is justified. There is a frank clinical evidence that removal of microscopically involved nodes gives higher cure rates than of those massively involved. Even removal of uninvolved nodes has no influence on cure rates.

The lymph node dissection has two aspects:

1. curative approach - elimination of tumor mass and sources of further metastatic spread
2. staging procedure and prognostic value

The curative value of removal of metastatically involved nodes is limited - the disease in most instances is a systemic one. This allows the start of systemic therapy.
In the absence of clinically suspicious nodes, there is disagreement as to the indications for elective node dissection. The literature data indicate that survival is longer in patients with clinically negative nodes than in those with clinically positive ones. There is a justifiable concern regarding elective node dissection due to the high incidence of postoperative complications (specially in inguinal and neck areas). The survival is considerably longer following regional node dissection for clinically negative and pathologically positive nodes (micrometastases), than for clinically positive nodes. It is highly desirable to more accurately predict the presence of occult regional lymph node metastases preoperatively. The clinical experience and knowledge of prognostic factors should be a guide for indication for elective lymphangectomy.
This procedure could define a group of patients with a worse prognostic factors who may benefit from systemic treatment.
The indications for lymphangectomy are different according to histology, sise and location of primary tumor, clinical stage of disease and various prognostic factors. It should be staded that the greater the tumor diameter, the deeper the invasion - the highest percentage of lymph node involvement and the worse is the prognosis.

The en bloc excision and incontinuity node dissection should be performed if the tumor is located near the regional lymph nodes. If the lesion is located at a considerable distance from the enlarged regional nodes the discontinuous node dissection should accompany the excision of the primary. The operative technique should follows the principles of oncological surgery - removal of the primary tumor and regional lymph nodes with a resonable margin of surrounding healthy tissues.

REFERENCES

Fisher, B., E.R. Fisher (1966). Transmigration of lymph nodes by tumor cells. Science, 152, 1397-1398.

Frazier, T.G. et al. (1977). Prognosis and treatment in minimal breast cancer. Am.Journal of Surg., 133, 697.

Goldman, L.I., W.H. Clark, E.A.Bernardino, A.M.Ainsworth (1977). The accuracy of predicting lymph node metastases in malignant melanoma by clinical examination and microstaging. Ann.Surg., 184, 537-542.

Holmes, E.C. et al. (1977). A rational approach to the surgical management of melanoma. Ann.Surg., 186, 481-490.

Pendergrast, W.F. et al. (1976). Regional lymphadenectomy and tumor immunity. Surg.Gynec.Obstet., 142, 385-390.

Pressman, J.J., R.F.Dunn, M.Burz (1967). Lymph node ultrastructure related to direct lymphaticorenous communication. Surg.Gynec. Obstet., 124, 963-973.

Silver, R.T., R.C. Young, and J.F. Hdland (1977). Some new aspecta of modern cancer chemotherapy. Am.Journal of Medicine, 63, 722-787.

Sträuli P. (1971). La barrière lymphatique, in Saegesser F., J. Pettavel (Ed.). Oncologie chirurgicale, Masson and Cie, Paris, pp. 159-175.

Cancer of the Breast: Coadjuvant Chemotherapy With Two Drugs

E. Caceres, M. Moran, M. Lingan, M. Cotrina, L. Leon and F. Tejada*

Instituto Nacional de Enfermedades Neoplasicas, Lima, Peru
**Comprehensive Cancer Center for the State of Florida*
Supported in part by PAHO and National Cancer Institute (USA) Latin American Program

The concept of breast cancer as a systemic disease is not new. In 1896 Beatson suggested that "the tumor in the breast is only a local manifestation of a blood affection" concept which probably explains why the recurrence and survival rates for breast carcinoma have changed relatively little over the past few decades.

Surgery and radiation therapy are both regional approaches to what frequently is a systemic disease. It is not surprising, therefore, that enlarging the scope of surgery or adding one regional modality (radiation therapy) to another regional modality (surgery) fails to increase the overall survival time.

Recent experience with prolonged systemic chemotherapy after operation suggests that early recurrence can be delayed, suggesting that patients with primary cancer of the breast with regional axillary lymph nodes involvement have hematogenously disseminated disease at the time of mastectomy.

This report presents our experience with adjuvant chemotherapyin 122 patients with stage II breast carcinoma followed for at least 2 years and treated with a two drug regimen given for 2 years. The aim of this study was to use a combination of drugs that it is within the capability of any hospital in Peru.

MATERIAL AND METHODS

Therapy consisted of cyclic administration of Cyclophosphamide (100 mg/sq.m. orally from the 1st to the 14th day) and 5-Flourouracil (600 mg/sq.m. intravenously on the first and eight days). Each cycle of Cyclophosphamide and 5-Flourouracil was repeated at 4 weeks intervals. Patients were treated for two years period.

Dosage was modified according to the degree of depression of the white blood cell count and of platelet count.

Follow-up studies included physical examination, blood cell count, liver function, chest and bone roentgenograms and liver or bone scan when necessary.

Stratification was done according to age, 49 or younger and 50 to 75 years, premenopausal and post-menopausal status and number of positive axillary nodes (one to three and four or more). All patients in the current protocol were compared with comparable group of historical control patients from our Institution who had received radical surgery and no chemotherapy or radiotherapy had been administered. Patients were informed of the investigational nature of the study and a written consent was obtained from each patient.

The end point of study was the first evidence of treatment failure, represented by the development of a local recurrence on the chest wall, and regional lymph node area (axilla, supraclavicular or internal mammary chain) or distant metastases.

TOXICITY

Adjuvant chemotherapy was fairly well tolerated. This statement is supported by the fact that therapy has been discontinued in only 1 patient (for psychological reasons). Furthermore, patients have received 89 per cent of the calculated doses of Cyclophosphamide and 5-Flourouracil.

The incidence of side effects from chemotherapy, calculated from a positive response for any symptom at any time after the patient entered the study, were the following.

Most patients eventually had some fatigue and anorexia during the 14 days of chemotherapy, but all were able to continue working either outside or within the home.

The next most frequent complaint was nausea, 69 per cent. However, this symptom rarely persisted more than four or five days after the first day of treatment. When vomiting (47.7 per cent) occurred, it was within a few hours after drug injection and did not persisted. Stomatitis was rarely present (2.2 per cent). Diarrhea, occasional with abdominal pain, occurred in the 12.5 per cent. Alopecia was occurred in 69 of 12.2 patients (56.6 per cent) and was severe enough for patients to wear a wig. However, after the first six months of treatment the hair usually grew back. Cystitis, without hematuria was the complaint in patients or 4.5 per cent. Amenorrhea, in most permanent, occurred in 55.3 per cent of the pre-menopausal women.

Myelosuppression of the Wbc between 3000 to $3900/m^3$ were present in 22.7 per cent, between 2000 to 2900 in 25 per cent and bellow 2000 in 14.7 per cent. Thrombocytopenia was not a problem and rarely was < 100.000.

One patient developed a second primary, a transcicional carcinoma of the bladder 21 months after the onset of therapy.

The lack of statistical difference is in the post-menopausal group and also the pre-menopausal group with more than four lymph nodes, but there is significant disease free interval in pre-menopausal patients with 1 to 3 positive lymph nodes compared to historical control. Probably all pre-menopausal patients are helped by adjuvant chemotherapy but in our population of patients, the majority of them (>50%) had over 50 per cent of lymph nodes involved with tumor (average 12) indicating that tumor burden was already large for good chemotherapy control.

It may well be that the proper drug combination has not yet been obtained in either the pre or post menopausal groups.

In view of the lack of response of the post-menopausal group, perhaps we need in this group to use some very aggressive and controversial multidrug regimens or more effective combination such as chemotherapy or to extend the period of treatment over a longer classic period of one or two years and perhaps to consider the addition of immuno stimulants as Levimasole - BCG or Cori Bacterium Parvum or the use of antiestrogenics.

Blood Gas Changes in Hepatic Artery Ligation

R. E. Madden, H. de Blasi and J. Zinns

New York Medical College, 1249 Fifth Avenue, New York, New York 10029, U.S.A.

ABSTRACT

Hepatic artery ligation (HAL) is widely used in the treatment of metastatic and primary liver tumors. Significant morbidity has been frequently encountered. In this study, alterations in oxygen content with HAL have been used in selecting those patients who will safely tolerate the procedure. Six patients with liver metastases were studied intraoperatively. PO_2 and O_2 saturations were simultaneously determined in arterial (HA) portal (PV) and hepatic vein (HV1) blood. Hepatic vein blood was then restudied after clamping the hepatic artery (HV2). Oxygen content was calculated in blood from each of the three sample sites by standard methods. Percent relative drop in oxygen content of the liver (Δ%) was then estimated by the formula. Four patients underwent HAL. In these patients, Δ%s were 11, 19.9, 13 and 15. Two patients were denied HAL because of low HV2 oxygen saturation after clamping. Δ %s in these were 26 and 36. There were no surgical mortalities and no significant morbidity due to HAL in this series. Intraoperative blood gas determinations may be of value in selecting those patients who will tolerate HAL and those in which an alternate strategy should be used.

Keywords: Artery, ligation, blood gases, metastases, liver, safety.

INTRODUCTION

Hepatic artery ligation has been used to palliate metastatic disease to the liver.[1,2,3,4] The basis for this treatment is the belief that hepatic tumors probably derive their nutrition from the arterial circulation.[5,6] In addition, angiographic and histologic evidence of induced tumor necrosis secondary to interruption of hepatic arterial flow has been reported.[7,8,9]

The morbidity, mortality, and therapeutic benefits have been recorded previously.[1,2,10] Notwithstanding the anatomic variation and adaptability of the blood supply to the liver,[5,11,12] the morbidity following common hepatic artery ligation depends on the subsequent alteration in oxygen availability to the liver after ligation. In some cases a drastic drop might occur and prove significantly injurious to the normal hepatic parenchyma. In many others, however, common hepatic artery ligation may be safely employed.[1,2,3]

We have developed a facile method for quick estimation of the safety of common hepatic artery ligation. This involves the calculation of the percent drop in blood oxygen content produced by temporary clamping of the artery. An operative decision is then made to ligate the artery, or, if deemed unsafe, to employ other therapeutic approaches.

Material and Methods

Six patients with hepatic metastases were studied. Five patients had liver disease secondary to colonic carcinoma, while one patient had metastatic breast disease. There were four males and two females in the group. The youngest patient was 54 years old, while the oldest was 76 years old.

Pre-operative diagnosis of hepatic metastasis was made by clinical assessment, liver enzymes, liver scan, and angiographic data. In all cases the diagnosis was confirmed intraoperatively.

On the morning of surgery the patients were first taken to the angiographic suite where a catheter was placed percutaneously in the femoral vein and, under fluoroscopic control, into an hepatic vein. After securing the catheter in place, the patients were transferred to the operating room.

Following operative confirmation of the metastatic disease, samples of blood were taken from the hepatic artery (HA), portal vein (PV) and the hepatic vein (HV_1). The oxygen tension and saturation of these specimens were determined. A vascular clamp was then applied to the common hepatic artery and its flow temporarily occluded. The liver was observed over the next fifteen to twenty minutes and changes in color noted. In two patients, the liver became cyanotic, while in the others no significant color changes occurred. After this period of arterial occlusion, another sample of hepatic venous blood (HV_2) was obtained and its oxygen tension and saturation determined.

It was decided empirically that if the hepatic venous saturation dropped below fifty-five percent or the oxygen partial pressure below 30 mm of mercury, hepatic artery ligation would be abandoned. The mitochondrial oxygen gradient could be too low below these levels and jeopardize hepatocyte viability.[13,14,15] In two out of the six patients, this proved to be the case.

RESULTS

The two patients who developed an hepatic venous saturation below fifty-five percent were those two in whom cyanosis of the liver was noted (Case 1 and 5 - Table 1). The other four patients underwent hepatic artery ligation for their metastatic disease (Table 2).

TABLE 1 Two Patients Developing Liver Cyanosis

	HV1 SATURATION	HV_1	HV2 SATURATION	HV_2	PERCENT OXYGEN CONTENT CHANGE
Case 1	81%	[16.4]	48%	[9.71]	26.2%
Case 5	68%	[12.4]	43%	[7.80]	36.2%

Definitions: HA=Hepatic Artery; PV=Portal vein; HV_1=Hepatic vein, pre-clamp; HV_2=Hepatic vein, post-clamp; [ ]=Oxygen content/100 ml.

TABLE 2 All Patients Studied, Hepatic Artery Clamping

CASE	AGE	DIAGNOSIS	OXYGEN CONTENT CHANGE	HEPATIC ARTERY LIGATION
1	64	Rectosigmoid carcinoma	26.2%	No
2	76	Carcinoma left colon	19.9%	Yes
3	67	Carcinoma left colon	11.2%	Yes
4	62	Rectosigmoid carcinoma	15.1%	Yes
5	54	Breast carcinoma	36.2%	No
6	64	Carcinoma left colon	10.7%	Yes

There were no surgical mortalities and no significant morbidity in this group.

Post-operatively, the six cases were reviewed and alterations in oxygen content were calculated. Oxygen content was calculated for each blood sample using the formula:

$$\text{Content} = 1.34 \times \text{Hgb} \times \text{Sat } \% \ O_2 = 0.0031 \times \text{partial pressure } O_2 \qquad (1)$$

The precent drop in oxygen content to the liver was then based on the premise that in the original pre-ligation state the hepatic oxygen content was equal to the sum of that delivered by the hepatic artery and portal vein minus the content of the hepatic vein: $HA + PV - HV_1$

In the post-ligation state, hepatic oxygen content was equal to that delivered by the portal vein minus the amount calculated in the hepatic vein, $PV - HV_2$. The ratio of the status of the post-ligation state to that of the pre-ligation state x 100 is defined as the percent change in oxygen content.

$$\frac{PV - HV_2}{HA + PV - HV_1} \times 100 = \% \text{ change in oxygen content.} \qquad (2)$$

Oxygen availability, which would necessitate flow measurements of the major vessels, and quantitation of arterial collateralization, were not determined. The hepatic vein oxygen saturation and tension signify tolerable oxygen gradients since these levels are greater than cellular levels.[13,11,14,15]

In the four patients who underwent and tolerated common hepatic artery ligation, the percent changes in oxygen content were 19.9%, 11.2%, 15.1%, and 10.7%. Two patients were denied hepatic artery ligation because of low venous saturation and cyanotic livers. The percent changes in oxygen content in these cases was 26.2% and 36.2%, representing significant alterations. (Table 2)

Discussion

It is recognized that the blood gas status at any time in a particular hepatocyte or tumor cell strongly depends upon its own metabolism as well as upon the nature of the microcirculation.[14] Nonetheless, certain trends in overall blood gas alterations can be interpreted as able to portend these effects. With this in mind, if one were able to calculate the oxygen content supplied to the liver, and then measure again the amount after ligating the common hepatic artery, the safety of the procedure might be predicted. If the difference in oxygen content is relatively small. i.e. less than 20%, assuring continued normal hepatic parenchymal supply, then interrupting the hepatic artery is a safe procedure.[16] On the other hand, if the drop in oxygen content to the liver is drastic enough to jeopardize normal cellular function, then the risk of the procedure must be questioned. The determination of blood gases intraoperatively appears to be useful in selected cases in which arterial interruption is contemplated as a mode of therapy.

REFERENCES

1. Fortner, J., R. J. Mulcure, A. Solis, R. C. Watson, R. B. Golbey. Treatment of Primary and Secondary Liver Cancer by Hepatic Artery Ligation and Infusion Chemotherapy. Annals of Surgery, 178: 2, 162, 1973.

2. Lien, W., N. Ackerman. The Blood Supply of Experimental Liver Metastases. Surgery, Vol. 68, No. 2, 334, 1970.

3. Mori, W., M. Masuda and T. Miyanaga. Hepatic Artery Ligation and Tumor Necrosis in the Liver. Surgery, Vol. 59, No. 3, 359, 1966.

4. Nagasue, N., K. Inokuchi, M. Kobayashi, Y. Ogawa, A. Iwaki, and H. Yukaya. Hepatic Dearterialization for Nonresectable Primary and Secondary Tumors of the Liver. Cancer, 38: 2593, 1976.

5. Healey, J. Jr. Vascular Patterns in Human Metastatic Liver Tumors. Surgery, Gynecology & Obstetrics, 120: 6, 1187, 1965.

6. Krogh, A. The Number and Distribution of Capillaries in Muscles with Calculation of the Oxygen Pressure Head Necessary for Supplying the Tissue. The Journal of Physiology, 52: 409-415, 1919.

7. Madding, G. F., P. A. Kennedy, S. Sogemeier. Hepatic Artery Ligation for Metastatic Tumor in the Liver. The American Journal of Surgery, Vol. 120: 95, 1970.

8. McDermott, W. Jr., T. Hensle. Metastatic Carcinoid to the Liver Treated by Hepatic Dearterialization. Annals of Surgery, 180: 3, 305, 1973.

9. Michaels, N. Newer Anatomy of Liver-Variant Blood Supply and Colateral Circulation, JAMA, 172: 2, 125, 1960.

10. Madding, G. F., P. Kennedy. Hepatic Artery Ligation. Surgical Clinics of North America, Vol. 52: 3, 1972.

11. Imamura, M., T. Suzuki, A. Nakase, and I. Honjo. Hemodynamic Changes in the Liver of the Rabbit After Hepatic Dearterialization. Surgery, Gynecology & Obstetrics, 140: 412, 1975.

12. McDermott, W. Jr., A. L. Parris, M. R. Clouse, W. A. Meissner. Dearterialization of the Liver for Metastatic Cancer: Clinical, Angiographic and Pathologic Observations. Annals of Surgery, Vol. 187, No. 1, 1978.

13. Davis, P. W., and D. W. Bronk. Oxygen Tension in Mammalian Brain. Federation Proceedings, 16: 689-692, 1957.

14. Kety, S. S. Determinates of Tissue Oxygen Tension. Federation Proceedings, 16: 666-670, 1957.

15. Sparks, F. C., M. B. Mosher, W. C. Hallauer, M. J. Silverstein, D. Rangel, J. E. Passaro and D. L. Morton. Hepatic Artery Ligation and Post operative Chemotherapy for Hepatic Metastases. Cancer 35: 1074, 1975.

16. Van Liew, H. D. Tissue Gas Tensions by Microtonometry: Results in Liver and Fat. Journal of Applied Physiology, 17: 359-363, 1962.

Eight Years' Experience With The Surgical Management of 321 Patients With Liver Tumors

J. G. Fortner, D. K. Kim, M. K. Barrett, S. Iwatsuki, D. Papachristou, C. McLaughlin and B. J. Maclean

Gastric & Mixed Tumor Service, Department of Surgery, Memorial Sloan-Kettering Cancer Center, New York, N.Y., U.S.A.

ABSTRACT

Three hundred twenty-one individuals with liver tumors have undergone exploratory laparotomy during the past eight years. Surgical procedures included hepatic lobectomy or segmentectomy (108 patients), hepatic vascular ligation and/or cannulation (134 patients), hepatic wedge resection (10 patients), unroofing of hepatic cyst (5 patients), isolation-chemoperfusion of the liver (4 patients), liver transplantation (8 patients), and liver biopsy only (52 patients). Ninety percent of these procedures were performed for malignant disease. The resectability rate for malignant disease was 33%.

This study focuses on the two most frequently performed operations- hepatic resection and vascular ligation and/or cannulation. One hundred eight individuals underwent resection : extended hepatic lobectomy (27), lobectomy (57), and segmentectomy (24). Primary liver cancer was present in 36, metastatic colorectal cancer in 25, miscellaneous other cancers in 27, and benign tumor, principally giant hemangioma, in 20. Forty-one resections utilized the isolation-perfusion technique. Standard resections employ an abdominal approach except when resection of the diaphragm is included. The 30-day operative mortality rate was 9% after resection. Three-year actuarial survival for all patients with malignant tumor undergoing resection was 46% being 81% for curative resection but only 18% for palliative resection.

Vascular cannulation and/or ligation was performed when there was diffuse hepatic involvement. Sites of primary tumor in this treatment group were liver in 37, colorectum in 55, and miscellaneous in 42. Individuals with vascular tumors, as determined by preoperative angiography, usually underwent hepatic artery ligation and cannulation (HALC); those with hypovascular tumors generally underwent HALC and portal vein cannulation. Eight of the nine postoperative deaths were early in the series and resulted from hepatorenal failure. This complication has been avoided in more recent cases by performing cannulation alone without ligation if tumor involvement encompasses 70% or more of the liver. Postoperative intrahepatic chemotherapy was administered according to one of three protocols for each of the primary site tumor groups. Individuals receiving at least two cycles of intrahepatic treatment had an improved one-year actuarial survival rate (33%) over those receiving fewer treatments (0%).

KEYWORDS

Hepatic resection
 Curative
 Major
 Palliative
 Wedge
Hepatic artery ligation/cannulation
Portal vein cannulation

This paper focuses on the two hepatic surgical procedures most frequently performed during the past eight years - hepatic resection and vascular ligation and/or cannulation. One hundred eight individuals underwent hepatic resection (Fortner and others, 1978). Eighty-eight (81%) of these resections were for malignant tumors. Of 36 primary liver cancers, 26 were hepatocellular, four were mixed hepatocellular and cholangiocarcinoma, three were sarcoma, one was cholangiocarcinoma, one was carcinoid, and one was unclassified. Additionally, there were five children with hepatoblastoma who were evaluated as a separate group. The majority of the metastatic tumors, 25 of 40, were colorectal in origin. Fifteen were of other types: melanoma in three, sarcoma in three, breast in two, carcinoid of the rectum in two, embryonal carcinoma in two, islet cell carcinoma of the pancreas in one, adenocarcinoma of the kidney in one, and adenocarcinoma of the stomach in one. There were four patients with primary gall bladder cancer and three with major bile duct cancer. Twenty of the resections were performed for benign disease: giant hemangioma of the liver in seven, focal nodular hyperplasia in five, cyst in three, hamartoma in three, adenoma in one, and Carroli's disease in one.

The thirty-day operative mortality rate for hepatic resection was 9% overall with a rate of only 4% for the standard type of resection. The operative mortality rate with the isolation-perfusion technique was 17% reflecting the advanced stage of disease in many individuals who underwent resection prior to 1975 (Fortner and others, 1974). The operative mortality rate with extended right hepatic lobectomy was 14%(4/27), right lobectomy 3%(1/36), left lobectomy 24%(5/21), and segmentectomy 0%(0/24). The rates were similar for curative and palliative resection being 10% and 12.5%, respectively. Cause of death for the ten patients dying within 30 days of surgery is shown in Table 1.

TABLE 1 30-Day Operative Mortality After Hepatic Resection

Cause of death	Number of patients
Coagulopathy with hepatic failure	4
Portal vein thrombosis (indwelling catheter)	1
Warm ischemia(40 min); postnecrotic cirrhosis	1
Unrecognized operative hemorrhage	1
Uncontrollable operative hemorrhage	1
GI bleeding, hepatorenal syndrome	1
Pulmonary embolus	1
Total	10

Several postoperative complications often occurred in the same patient. The most common was subphrenic abscess which developed in one-fifth of the patients. This has been virtually eliminated during the past two years following the introduction of a closed system of drainage.

Forty-six percent of the resections for malignant disease were termed curative. A resection was termed palliative if there was regional spread (23 patients) or distant metastasis (19 patients). The most common type of regional spread was vascular, either invasion of the intrahepatic vessels or of the vena cava (Fortner, Kallum, and Kim, 1977). Lymph node metastasis was the most frequent type of extrahepatic metastasis.

Actuarial survival rates were calculated for the first three postoperative years. The ten thirty-day operative deaths are excluded. Of the 36 individuals undergoing curative resection, 100% were alive at one year and 81% at three years(Table 2).

TABLE 2 Actuarial Survival in Individuals Undergoing Hepatic Resection

		Actuarial survival	
Curative resection	Number of patients	One year	Three years
Liver	13	100%	88%
Colorectal	17	100%	72%
Other	6	100%	83%
Total	36	100%	81%
Palliative resection			
Liver	16	78%	31%
Colorectal	6	56%	0%
Other	20	70%	0%
Total	42	71%	18%

Primary liver cancer had the highest survival rate, 88%, at three years. Approximately two-thirds of patients lived one year after palliative resection and 18% lived three years. Survival rates were similar for those with regional spread and with distant metastasis being 17% and 21% at three years.

The three-year survival rate was 48% for individuals with metastatic colorectal cancer undergoing either curative or palliative resection. This can be compared with a report by Attiyeh, Wanebo, and Stearns (1978) of 19 patients treated over a 26-year period who underwent wedge resection for metastatic colorectal cancer to the liver. The median size of the metastasis was 1.5 cm with 84% being solitary. This compares with a median size of 6.5 cm in this series, with 52% being solitary. The three-year actuarial survival rate was not significantly different for the two procedures: 56% after wedge resection and 48% after major hepatic resection. Thus, lobectomy is not an obligatory procedure. Resection of the metastasis with a margin of normal liver tissue can be a satisfactory procedure. The extent of surgical resection depends on the size and location of the tumor. A large mass of metastatic cancer need not preclude cure.

Survival rates for patients with primary hepatoma undergoing curative resection were high. Even patients undergoing a palliative resection had a good result. These results reflect, in part, the type of cancer being treated. Cirrhosis was considered a contraindication for major hepatic resection with only three such patients in this series. The survival rate for individuals with metastatic colorectal cancer undergoing curative resection also was excellent. Palliative resection for colorectal metastasis appears not to be indicated as no individuals have survived two years. The number of resections performed in individuals with the various other primary cancers is too small to provide meaningful survival figures.

Extensive liver involvement by primary or metastatic tumor precluded hepatic resection in 134 patients who underwent ligation and/or cannulation of the hepatic vessels. The primary tumor arose in the liver in 37 patients, in the colorectal area in 55, and at miscellaneous other sites in 42. The prognosis after hepatic artery ligation has been shown to correlate with the degree of tumor vascularity: patients with hypervascular lesions demonstrated the longest survival (Kim and others, 1977). It appears that hypervascular tumors derive their blood supply from the hepatic artery while hypovascular tumors receive blood from the portal vein. Thirty-six livers were classified as Grade I, II, or III in order of increasing vascularity as judged by preoperative arteriography. Over two-thirds of hepatomas were Grade III lesions while only 14% of metastatic colorectal carcinomas were hypervascular. Thus, individuals with vascular tumors usually underwent hepatic artery ligation and cannulation. Those with hypovascular tumors generally were subjected to hepatic artery ligation and cannulation with portal vein cannulation. Of 134 patients, 78 received hepatic artery ligation and cannulation, 24 hepatic artery ligation and cannulation combined with portal vein cannulation, 15 hepatic artery ligation alone, 8 portal vein cannulation alone, 5 hepatic artery ligation alone, and 4, other combinations. Eight of the nine 30-day operative deaths were early in the series and resulted from hepatorenal failure (Kim and others, 1976). The best treatment in such cases is prevention since renal failure following hepatic artery ligation was associated with a 78% mortality rate. This complication has been avoided in more recent cases by performing cannulation alone without ligation if tumor involvement encompasses 70% or more of the liver. A program of mannitol diuresis has practically eliminated renal failure as a postligation complication in this series.

Intrahepatic chemotherapy was administered according to one of three protocols depending on the primary site of the tumor (Fortner, 1977; Fortner and Pahnke, 1976). Heparinized saline is injected in the catheter following each infusion. If the catheter becomes occluded by thrombi, treatment is continued by peripheral intravenous route. Adequate treatment arbitrarily was defined as the administration of at least two cycles of drugs during the first three postoperative months. Adequate intrahepatic chemotherapy was administered to 65% of patients undergoing hepatic artery or portal vein cannulation. The actuarial survival rates in individuals receiving adequate intrahepatic chemotherapy are shown in Table 3.

TABLE 3 Adequate Intrahepatic Chemotherapy After Ligation/Cannulation

Primary tumor	Number of patients	Actuarial survival One year	Actuarial survival Two years
Liver	20	39%	28%
Colorectal	32	32%	12%
Other	23	26%	16%
Total	77	33%	18%

One-third of patients lived one year and 18% a second year. Survival was highest in the primary liver tumor group both at one and two years. In addition to a prolongation of life, patients receiving intrahepatic infusion of chemotherapy experienced significant relief from symptoms attributed to liver enlargement such as epigastric pain and fullness and right shoulder pain. Most patients were ambulatory and, with a few exceptions, were able to live a useful life with their families. No individuals receiving inadequate chemotherapy have survived one year

An additional eleven patients received adequate doses of chemotherapy via the intravenous route. Their survival rates were somewhat poorer than those treated intrahepatically, being 27% at one and 8% at two years. A trial is underway at the present time in which individuals are randomized to receive chemotherapy via either the intrahepatic or the systemic route.

Ligation/cannulation proved most successful in individuals with primary liver tumors, expecially if adequate chemotherapy was given. Patients with metastatic colorectal or other metastatic neoplasms did more poorly perhaps because most of them had already been failures from systemic chemotherapy.

REFERENCES

Attiyeh, F., H. Wanebo, and M.W. Stearns, Jr. (1978). Hepatic resection for metastases from colorectal cancer. Dis Colon & Rectum, 21, 160-162.

Fortner, J.G. (1977). Current management of tumors of the liver. In G.F.Madding and P.A.Kennedy (Eds.), Symposium on Hepatic Surgery, Surgical Clinics of North America, Saunders, Philadelphia. pp.465-472.

Fortner, J.G., B.O.Kallum, and D.K.Kim (1977). Surgical management of hepatic vein occlusion by tumor. Arch Surg, 112, 727-728.

Fortner, J.G., D.K.Kim, B.J.Maclean, M.K.Barrett, S.Iwatsuki, A.D.Turnbull, W.S. Howland, and E.J.Beattie, Jr.(1978). Major hepatic resection for neoplasia. Personal experience in 108 patients. Ann Surg, 188,363-371.

Fortner, J.G. and L.D.Pahnke (1976). A new method for long-term intrahepatic chemotherapy. Surg Gynecol Obstet, 143, 979-980.

Fortner, J.G., M.H.Shiu, D.W.Kinne, D.K.Kim, E.B.Castro, R.C.Watson, W.S.Howland, and E.J.Beattie, Jr. (1974). Major hepatic resection using vascular isolation and hypothermic perfusion. Ann Surg, 180, 644-652.

Kim, D.K., R. Penneman, B.O.Kallum, M.Carillo, E.Scheiner, and J.G.Fortner (1976). Acute renal failure after ligation of the hepatic artery. Surg Gynecol Obstet, 143, 391-394.

Kim, D.K., R.C.Watson, L.D.Pahnke, and J.G.Fortner (1977). Tumor vascularity as a prognostic factor for hepatic tumors. Ann Surg, 185, 31-34.

Preoperative Chemotherapy of Gastrointestinal Tumors: A Feasibility Study

A. Ch. Avgoustis*, G. P. Stathopoulos, A. B. Polychronis* and A. N. Papaioannou***

**B' Surgical Unit, Evangelismos Medical Center, Athens, Greece*
***B' Department of Medicine, University of Athens, Ippokration General Hospital, Athens, Greece*

ABSTRACT

The feasibility of preoperative chemotherapy was studied in a series of patients with various epithelial tumors of the gastroesophageal junction, stomach, periampullary region, colon and rectum. The rationale for the administration of chemotherapy before rather than after operation is based on the assumption that by so doing, we may inhibit growth of existing micrometastases or the development of new ones in the perioperative period when immunosuppression and other tumor-enhancing factors are at work. Operation follows chemotherapy by two to three weeks when relative immunostimulation develops and this hopefully offsets at least partially the suppression of immunity of the patient undergoing surgery. In tumors of the colon and rectum, but not of other sites, preoperative chemotherapy as used, was effective in reducing the size of the primary by 50% or more and in improving the operative conditions due to reduction of blood loss during operation. There was no increase in mortality or morbidity and the entire operative venture is made faster and safer, but the efficacy of this policy remains to be tested further.

Preoperative chemotherapy; GI tumor; Adjuvant treatment

INTRODUCTION

It is now generally accepted that most, if not all patients with common solid tumors such as of the breast, colon and others, have disseminated disease at the time of diagnosis. This belief led to the formulation of the concept of micrometastases which are now considered the fundamental failure cause of the standard treatment for solid tumors in man (Frei III, 1977; Papaioannou and others, 1978). In the current clinical research effort various treatment schedules are employed for the treatment of the systemic component of the disease after operation. Unfortunately, recent results from completed clinical trials did not fulfill our expectations. All prospectively randomized and controlled studies have failed to show any worthwhile advantage of patients receiving postoperative adjuvant chemotherapy for cancer of the stomach (Rake and colleagues, 1976; Kingstone and coworkers, 1978) and colorectal cancer (Dwight and others, 1973; Grage and colleagues, 1977; Higgins and co-workers, 1976; Lawrence and others, 1975; Grossi and colleagues, 1977).

The incresingly more convincing evidence that various factors during the periooperative period may in some respects be harmful to the ultimate survival of the cancer patients led us to study a change from the presently popular policy of operating first and then give chemotherapy, if necessary, into starting with chemotherapy and then operate. The rationale for this change is based on well documented evidence for the presence of immunosuppression and other tumor-enhancing factors induced in the perioperative period which are likely to result in enhancement of existing or new micrometastases (Paparioannou and co-workers, 1978a). In theory, this could be prevented either by reducing the potential for growth of micrometastases or even by altogether eliminating them before operation. In our past experience with surgical complications of advanced neoplastic intrabdominal disease, we found that operations after chemotherapy were usually safe if performed 2 to 4 weeks following treatment. This experience concerned a substantial number of patients with colorectal or other gastrointestinal tumors in whom such intervention became imperative usually for emergencies. With this knowledge we began using preoperative chemotherapy in various schedules to progressively less advanced and more recently in operable tumors of the gastrointestinal tract.

The present report concerns the initial phase of application of this principle in patients with adenocarcinoma of the gastrointestinal tract whose prognosis leaves much to be desired at an international level.

PATIENTS AND METHODS

Preoperative chemotherapy was given to five patients with locally advanced carcinoma of the colon invading the uterus, abdominal wall, or iliac bone. Four other patients received preoperative chemotherapy for carcinoma of the rectum invading the urinary bladder, prostate or uterus. It was also given to a patient with obstructive jaundice from an inoperable carcinoma of the common bile duct, diagnosed by laparoscopy, to another with locally advanced carcinoma of the stomach and to two women with palpable carcinoma of the prepyloric region of the stomach extending to the pyloric canal. Various drugs and schedules were used in these patients brefore operation, without a set protocol. Following gradual modification in the type of agents and schedules used, we began a new prospective study in November, 1976 (Papaioannou and others, 1978), administering the following preoperative chemotherapy scheme:

Day 1 Oncovin I.4 mg/m^2 maximun 2 mg iv push at 20.00 hours

Day 2 Methotrexate 30 mg/m^2, 5-Fluorouracil 300 mg/m^2 and Adriamycin 20 mg/m^2 all iv push at 08.00 hours

Day 3 5-Fluorouracil 300 mg/m^2 iv push at 08.00 hours

Day 4 5-Fluorouracil 300 mg/m^2 and Adriamycin 20 mg/m^2 iv push at 08.00 hours

Not all of these agents are effective as single agents in metastatic gastrointestinal tumors. In fact, some other agents could have been selected for that purpose. The above combination, however, lacks substantial hematological toxicity and is not overtly immunosuppressive so that operation can be safely done within two weeks in most cases. On the other hand, this regimen is used primarily to affect the micrometastases and not the primary tumor. Since the sensitivity of microscopic foci is not necessarily that of gross disease (Freii III, 1977), we felt that a combination with phase specific and non-specific cycle dependent agents and at the same time not particularly toxic or immunosuppressive, would be more appropriate than a combination of agents that would have been chosen for metastatic disease.

One half hour before operation, we begin 500 cc of Ringer's solution with 1000 mg 5-Fluorouracil in IV drip that lasts for six hours. One half of this solution is given during operation (usually over a two hour period so that higher levels are achieved during operation) and the remaining over the ensuing four hours.

RESULTS

All patients with colorectal cancer appeared to have benefited from preoperative chemotherapy. Specifically, four patients with advanced colorectal tumors had uneventful postoperative course. The other patient with advanced carcinoma of the cecum invading the abdominal wall developed pancytopenia and septicemia which forced delay of the operation (right hemicolectomy with abdominal wall excision) for one month. Postoperatively the same patient bled into the abdomen and eventually died from sepsis due to rupture of her ileotransverse anastomosis. From the above patients, two developed urinary tract infection and one infection of the perineal wound. The two women with advanced carcinoma of the pylorus were submitted to radical gastrectomy and had uneventful postoperative course. Although limited, our experience with these gastric carcinoma cases was that tumor size reduction was either minimal or not achieved at all by preoperative chemotherapy. The patient with carcinoma of the esophagogastric junction probably had some response to preoperative chemotherapy since his dysphagia improved, but he died four weeks postoperatively from sepsis due to rupture of the anastomosis in the chest.

Local complications of the wound or intrabdominal complications other than those mentioned, e.g. abdominal wound infections, seromas, hematomas, dehiscences, postoperative hernias, intrabdominal abscesses, rupture of anastomoses, prolonged postoperative paralytic, ilens, or postoperative bowel obstructions were not observed. Systemic complications other than those mentioned, or thromboembolic episodes were likewise not recognized. In all patients with colorectal tumor an impressive reduction of tumor size was observed close to or more than 50% in all patients; in fact, the patient who developed pancytopenia had a tumor reduction close to 90%. The general impression was gained that operative difficulties not only were not increased as a result of preoperative chemotherapy, but if anything, were decreased because of reduced bleeding and greater ease of tissue plane dissection.

SIDE EFFECTS

One of the patients with carcinoma of the stomach developed a severe anaphylactoid reaction on the third day of his preoperative treatment and died on the 17th day without an operation (Bellenis and colleagues, 1978). The patient with common bile duct carcinoma had no response from the preoperative chemotherapy and died from liver failure ten days after chemotherapy without an operation. Almost all our patient receiving preoperative chemotherapy showed varied degress of transient alopecia, and of gastrointestinal disturbances.

DISCUSSION

The current clinical trials using postoperative adjuvant chemotherapy for solid tumors have shown that the long term results for gastric and colorectal tumors are disappointing. We have proposed that a probable cause of failure is the enhancement of micrometastases during and after operation, as a result of perioperative immunosuppression and other tumor-enahncing factors which are at work during this period and the delay in their treatment (Papaioannou and co-workers, 1978a).
The magnitude of the problem is not apparent since estimation of the total burden

of micrometastases is impossible with present technology; theoretically, however, it could be equal or even larger than that of the primary at the time of diagnosis (Frei III, 1977). It is quite possible that entrenchment of these microscopic foci during perioperative immunosuppression as a result of operative stress, anasthesia and various drugs (Riddle and Berenbaum, 1967; Park and others, 1971; Cochran and colleagues, 1972) but possibly also as a result of the preoperative anxiety or postoperative depression (Papaioannou, 1974) may be important for the ultimate outcome of these patients. On the other hand, chronic nutritional deficits of patients with neoplasms of the gastrointestinal tract may also affect their immunity. This is particularly important in patients with neoplasms of the large bowel and rectum, who as a group exhibit substantially lower immune responses than most patients with neoplasms at other sites (National Research Council, 1977; Wanebo and co-workers, 1978). It could then be conceivably harmful to operate first on the primary, when the conditions are favorable for enhancement of the existing micrometastases or the development of new ones.

The choice of chemotherapeutic agents as previously mentioned was made to affect a broad spectrum of tumor cells (in and out of cell-cycle) while interfering as little as possible with the immune and hematological status of the host so that operation may be done after the shortest possible interval. With this in mind we selected minimally immunosuppressive or myelotoxic both phase-specific and cycle specific agents, lacking overlapping toxicity. 5-Fluorouracil during and after operation for six hours is given in order to affect the potential of cancer cells forced into the circulation during operative manipulations and which is certain from previous work in man (Roberts, 1961) to increase considerably as soon as manipulations begin. Quantitation of cell-shedding to the best of our knowledge has not been attempted in man but its magnitude at least in one experimental system is as high as 1,000,000 cells per gram of tumor per 24 hours (Butler and Gullino, 1975). Although it is hoped that after preoperative chemotherapy a substantial part of the primary tumor is deprived of clonogenic cells capable of forming metastases, we felt that minimally immunosuppressive intraoperative chemotherapy may be of some value.

The timing of operation after chemotherapy is particularly important to the patient. It is well known that immunosuppression resulting from chemotherapy is followed by a rebound of immunological capability that reaches beyond the pre-chemotherapy level depending upon the combination used in about two weeks (Hersh and Oppenheim, 1967). We felt that this period of relative immunostimulation of the host may be ideal to perform the operation in order to offset some if not all of the operative immunosuppression. During this period of immunostimulation the white cell count goes back to normal, and there is a relative increase of peripheral lymphocyte numbers and of their functional activity that usually last for one more week (Harris and others, 1976).

There are two important observations from this preliminary study (a) that patients with carcinoma of the colon and rectum receiving preoperative chemotherapy had a gross reduction of their primary tumor, in some instances more than 50% of their original size. This is a substantial therapeutic gain since the main cell population affected by chemotherapy is that of actively dividing (clonogenic) cells among which are cells that are capable of forming metastases. This means that a 50% size reduction may result in probably more than 99% reduction in clonogenic cells (Fisher, 1977). (b) during operation, we observed substantial reduction of blood loss, that has been also observed during mastectomies of our patients who receive preoperative chemotherapy. This is an important event from the technical point of view since it facilitates operations and may, therefore, be considered an additional benefit of the policy. It was felt, however, that immediate benefit did not occur in patients with gastric and pancreatic cancer with the agents and schedules used. This does not preclude, however, the possibility that more effective drugs may not be used to greater advantage if given before operation in these patients as well. Serious untoward effects of preoperative chemotherapy on postoperative wound healing, infection or systemic complications were not observed. Both these observations were later confirmed during our prospective trial (Papaioannou and colleagues, 1978).

REFERENCES

Bellenis, I. P., N.K., Sphyras, and A.N. Papaioannou (1978). Post-chemotherapy anaphylactoid reactions. Presented at the annual meeting of The Mediterranean Society of Chemotherapy, Madrid, September 19.

Bonadonna, G., A. Rossi, P. Valagussa, A. Banfi, and U. Veronesi (1977). The C.M.F. program for operable breast cancer with positive axillary nodes. Updated analysis on the disease free interval, site of relapse and drug tolerance. Cancer, 39, 2904-2916.

Butler, T.P., and P.M. Gullino (1975). Quantitation of cell-shedding into efferent blood of mammary adenocarcinoma. Cancer Res., 35, 512-516.

Cochran, A.J., W.G.S. Spilg, R.M. Mackie, and C.E. Thomas (1972). Postoperative depression of tumor-directed cell-mediated immunity in patients with malignant disease. Br. Med. J., 11, 67-70.

Dwight, R.W., E.W. Humphrey, and G.A. Higgins (1973). FUDR as an adjuvant to surgery in cancer of the large bowel. J. Surg. Oncol. 5, 243-251.

Fisher, B., A. Glass, and C. Redmond (1977). L-phenylalanine mustard (L-PAM) in the management of primary breast cancer. An update on earlier findings and a comparison with those utilizing L-PAM plus 5-Fluorouracil. Cancer, 39, 2883-2903.

Grossi, C.E., W.I. Wolff, and T.F. Nealon (1977). Intraluminal fluorouracil chemotherapy adjunct to surgical procedures for resectable carcinoma of the colon and rectum. Surg. Gyn. + Obstet., 145, 549-554.

Harris, J., D. Sengar, T. Stewart, and D. Hyslop (1976). The effect of immunosuppressive chemotherapy on immune function in patients with malignant disease. Cancer, 37, 1058-1069.

Grage, J., G. Cornelle, K. Stravits, K. Jonas, and G. Metter (1975). Adjuvant therapy with 5-FU after surgical resection of colorectal cancer. Proc. Assoc. Cancer Res., 16, 258.

Hersh, J.B. Jr., and J.J. Oppenheim (1967). Inhibition of in vitro lymphocyte transformation during chemotherapy in man. Cancer Res., 27, 98-102.

Higgins, G.A., E. Humphrey, G.L. Juler, H.H. Leveen, J. McCoughan, and R.J. Keuhn (1976). Adjuvant chemotherapy in the surgical treatment of large bowel cancer. Cancer, 38, 1461-1467.

Kingstone, R.D., D.J. Ell, J. Powel, V.S. B rookes, J.A.M. Waterhouse, M.D. Hurst, and J.A. Smith (1978). The west midlands gastric carcinoma chemotherapy trial. Planning and results. Clin. Oncol., 4, 55-69.

Lawrence, W., J.J. Terz, J.S. Horsley, R.E. King, W. Lovett, R.W. Brown, B.W. Ruffner, and W. Regelson (1975). Chemotherapy as an adjuvant to surgery for colorectal cancer. Ann. Surg., 181, 616-623.

Lawrence, W., J.J. Terz, J.S. Horsley III, P.W. Brown, and C. Romero (1978). Chemotherapy as an adjuvant to surgery for colorectal cancer. Arch. Surg., 113, 164-180.

National Research Council (1977). Review of gastric and colorectal cancer. Report of a conference held on April 2 and 3, Edinburgh.

Papaioannou, A.N. (1974). The Etiology of Human Breast Cancer. Endocrine, Genetic, Viral, Immunologic and Other Considerations. Springer-Verlag, Berlin, New York.

Papaioannou A.N., J.N. Nomicos, G.P. Stathopoulos, A. Ch. Avgoustis, and J. Th. Olympitis (1978). Pre- and postoperative chemotherapy with or without anticoagulation for colorectal cancer. Presented at British Society for Surgical

Oncology, Cardiff, July 21.

Papaioannou, A.N., M.B. Scouros, A. Ch. Avgoustis, J. Th. Olympitis, and I. Bellenis (1978). The management of solid tumors: Is it time for changes in surgeons' attitudes. Scient. Exhibit S-121, Am. Col. Surgeons Congress, San Francisco, October.

Park, S.K., J.I. Brody, H.A. Wallace, and W.S. Blakemore (1971). Immunosuppressive effects of surgery. Lancet, 9, 53-54.

Rake, M.O., C.N. Mallinson, B.J. Cocking, M.T. Cynaski, C. Fox, A. Jackson, and B. Diffey (1976). Assessment of the value of cytotoxic therapy in the treatment of carcinoma of the stomach. Gut, 17, 832.

Riddle, P.R., and M.C. Berenbaum (1967). Postoperative depression of the lymphocyte response to phylohemmagglutinin. Lancet, I, 746-748.

Roberts, S.S. (1961). Spread by the vascular system. In W.H. Cole, G.D. McDonald, S.S. Roberts, H.W. Southwick (Ed.), Dissemination of Cancer; Prevention and Therapy. Appleton-Century Crofts, Inc., 1961, New York, New York.

Wanebo, H.J., C.M. Pinsky, E.J. Beattie Jr., and H.F. Oettgen (1978). Immunocompetence testing in patients with one of the four common operable cancers. A review. Clin. Bull., 8, 15-22.

The Ileocecal Bladder After Radical Cystectomy (A Study of 62 Cases)

M. Khafagy, M. N. El-Bolkainy, M. Bahgat, A. Osman and A. El-Said

National Cancer Institute, Faculty of Medicine, Cairo University, Egypt

ABSTRACT

62 patients underwent radical cystectomy and ileocecal bladder reconstruction for carcinoma of the bladder. The ileocecal region was used as an artificial bladder, its valve protecting the kidneys from back pressure exerted by the external urethral sphincter. The ureters were anastomosed to the ileal part and the cecum was anastomosed to the urethra or bladder neck. Nine patients died postoperatively. All patients except one had perfect control of micturition, during the day. Two female patients in whom the bladder neck was preserved were also fully continent at night. The remaining patients had a minor degree of nocturnal enuresis with only a few drops of urine voided involuntarily during sleep. All patients feel the desire to void when the bladder is full. Voiding cystograms in ten patients did not reveal reflux in the ureters. IVPs done up to 6 years postoperatively showed preservation of kidney function and configuration in the majority of patients. A balanced "vesico-urethral" unit could be achieved as evidenced by the normal flow curves recorded.

INTRODUCTION

The external urethral sphincter with its intact nerve supply is often preserved after radical cystectomy for carcinoma of the bladder. Thus the use of a loop of bowel to replace the bladder after total cystectomy usually gives the patient a continent bladder but results in ureteral reflux which leads to deterioration of the renal function. The problem of urinary diversion after radical cystectomy is difficulty of imitation of the normal ureterovesical mechanism. As an attempt to solve this problem we decided to use the ileocecal region as an artificial bladder (Khafagy,1975) because of its valve that would protect the kidneys from the back pressure exerted by the external urethral sphincter. Both ureters were anastomosed to the ileum and the cecum was anastomosed to the urethra or bladder neck. Recently, Sevaggi and associates (1972) demonstrated the competence of the ileocolic valve of dogs in preventing ureteral reflux.

MATERIAL

In the period from 1972 to 1977, 62 patients with carcinoma of the urinary bladder underwent radical cystectomy and ileocecal bladder reconstruction. The procedure was carried out in one stage in 53 patients and in two stages in nine. 52 were male and 10 were female. Their age ranged from 30 to 65 years with a median of 45 years. The clinical staging of the disease is presented in Table 1. Most of the patients had T_3 lesions (83.6%). The lower border of the tumor was determined by cystoscopy.

TABLE 1. Clinical Stages (55 cases)

Stage	NO.	Per Cent
T_1	4	7.2
T_2	4	7.2
T_3	46	83.6
T_4	1	1.8
Total	55	

In 7 cases, operated elsewhere, information on stage was unknown. Colonic preparation was started 5 days preoperatively with plain water enemas, saline catharsis and kanamycin orally. A preliminary report of the first 10 patients was previously published (Khafagy, 1975).

SURGICAL TECHNIQUE

The patients were explored through a long right paramedian incision, extending from the symphysis pubis to about 5 cm. above the umbilicus. Dissection of the iliac lymph nodes was done on both sides. After the posterior dissection and division of the lateral and posterior ligaments of the bladder, a vaginal clamp was applied about one inch below the lower border of the tumor and the specimen was transected. An adequate safety margin was assured by frozen section examination.

The distal level of transection of the specimen as outlined in Table 2 was dictated by the lower border of the tumor.

TABLE 2. Level of Transection of the Cystectomy Specimen

Level	NO.	Per Cent
2 cm. above bladder neck	1	1.6
Bladder neck	24	38.7
Middle of the prostatic urethra	19	30.6
Lower third of the prostatic urethra	8	12.9
Membranous urethra	3	4.8
Previous cystoprostatectomy	7	11.3
Total	62	

A previous radical cystoprostatectomy with urinary diversion, was done elsewhere in seven patients, ureterocutaneous in five, ileal conduit in two and rectal bladder was done in two other cases. The operation was done in two stages intentionally in two patients because of the poor general condition of the patients. Radical cystectomy and ureterocutaneous was performed as a first stage followed six months later with ileocecal bladder reconstruction.

The ileocecal region was mobilised and the segment was isolated, based on the ileocolic artery (Fig. 1,A and B). The ileal part was

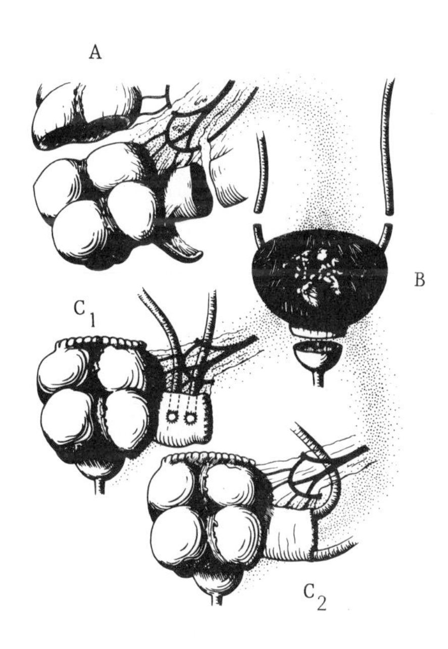

Fig. 1. Diagram of ileocecal bladder reconstruction. A, isolation of ileocecal segment. B, cystectomy for carcinoma of bladder. C_1, end-to-side anastomosis of ureters to ileum. C_2, end-to-end anastomosis of ureters to ileum.

about 10 cm. and the cecal part was from 15 to 20 cm. long. The ileum was then anastomosed to the ascending colon to restore continuity of the bowel. Both ureteral ends were anastomosed to the ileal part of the new bladder, either end-to-end or end-to-side (Fig. 1, C_1 and C_2). The cut end of the ascending colon in the ileocecal segment was closed in 2 layers using 3-zero chromic catgut sutures. The appendix was removed and a Foley catheter was inserted through its lumen and brought through the abdominal wall as a cecostomy. The lower part of the cecum was anastomosed to the urethra or to the bladder neck, using 1 layer of interrupted 1-zero chromic catgut sutures. Two polyethylene tubes were inserted on the side wall of the pelvis and connected to low grade suction for drainage. A Foley catheter was inserted through the urethra into the new bladder. The estimated blood loss was about 1 to 2l. during the procedure. The average operating time was from 4 to 4.5 hours.

RESULTS

Convalescence was uneventful in 47 patients. Nine patients died in the postoperative period, (Table 3). Advanced bilharzial fibrosis of the liver was usually associated with bilharzial bladder cancer. Liver failure with severe jaundice and coma was the cause of death in two patients.

TABLE 3. Postoperative Mortality

Cause	NO.
Peritonitis	2
Liver failure	2
Heart failure	2
Septic shock	2
Secondary haemorrhage	1
Total (14.5%)	9

Six patients developed 9 complications (Table 4). Persistent suprapubic fistula for more than 6 months occurred in two patients. One of them healed one year later after two repeated transurethral resection of the new bladder neck. One patient developed a vesicovaginal fistula in the postoperative period which was closed abdominally three months later.

TABLE 4. Postoperative Morbidity

Type	NO.
Wound Sepsis	5
Persistent suprapubic urinary fistula	2
Vesicovaginal fistula	1
Hematemesis (stress ulcer)	1

Pathological examination of the specimens revealed that the majority of the cancers were of the P_3 category (75.8%). Superficial tumors (P_1 & P_2) accounted for only 9.7% of cases, (Table 5). Pelvic lymph node metastases, mostly in the obturator group were encountered in six patients.

TABLE 5. Pathological Staging

Stage	NO.	Per Cent
Pis	-	-
Pla	-	-
Plb	1	1.6
P2	5	8.1
P3	47	75.8
P4	2	3.2
Previous cystectomy	7	11.3
Total	62	

Follow up of the patients ranged from 6 months to 6 years. Eight of them were operated upon than 5 years ago.

Fourteen of 44 patients who underwent one stage radical cystectomy and ileocecal bladder reconstruction developed lateral pelvic wall recurrence (Table 6). Most of these patients died within the first year of follow up. It was found that a safety margin of one inch below the lower border of the cancer was adequate as evidenced by absence of local recurrence at the new "bladder" neck or in the urethra.

TABLE 6. Recurrence after radical cystectomy and ileocecal bladder (one stage).

Site	NO.	Per Cent
Local at the cecourethral region	-	-
Pelvic wall	14	31.8
Systemic	5	11.4
NED	21	47.7
Untraced	4	9.1
Total	44	

Although continence could be achieved in the majority of patients, 90.6% (Table 7), nocturnal enuresis was encountered in all but two. It is worth noting that complete continence was observed among two female patients in whom the level of transection of the specimen was at the bladder neck.

TABLE 7. Urinary Continence after Ileocecal Bladder.

Status	NO.	Per Cent
Diurnal and nocturnal continence	2	3.8
D. continence and N. incontinence	46	86.8
D. and N. incontinence	1	1.9
Lost to follow up	4	7.5
Total	53	

During the first few months after surgery, most of the patients had extreme frequency of micturition i.e. every 10 to 15 minutes, which improve gradually by teaching them perineal exercises. At the end of one year, a reasonable rate of voiding was obtained, every 2 to 4 hours or more, in 67.8 per cent (Table 8). Furthermore the five patients who survived five years or more could control urine for periods from four to eight hours.

TABLE 8. Frequency of micturition after one year

Rate of voiding	NO.	Per Cent
D. and N. incontinence	1	3.6
Every half hour	1	3.6
Every 1 - hour	6	21.4
Every 2 - hours	6	21.4
Every 4 - Hours	13	46.4
Persistent urinary fistula	1	3.6
Total	28	

During follow up, eigth patients developed stricture at the cecourethral anastomosis, five of them followed a second stage ileocecal bladder reconstruction. Seven were successfully treated; one by intermittent dilatation and 6 by transurethral resection.

Urographic studies were done every 6 months after the operation. Functional evaluation of the results of this procedure was restricted to the 30 patients who survived the procedure free of disease. In 56 of 59 renal units, the follow up configuration of the pelvicalyceal system was unchanged or improved compared to the preoperative status (Fig.2). Deterioration was observed in three kidneys due to the development of stricture at the ureteroileal anastomosis (Table 9).

TABLE 9. Radiological appearance of 59 renal units

Status	NO.	Per Cent
Unchanged	54	91.5
Improved	2	3.4
Deterioration	3	5.1
Total	59	

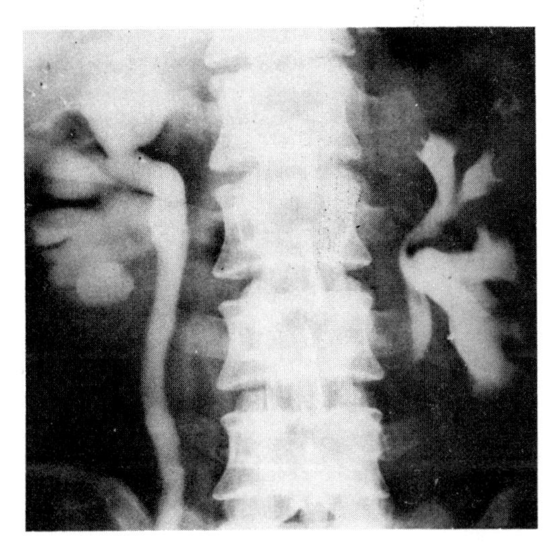

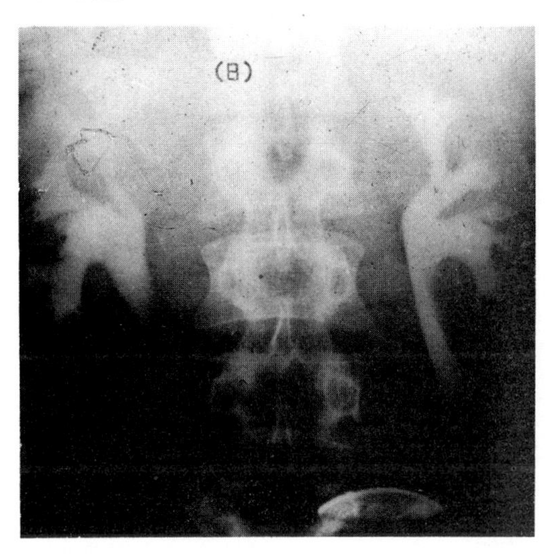

Fig. 2. I.V.P. A, preoperative B, 6 ys. postoperative. Demonstrates improved appearance.

The capacity of the new bladder ranges from 300 to 350 ml. Voiding cystograms done in 10 patients, did not reveal any ureteral reflux (Fig. 3). Pyelonephritis was observed in only 4 patients in this series.

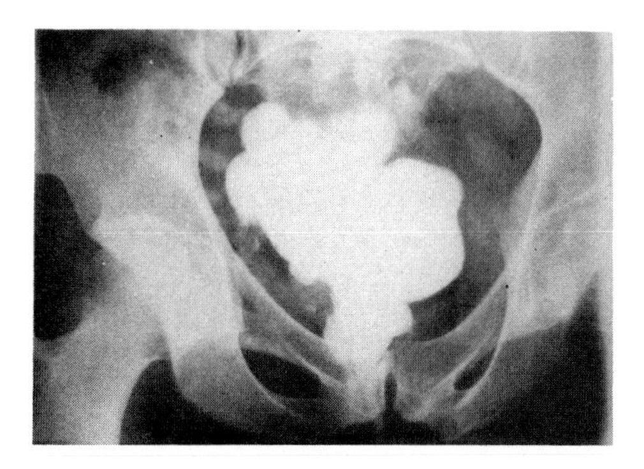

Fig.3. Voiding cystogram of ileocecal bladder shows absence of reflux.

Five of the patients who survived for more than one year were subjected to a urodynamic assessment by recording their flow rate, voiding and intra-abdominal pressures using a Disa urodynamic system (Disa, Electronics, Denmark). Excellent flow curves were obtainable after this procedure. Increasing the intro-abdominal pressure was the drive for voiding in these patients, as evidenced by simultaneous recording of intra-abdominal and "intravesical" pressures.

DISCUSSION

This operation proved to be technically feasible without increased mortality or morbidity. The operation is applicable for a large number of bilharzial bladder cancer patients, since the trigone is rarely affected by the neoplasm (5.7 per cent) and the prostate is involved in only 6 per cent (El-Bolkainy,1972). The incidence of postoperative mortality of 14.5 per cent in this series is comparable to the figures in other series for radical cystectomy and ileal conduit and rectal bladder which ranges from 12 to 14.6 per cent (Whitmore,1977 and Ghoneim,1972).

Based on the analysis of preliminary survival data, patients undergoing this procedure were not at higher risk for development of recurrence or metastases as compared to the standard radical procedure.

In all patients, the cecum was anastomosed without tension to the urethra or bladder neck. No problems were encountered in anastomosing the ileum to the ascending colon. Persistent diarrhoea owing to loss of the ileocecal valve was not encountered.

The renal function was preserved in the majority of cases. This is due to the efficiency of the ileocecal valve in reflux prevention. There was no need for plication of the ileocecal region to increase the efficiency of the ileocecal sphincter. Voiding cystogram showed that the ileocecal sphincter prevented ureteral reflux, nearly imitating the ureterovesical mechanism. Kidney function and configuration of the pyelocalyceal system were preserved even after more than five years of follow up.

Moreover a balanced vesicourethral unit could be achieved as evidenced by dynamic studies. Our findings that the intra-abdominal pressure is the drive for voiding in these patients are similar to the results reported by Gleason 1972, following augmentation cystoplasty.

All patients were fully satisfied with the continence obtained especially during the second and third year of follow up. Most of them were gainfully employed in the fields as farmers.

Ileocecal bladder reconstruction is an attempt to overcome the inevitable social inconveniences and functional losses resulting from cystectomy and standard diversion procedures.

REFERENCES

El-Bolkainy, M.N., M.A Ghoneim and M.A. Mansour (1972). Carcinoma of the bilharzial bladder in Egypt: Clinical and pathological features. Brit. J. Urol., 44, 561-570.

Ghoneim,M.A., M.A. Mansour and M.N. El-Bolkainy (1972). Radical cystectomy for carcinoma of the bilharzial bladder. Brit.J. Urol., 44, 461-466.

Gleason, D.M., R.F. Gittes, M.R. Bottaccini and J.C. Byrne (1972). Energy balance of voiding after cecal cystoplasty. J. Urol., 108, 259-264.
Khafagy, M., M.N. El-Bolkainy, R.S. Barsoum and S. El-Tatawy (1975). The ileocecal bladder: A new method for urinary diversion after radical cystectomy (A preliminary report). J. Urol., 113, 314-316.
Selvaggi, F.P., P. Zaini and J.D. Battenberg (1972). Use of the canine ileocolic valve to prevent reflux. J. Urol., 107, 372-376.
Whitmore, W.F, Jr., M.A. Batata, M.A. Ghoneim, H. Grabstald and A. Unal (1977). Radical cystectomy with or without prior irradiation in the treatment of bladder cancer. J. Urol., 118, 184-187.

Radical Cystectomy for Carcinoma of the Bilharzial Bladder

Mohamed A. Ghoneim

Department of Urology, Mansoura University, Mansoura, Egypt

Abstract: One hundred and sixty-two patients with carcinoma of the bilharzial bladder were treated by radical cystectomy. The postoperative mortality was 12.9% and the five-years survival rate was 38.9%. Correlations of the survival figures relative to the various pathologic features of the tumor revealed that squamous and transitional tumors have essentially the same prognosis, and that the most important prognostic parameter was the tumor grade. In most cases, treatment failures were due to local recurrences which developed within one year after operation.

Introduction: In Egypt, carcinoma of the bilharzial bladder presents certain particular features (Makar, 1955; El-Sebai, 1961 and El-Bolkainy, 1972). Most of the cases are seen in an advanced stage, and the tumors are usually of the squamous cell type and are not radioresponsive. Radical cystectomy with urinary diversion appears to offer the best prospects for cure in most of the cases. The aim of this article is to describe our technique for radical cystectomy and present the end results of treatment using this modality.

Material: Between 1968 and 1973, 190 patients with clinically "resectable" bladder tumors were admitted to our department. Distribution of age and sex is given in Table 1.

Table 1. Age and Sex Distribution

Sex	-30	30-	35-	40-	45-	50-	55+
Males	7	7	17	46	29	19	14
Females	6	8	8	14	7	6	2
Total	13	15	25	60	36	25	16

Of the 190 patients, 162 underwent radical cystectomy with some form of urinary diversion. Segmental resection was performed in 19 cases. Six patients were considered inoperable during surgical exploration because of positive lymph nodes above the pelvic brim. Three patients refused the operation.

In general the indications for cystectomy were (1) Superficial tumors unsuitable for transurethral resection due to their size,and(2) Infiltrating tumors unsuitable

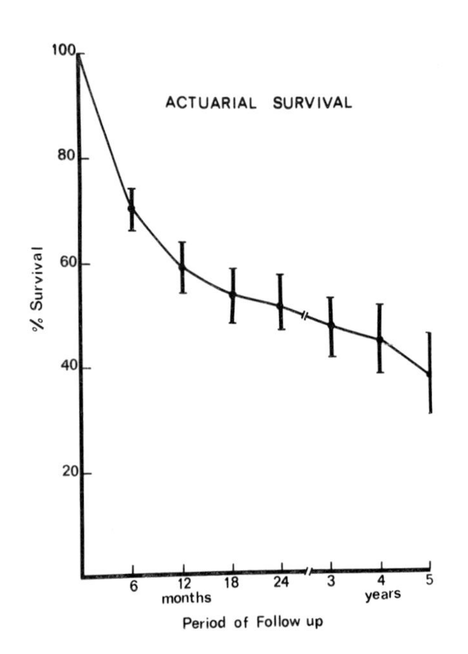

Fig.I Actuarial survival in general, Vertical bars represent the stan-ard error (P= 0.05)

Survival was also correlated to the histology, grade and pathological stage of the tumors. There was no significant difference in survival among patients with squamous and transitional cancers (37% and 41% respectively). Patients with low grade tumours had a five year survival rate of 52.3%, while survival with high grade ones was 25.8% only; a statistically significant difference. Cases with superficial tumours (P1 and P2) had a better survival rate than those with infiltrating ones (P3 and P4) (54.3% and 33.5% respectively).
The regional lymph nodes were involved in 28 cases (17.2%). Cases with negative nodes had a significantly better survival rate than those with positive ones. Nevertheless, the radical operation could provide a 20% survival for patients with positive nodes.
Table IV shows the distribution of treatment failures (recurrences and/or metastasis) according to their site and time of onset. Most of the recurrences were local in the pelvis and developed within twelve months after initiation of treatment.

Table IV. Treatment failures distribution according to onset and locality.

Period (years)	Regional	General	Undetermined	Total	Per Cent
0-	36	2	8	46	60.5
1-	8	2	...	10	13.5
2-	5	...	1	6	7.8
3-	1	1	...	2	2.6
4-	6	...	1	7	9.2
5+	2	1	2	5	6.5

for segmental resection in view of their size and/or location.

Methods: The extent of the radical operation is to remove the bladder with its perivesical fat, peritoneal covering, the prostate, and the seminal vesicles, together with the distal common iliac, internal iliac, and external iliac lymph nodes. In the female, the bladder, urethra, uterus, and upper two thirds of the vagina with the pelvic cellular tissue and the aformentioned lymph nodes are removed. The technical details of this proceduse were the subject of a previous publication (Ghoneim, 1972).
The various methods used for urinary diversion were rectal bladder in 97 cases (Mauclaire 89 and Gersuny-Lously 8), ureterocutaneous 45, ureterocolonic 7, and ileal loop conduit (Bricker) 13.

Results: Twenty one patients died in hospital after initiation of treatment, a postoperative mortality rate of 12.9%. The various causes of post-operative mortality were septic shock in 9 patients, fecal fistula 3, urinary leakage 3, pulmonary edema 1, hepatorenal failure 2, and deep vein thrombosis and embolism 1. The most important single cause was septic complications.
The distribution of cases according to clinical and pathologic stage is presented in Table II.

Table II. Clinical and pathologic stage

Stage	Clinical Number	Pathologic P_1	P_2	P_3	P_4
T1	2	I	I	..	..
T2	30	..	7	21	2
T3	129	..	22	85	1

It is clear that most of the cases present initially with an advanced stage. Moreover there was a tendency of underestimating the pathologic extent of the disease, particularly at the T2 stage.
The tumor histology and grade are given in Table III.

Table III. Tumor histology and grade.

Histology	Grade Low	Grade High	Total
Squamous	71	40	111
Transitional	9	33	44
Adenocarcinoma	1	6	7
Sarcoma	..	..	2
Total			162

Squamous cell tumors account for 66.7% of the cases; of these, 62% are of well differentiated variety. Transitional cell carcinoma is less commom and represents 26% of cases, and adenocarcinoma is rare, constituting 5% only.
Survival was calculated using the actuarial technique (Fig.1). Postoperative mortalities were included in the computation of the curves. Eighty seven patients were followed up for five years or more. Follow-up data were unavailable in 4 cases only. Survival was defined as living without clinical evidence of disease. The five-year survival was 38.9%.

Discussion; The general technical principles employed are similar to those previously described by Paquin qnd Mqrshall (1956) qnd El-Sebai (1961). The operative mortality of 12.9% compares favorably with other reports(Whitmore,1962; Sakati,1966; Stones,1966). However, recent articles indicate that these figures can be remarkably reduced by staging of the operative procedure (Crimes,1972). Rectal bladder with a terminal left iliac colostomy was the most common method utilized for urinary diversion. The technique, its justifications and the results were reported in a series of previous publications (Ghoneim,1970 & 1974). Analysis of survival figures relative to the various pathologic features of the tumor revealed that squamous and transitional tumors have essentially the same prognosis. It was also demonstrated that the tumor grade is the most critical prognostic index; a fact which was also emphasized by Thompson (1960). The majority of treatment failures were due to local recurrences in the pelvis. Moreover, most of them developed within twelve months after initiation of treatment. These findings strongly suggest that surgical excision alone, despite being radical was inadequate to deal with the extent of the "local pathology". An adjuvant line of treatment directed locally to the pelvis may improve survival. Preoperative radiotherapy appears to provide the logical tool. On these bases, a prospective randomized study comparing radical cystectomy and 2,000 rad preoperative radiation dose given in five daily fractions followed by cystectomy, is the subject of a current clinical trial.

References:

(1) Crimes,J.H.,Hart,J.M.,Glenn,J.F.,and Anderson,E.E. (1972). Staged approach to invasive vesical malignancy. J. Urology,108,872-875.

(2) El-Bolkainy,M.N., Ghoneim,M.A., and Mansour,M.A.(1972). Carcinoma of the bilharzial bladder in Egypt. Brit. J. Urology,44,561-570.

(3) El-Sebai,I.(1961). Cancer of the bladder in Egypt. Kasr El-Aini J.Surgery, 2, 182-241.

(4) Ghoneim,M.A.(1970). The recto-sigmoid bladder for urinary diversion. Brit.J. Urology, 42,429-433.

(5) Ghoneim,M.A., Mansour,M.A. and El-Bolkainy,M.N.(1974). Radical cystectomy for carcinoma of the bilharzial bladder. Brit. J. Urology,44,461-466.

(6) Ghoneim,M.A., and Ashamalla,A.(1974). Further experience with the recto-sigmoid bladder for urinary diversion. Brit. J. Urology, 46,511-519.

(7) Makar,N. "Urological aspects of bilharziasis in Egypt". S.O.P.Press, Cairo 1955, pp.51-83.

(8) Paquin,A,J. and Marshall,V.F.(1956). A technique for radical cystectomy. Cancer,9,585-595.

(9) Sakati,I.A., and Marshall,V.F.(1966). Postoperative fatalities in urology. J. Urology, 95, 412-414.

(10) Stones,J.H., and Hodges,C.V. (1966). Radical cystectomy for invasive bladder cancer. J. Urology, 96, 207-210.

(11) Thompson,G.I.,(1960). Prognosis in vesical neoplasm. J.A.M.A.,172,28-33.

(12) Whitmore,W.F., and Marshall,V.F.(1962). Radical total cystectomy for cancer of the bladder. J. Urology, 87,853-868.

(28) Prout,G.R.,Slack,N.H., and Bross,I.D.(1971): Pre-operative irradiation as an adjuvant in the surgical management of invasive bladder carcinoma. J. Urology. 105, 223-231.

(29) Putten,L.M.(1968): Tumour reoxygenation during fractionated radiotherapy; studies with transplantable mouse osteosarcoma. European Journal of Cancer. 4, 173-182.

(30) Werf-Messing, V.D.(1973): Carcinoma of the bladder treated by preoperative irradiation followed by cystectomy. Cancer, 32, 1084-1088.

(31) Whitmore,W.W., Batata,M.A., Ghoneim,M.A., Grabstold,H. and Unal,A.(1977): Radical cystectomy with or without prior irradiation in the treatment of bladder cancer. J. Urology,118, 184-187.

Index

The page numbers refer to the first page of the article in which the index term appears.